SPIDERS OF NEW ZEALAND

Dedicated to our children

SPIDERS OF NEW ZEALAND

AND THEIR WORLD-WIDE KIN

Ray Forster and **Lyn Forster**

With a Foreword by
Norman I. Platnick

University of Otago Press

in association with

Otago Museum

Published by University of Otago Press
PO Box 56/56 Union Street, Dunedin, New Zealand
Fax: 64 3 479 8385
Email: university.press@stonebow.otago.ac.nz

Published in association with Otago Museum

© R.R. Forster and L.M. Forster 1999
First published 1999
ISBN 1 877133 79 5

Printed through Condor Production Ltd, Hong Kong

CONTENTS

Foreword vii
Acknowledgements 1
Introducing spiders 3

ONE **Structure and behaviour of spiders 9**

TWO **The life of a spider 33**

THREE **Spider relatives 47**
Mites: Order Acari 48
False scorpions: Order Chelonethi 49
Harvestmen: Order Opiliones 52

FOUR **Trapdoor spiders and their kin: Mygalomorphae 59**
Hexathelidae 61
Idiopidae 64
Nemesiidae 66
Migidae 67

FIVE **Living fossils: Araneomorphae 69**
Hypochilidae 72
Austrochilidae 73
Gradungulidae 73

SIX **Free-living spiders 82**
Wolf spiders: Lycosidae 82
Nurseryweb spiders: Pisauridae 87

SEVEN **Crab spiders 95**
Giant crab spiders: Sparassidae 95
Small crab spiders: Thomisidae 95

EIGHT **Hunting spiders 105**
Lynx spiders: Oxyopidae 105
Hopping spiders: Clubionidae 106
Fleet-footed spiders: Corinnidae 106
Prowling spiders: Miturgidae 107
Stealthy spiders: Gnaphosidae 109
White-tailed spiders: Lamponidae 111
Pirate spiders: Mimetidae 112
Shield spiders: Malkaridae 113
Scuttling spiders: Cycloctenidae 114

NINE **Jumping spiders: Salticidae 117**

TEN **Six-eyed spiders 135**
Orsolobidae 136
Dysderidae 139
Segestriidae 141
Periegopidae 143
Scytodidae 144
Oonopidae 144

ELEVEN **Orbweb spiders** 145
Araneidae 156
Tetragnathidae 164
Nanometidae 166
Ray spiders: Theridiosomatidae 168
Uloboridae 169
Deinopidae 169

TWELVE **Spaceweb spiders** 171
Cobweb spiders: Theridiidae 171
Stiphidiidae 183
Pholcidae 187
Linyphiidae 189
Synotaxidae 192
Cyatholipidae 193
Neolanidae 195

THIRTEEN **Midget spiders** 197
Anapidae 197
Micropholcommatidae 199
Mysmenidae 200
The Archaeid group 202
Holarchaeidae 203
Mecysmaucheniidae 203
Pararchaeidae 204
Erigoninae (Linyphiidae) 205
Hahniidae 207
Phoroncidiinae (Theridiidae) 208

FOURTEEN **Seashore spiders** 209
Desidae 210
Anyphaenidae 212
Agelenidae 213

FIFTEEN **Hackled-silk spiders** 221
Dictynidae 222
Agelenidae 223
Amaurobiidae 223
Nicodamidae 227
Oecobidae 228
Amphinectidae 229

SIXTEEN **Four families** 233
Zodariidae 233
Ctenidae 234
Psechridae 234
Huttoniidae 235

SEVENTEEN **Harmful spiders** 237

EIGHTEEN **How to find and study spiders** 247

APPENDIX I **World list of spider families** 255
APPENDIX II **Historical notes on early arachnologists** 256
Select bibliography 258
Index 265

FOREWORD

The scientific study of spiders began in Europe and until relatively recently the vast majority of arachnologists have worked either in Europe or North America. It isn't surprising, therefore, that the higher classification of spiders, as worked out a century ago by Europeans, has worked best for north temperate groups, or that most of the classical information on the anatomy, physiology, behaviour, and ecology of spiders has been based on north temperate groups.

Recent decades have seen dramatic changes, however. Serious studies of tropical and south temperate animals have resulted in remarkable improvements in our understanding of spider interrelationships, phylogeny, and biogeography, as well as sizeable additions to our knowledge of how spiders live. The authors of this book, Ray and Lyn Forster, have long been in the forefront of these advances, in many cases pioneering in New Zealand discoveries that have later been found to apply as well to taxa in other parts of Australasia, in southern South America, and in southern Africa.

In this book, two lifetimes of field investigations and laboratory study have come together to paint an intimate portrait of a fascinating fauna, in a way that could only have been achieved through detailed research done on the spot, over a protracted period. The Forsters, while spinning an eminently readable and enjoyable tale encompassing the natural history of a major proportion of the New Zealand biota, make clear how much still remains to be learned. Clearly, we live on a still little-known planet. With this book in hand, budding naturalists in New Zealand and elsewhere will be well prepared to proceed with vital explorations on those new frontiers. If they do so with as much panache and distinction as have these authors, their success is assured!

NORMAN I. PLATNICK

Peter J. Solomon Family Curator, Department of Entomology,
American Museum of Natural History

Adjunct Professor, Department of Biology, City College,
City University of New York

Adjunct Professor, Department of Entomology, Cornell University, USA

ACKNOWLEDGEMENTS

We wish to express our grateful thanks to the many people who have assisted us in the gathering together of spider knowledge over the years, notably Norman Platnick, Robert Raven, Jon Coddington, Charles Griswold and Jerome Rovner. We owe Norman Platnick a special debt of gratitude for his helpful comments on the manuscript. We have been encouraged by the considerable support for this book by Mike Fitzgerald, John and the late Frances Murphy, Grace Hall, Tony Harris, Brian Patrick, Cor Vink, Andrew McLachlan, Ian Millar and Phil Sirvid. We have enjoyed and benefited from correspondence and discussions with other spider enthusiasts from all over the world and we owe them all a great deal.

Several illustrations have been generously supplied by Robert Raven, David Blest, Brent Opell, Ian Millar, Andrew McLachlan, Wayne Harris, David Court, Jerome Rovner and the late Frances Murphy (courtesy of John Murphy), and we express our appreciation to them.

We thank our children, Douglas, Marjorie, Malcolm, and Christine for their support and tolerance. Douglas and Malcolm regularly collected species from high altitude sites.

In the main, the black and white drawings, copied from transparencies of colour slides, have been splendidly executed by Barry Weston and we extend our thanks to him. Most of the photographs were taken by R.R. Forster while some additional black and white drawings were undertaken by L.M. Forster.

Finally we thank our publishers, University of Otago Press – and especially Wendy Harrex, Fiona Moffat, Linda Pears and John Cottle – for their part in bringing this project to a satisfying conclusion.

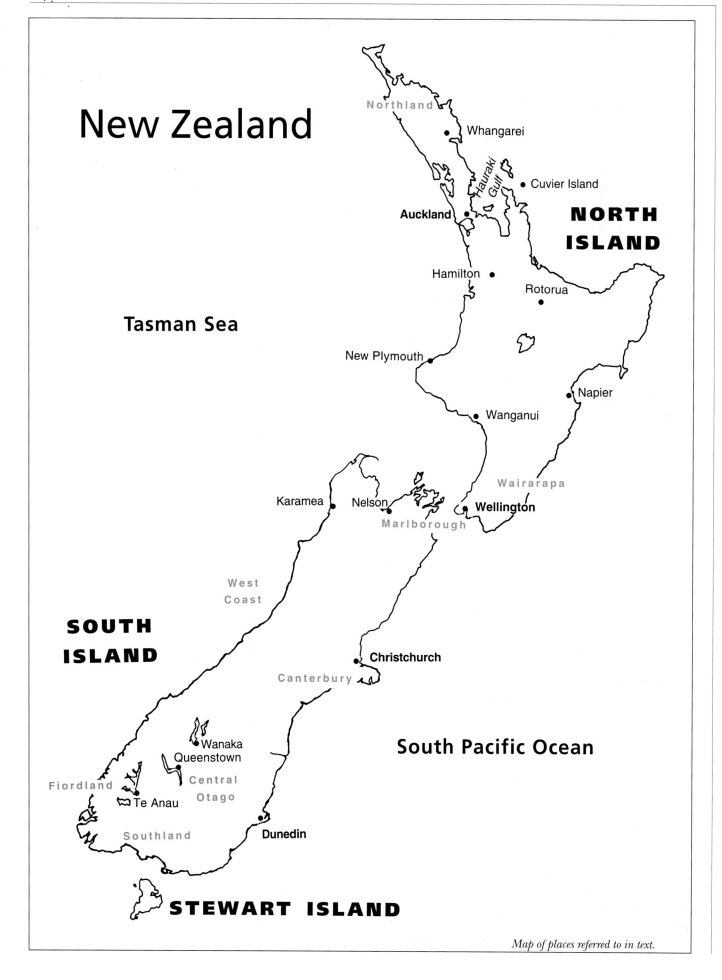

Map of places referred to in text.

INTRODUCING SPIDERS

Sometimes, on a brisk early morning, the sparkle of a dew-laden web catches your eye. Such a creation is a photographer's delight and brings pleasure to all those who happen to see it. But the spiders which fashion this remarkable silk snare are less well known and few people are lucky enough to witness their special skills of construction. How do spiders go about this task?

As we walk through the bushes a green jumping spider peers from beneath a leaf and turns its head to watch us passing by. Is it our imagination or is that spider really looking us up and down?

Out in the garden a flower bud on a branch attracts our attention. It seems ready to burst into bloom. But wait – did it move? Is what we now see a bud, or a spider wandering away?

Now and again we notice holes in the ground and, if we look more closely, we see that some of them are lined with silk. What lives in these holes?

Perhaps when we are strolling along the seashore, a quick flurry in the sand catches our eye, but we see nothing scurrying away. Was it just an illusion?

As you read this book we hope you will come to realise, as we did, that spiders are remarkable little animals often with strange features and astonishing behaviours, and that very few are to be feared. You will see that they have an important place in the nature of things and that they should be cherished as wild creatures in the same way as we cherish our birds and our plants.

How many different kinds of spiders are there?

When we wrote *New Zealand Spiders: An Introduction* in 1973 the formal stages of the description of the New Zealand fauna were in their infancy and less than one-quarter of the probable number of New Zealand species had been recorded – that is, some 500 of an estimated fauna of 2,500 species. During the next twenty-five years a further 800 species have been named, bringing our known fauna to 1,300. In addition, 127 new genera and 16 new families have been established although there are still a number of common families to be revised. Naming these groups requires the preparation of detailed descriptions and illustrations of all new species as well as other associated work, and these names only become accepted when this work is published.

How spiders are named

The two names given to a spider (or any animal or plant) are the name of the **species** (in this case, a particular kind of spider) and the name of the **genus** (which links this particular kind of spider with other similar members of this group). The name of the genus is placed first, the species name second. For example, *Latrodectus katipo* is the name given to New Zealand's native poisonous spider, which means that there are a number of species of the genus *Latrodectus*, one of which is the *katipo*. To signify that this is a scientific name it is printed in *italics* in the text (but roman in captions). In this case the common name of this spider is also katipo, the 'species' part of its scientific name.

Although the same generic name can be used for both a plant and an animal it must be unique within either of these two kingdoms. This means that before establishing a new generic name, it is necessary to ensure, on a worldwide basis, that the name chosen has never been used before. Unfortunately a name is repeated from time to time and this is one of the causes for those confusing changes in scientific names. For example, the name *Pounamua* given to a group of native six-eyed spiders was found to have been used for a group of extinct moa birds some years earlier. To avoid confusion, the generic name for these spiders was changed to *Pounamuella*. Above the level of genus, a larger grouping of animals having certain features in common is called the **family** which, in animals, can be recognised by the ending **-idae**. In referring to these spider groups, we often do so adjectivally, e.g. the family Araneidae may be described as araneids, or the family Theridiidae as theridiids. Subfamily names end in **-inae**.

One of the difficulties in classifying New Zealand spiders has been that many of our spiders do not fit readily into the more well-known families of the Northern Hemisphere. These, of course, were established long ago when spiders of the southern continents were little known. This has necessitated the establishment, over the last few decades, of more than twenty new families to encompass the southern spiders. The reason is that various family characters of these southern spiders differ from those of the Northern Hemisphere. While our knowledge of the southern fauna has greatly increased, much more work is needed before a clear understanding of their evolutionary progression is reached. To this end there has to be universal agreement on the structure of a higher classification which reflects their evolution and relationships.

Dividing spiders into groups.

In the global scheme of things, all known spiders are grouped together under the Order **Araneae**. Two suborders have been established. The name of one of these groups, **Mesothelae,** to which *Liphistius* and *Heptathela* (below) belong, means that these spiders have spinnerets in the mid-ventral abdominal position. These and other anatomical features are explained in chapter 1. Like mygalomorphs, however, the chelicerae are paraxial and there are four booklungs, but the most distinguishing feature, which separates the Mesothelae from all other spiders, is the segmented abdomen. This particular feature is believed to link spiders with very ancient segmented invertebrates. There are forty or so species of *Liphistius* found only in Eastern Asia where they live, some in caves, some on hills and some in the jungle. They make silk-lined burrows

Liphistiomorph spiders such as Heptathela kimurai *from Gifu, Japan are representative of the earliest fossil spiders and are characterised by their clearly segmented abdomens. Although fossil evidence shows that they were once widespread they are now restricted to some islands north of Australia, China and Japan.*

which are closed with a lid. Moreover, they are the most primitive living spiders in the world today and ones which have never found their way to New Zealand shores, or indeed, to any southern country

The second suborder is **Opisthothelae** which means that the spiders in this group have spinnerets at the end of the abdomen. This is in turn sub-divided into the **Mygalomorphae** and **Araneomorphae**. Most mygalomorphs are large bulky spiders restricted to firm substrates. They strike downwards at their prey with paraxial chelicerae and have two pairs of booklungs but no tracheae. Araneomorphae, on the other hand, transfix and carry their prey with diaxial fangs, can move about more freely and most of them have one pair of booklungs and tracheae. Moreover, araneomorphs have developed novel forms of prey capture, including the use of silk snares, as well as elaborate courtship and mating procedures. These matters are discussed in detail in the following chapters.

The diagram below is a simple representation of spider groups as we know them today. Nevertheless, even now sub-groups are seldom clear-cut and arachnologists the world over continue to ponder on the criteria used as new information and alternative views come to the fore.

A simplified version of a higher classification (after Forster et al, 1987)

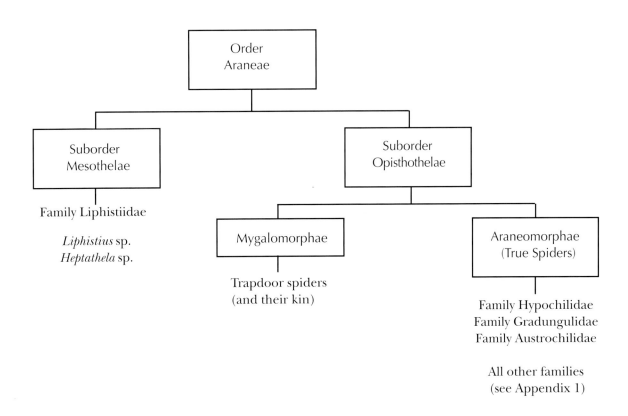

A diagrammatic representation showing the main subgroupings of spiders today. Liphistius *and* Heptathela *are separated from all other spiders while mygalomorphs are seen as 'primitive' but still surviving strongly today. The Araneomorphae are known as True Spiders because the advent of diaxial chelicerae, tracheae and greater use of silk paved the way for their diversification and adaptation to a vast range of habitats and lifestyles.*

Newcomers to these shores

Of the 107 families listed in Appendix 1, 55 are found in New Zealand. Fifty of these families include endemic species, but five introduced families, Lamponidae, Dysderidae, Pholcidae, Corinnidae, Sparassidae, and one subfamily, the Erigoninae, have become well established. These six groups are represented in New Zealand by one or more introduced species but all are now a permanent part of our fauna. The Lamponidae are represented by the Australian white-tailed spiders *Lampona cylindrata* and *Lampona murina*, and Dysderidae by the six-eyed *Dysdera crocota* which was probably introduced from England many years ago. Daddy-long-legs is the common name for the pholcid representatives here (*Pholcus* spp.), known for their predilection for living inside warm dry habitations while the fast-moving *Supunna*, also from Australia, belongs to the Corinnidae. One of the best known introductions, however, is undoubtedly the now common Avondale spider from the Auckland suburb of the same name. This Australian Triantelope is called *Delena cancerides*, a member of the Sparassidae. Undoubtedly, however, the tiny money spiders, Erigoninae – a subfamily of the Linyphiidae – are the most successful immigrants both in species and in numbers, but being so small are less likely to be recognised.

The new settlers, however, have included several species which belong to more cosmopolitan families already represented by our endemic fauna. Australian spiders, such as *Eriophora pustulosa* (Araneidae), *Hemicloea rogenhoferi* (Gnaphosidae), *Badumna longinqua* (Desidae) and *Helpis minitabunda* (Salticidae) come to mind while a more recent immigrant, the redback spider, *Latrodectus hasselti* (Theridiidae) may well become a permanent resident. The presence of the European *Meta segmentata* (Tetragnathidae) has recently been noted.

While several of those spiders which invade our homes and pastures have been established in this country for a very long time, many regular visitors fail to become residents. One such spider is *Heteropoda venatoria* (Sparassidae), better known as the banana spider, so called because it frequents the banana groves in its native habitat. Amongst the more transient visitors have been live females of *Oecobius* (Oecobiidae) and *Scytodes* (Scytodidae), the former discovered in a house near Auckland and the latter on a wall at a school in Matamata (east of Hamilton in the North Island). Both species have been introduced into other countries and could well become established here, at least in the North Island. The Ogre-faced spider *Deinopis* (Deinopidae) has been found twice, a male and a female, living near overseas shipping wharves in the North Island and it is assumed they arrived with cargo from Australia. This is another family that could become established. From time to time various species of tarantulas (opposite) from different families are found on ships or in unloaded cargo in our ports. Despite their somewhat fearsome reputation, these spiders are often kept as pets in both England and the United States.

At present two other families listed in Appendix 1, Neolanidae and Huttoniidae, are known only from New Zealand but they are quite likely to be found in Australia.

Value of collections

Clearly one of the most important aspects of naming species lies in the availability of collections of plants and animals held, in many instances, by museums. These collections can only be acquired by the gathering of specimens from all over the country, a process which usually takes many years. This unsorted material is studied by specialist taxonomists whose task it is to examine and describe the specimens and to assign them to their appropriate group or, if necessary, to new groups.

Today, the value of naming species is recognised not merely because it enables us to identify them but also because each species possesses an enormous amount of genetic information. Moreover, this country has made a global com-

A mygalomorph *Aphonopelma chalcodes, from Arizona, and often kept as a pet in the United States, is typical of the large spiders usually called 'tarantulas'. Other uses of this name, including the earlier use of 'tarentula' for some Italian wolf spiders has led to confusion. These large mygalomorphs are not found in New Zealand*

mitment to conserving biodiversity, which means that we must endeavour to conserve representative ecosystems. We can only be sure of doing this if we can name all those plants and animals that make up an ecosystem.

Establishing relationships

After more than two hundred years of serious study, we might expect that at least the broader grouping of spiders into families would be settled. Unfortunately this is not so. Some authorities consider that family groupings should be based on convenience, so that animals which look alike are placed together despite the fact that they may have evolved from quite different ancestors. If the sole purpose of a classification is to place a specimen into a convenient pigeonhole, this method has some appeal, but today it is widely accepted that classifications should reflect the true relationships of living organisms and, if possible, their course of evolution. This needs to be emphasised because spider classifications are in a state of flux during the transition from a system based on superficial similarities to one based on true ancestry.

Today, at the turn of the century, workers in many countries are publishing new spider classifications based on phylogenetic relationships. Traditional methods of classification are supplemented by various new techniques such as cladistics, DNA molecular analyses, as well as the assessment of many 'hard-to-see' characters by means of electron microscopy and scanning micrographs.

While new categories at the species level tend to have local relevance only, the establishment of the higher taxonomic categories of genus and family have significance in other countries and, in particular, demonstrate the ancient relationship of the New Zealand fauna with that of the southern continents and islands. Many genera and families established on the basis of New Zealand spiders are found in some of the other southern lands, in particular Australia but also in Chile and South Africa. These relationships demonstrate that elements of the fauna have persisted in these lands after the break-up of the southern land mass, Gondwana, millions of years ago. To emphasise the global significance of these relationships, a number of monographs co-authored by Platnick *et al* have been published by the American Museum of Natural History, New York.

But why study spiders?

We could just as easily ask why you collect stamps, go fishing, watch videos, or play games on a computer. For us, the short answer is much the same – because we like doing it. We should add, though, that spiders are fascinating to watch, and they present a challenge to us in trying to determine their relationships and how they might have evolved. Moreover, spiders and other invertebrates colonised the world long before humans did – it could be said that they prepared this earth for the arrival of higher animals. A great many people have an interest in birds, for example, but how many of them realise that but for invertebrates, most birds would not survive? Even the native New Zealand bellbird, known as a nectar feeder, supplements its diet from time to time with insects and spiders, often catching moths and beetles on the wing. If it were not for spiders, our vegetation and crops would be even more severely depleted by insects and other pests. In his book, *The World of Spiders*, Bristowe calculated that, in Britain the weight of insects eaten by spiders every year was greater than the weight of the human inhabitants of that country.

Such studies also have a practical basis. Interrelationships between spider species in different countries may reveal or confirm past geological, geographical or ecological links. Such 'bioprospecting' can be useful in determining whether a particular rock or mineral found in one locality might be found in another, given that two related groups or species of invertebrates once co-existed in the same geological timeframe. Moreover, understanding the history and distribution of spider species is a valuable tool in the management of conservation priorities.

Spider venoms are being assessed by drug companies in some parts of the world as potentially useful in the treatment of certain debilitating diseases. The biodegradable nature of venoms has also led to investigations of their usefulness as insecticidal agents. Spider silks are exceptionally strong and lightweight, so their potential for the manufacture of parachutes and bulletproof vests is under scrutiny. Today, there is increasing awareness of the need to conserve as many species as possible, for not only do they form part of the living world today, but their role in the future should not be underestimated.

The task of taxonomists – those scientists whose task is to identify and name the plants and animals, both living and fossil – is to provide clear information about structural characters which enable others to identify them and to further develop an overall classification which reflects their evolution. The scientific community which carries out this task is worldwide in its composition and descriptions of the flora and fauna of each country are subject to international scrutiny.

To our readers

We hope this book will be useful to those who have an interest in natural history as well as those who wish to further their knowledge of spiders. With this in mind, we have grouped families in several different ways – e.g., seashore spiders (habitat) or midget spiders (size) and so on. While very few spiders are harmful, we have written about those that are and hope our readers will realise that hundreds of spiders pose no threat at all. We also include a chapter on how to find and study spiders so that anyone wishing to further their knowledge can make a start. Of course, people soon work out their own specialised methods to observe particular aspects which interest them.

To avoid repetition in the text, we have provided short historical notes (Appendix 2) about people who have contributed in many ways to the early knowledge of spiders in this country. The bibliography includes those who are mentioned – or whose studies have been referred to – in the text, and many others whose work is relevant to a further understanding of the topics we have discussed.

CHAPTER ONE

STRUCTURE AND BEHAVIOUR OF SPIDERS

While much pleasure and enjoyment may be gained from the simple observation of spiders living their normal lives, it is useful to have some background knowledge so the significance of their structure and typical behaviour can be appreciated. Without this knowledge, behavioural features, which might be readily recognised in larger animals, such as birds and mammals, are easily overlooked in spiders and much of the interest in studying them is lost. Fortunately several of the books and papers in the bibliography provide this information in some depth, and as they are readily available in libraries only a brief summary is given below.

To begin with, spiders are *invertebrates*, that is to say, animals which do not have a backbone. Instead, their soft bodies are enclosed in a hard chitinous covering called an *exoskeleton*. It is this exoskeleton which gives the spider its support and shape, its capacity to move about and spin silk, its ability to withstand water loss as well as a measure of protection from its enemies.

Parts of the body

Main features

The body of a spider (Figs 1.1 & 1.2) is clearly divided into two portions, the *cephalothorax* and the *abdomen* which are joined by a narrow waist called the *pedicel*. Cephalothorax is the name for the combined head and thorax which are always joined to form a single unit. The upper surface of this region is covered by a plate called the *carapace* which has a U-shaped groove marking

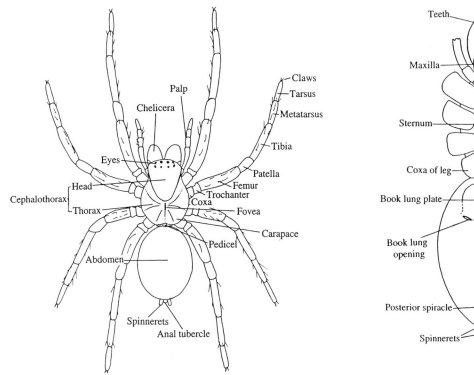

Fig. 1.1 (below left): Dorsal view showing the main external features of a typical spider.

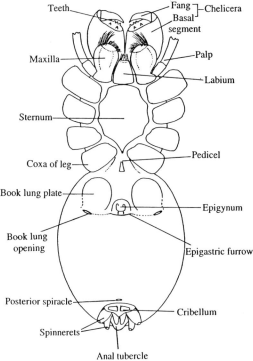

Fig. 1.2 (below right): Ventral view of a typical spider (legs omitted).

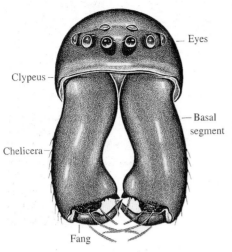

Fig. 1.3 Front view of a spider showing eyes, clypeus (area below the eyes) and (diaxial) chelicerae.

the end of the head. Behind the groove, in the middle of the thorax, there is usually a slight hollow called the *fovea*, and this marks the place where the internal muscles are attached. On the cephalothorax are the *mouthparts, chelicerae, pedipalps, legs* and *eyes*, while the abdomen holds the *heart, respiratory* and *reproductive* systems, as well as the *spinnerets* and *glands* which produce the *silk*. Most of the under surface of the cephalothorax is covered by a large plate, the *sternum*, which in turn has a small plate called the *labium*, hinged or fused in front. The *nervous* and *digestive* systems extend throughout the body and are discussed below.

The front of the spider

The space between the front row of eyes and the forward edge of the carapace is called the *clypeus* (Fig. 1.3). The mouthparts consist of a labium which is flanked on either side by a small flexible plate, the *maxilla* (plural maxillae), to each of which is attached a pedipalp (see Fig. 1.2). The maxillae are situated below the opening of the mouth and move freely, so they can act as chewing devices during feeding. The chelicerae are situated in front of the mouth below the anterior margin of the carapace and are easily seen when the spider is viewed from the front.

The function of the chelicerae

Each chelicera consists of a stout *basal segment* and a moveable spine-like *fang* (see Fig. 1.3). The fangs are used to pierce prey and, except in a very few spiders (e.g., Uloboridae), to deliver the poison which subdues the prey. The chelicerae are oriented in two different ways. In more primitive spiders (e.g., Mygalomorphae), the basal segment and fangs are parallel to each other (*paraxial*, Fig. 1.4) and to use them the spider has to raise its head and strike downwards. As a result prey is pinned to the substrate and generally has to be dragged backwards. In advanced or True Spiders (e.g., Araneomorphae) the basal segment and fangs are directed inwards (*diaxial*, see Fig. 1.3) and strike towards each other. Consequently prey is firmly transfixed and the spider is able not only to carry its meal away, but also to use its fangs for mastication. This difference in cheliceral orientation and function is a major distinguishing feature between the primitive group of spiders, Mygalomorphae, and the more advanced or True Spiders, Araneomorphae.

When not in use, the fang rests in a shallow groove on the basal segment which usually has a row of stout teeth on each side (Fig. 1.5a). Below this groove are a number of pores through which fluids are released when prey is being chewed. The poison is stored in glands surrounded by muscle fibres which contract as the spider bites. It is then forced through a narrow duct to a slit behind the tip of the fang (Fig. 1.5b), a position which prevents the opening being blocked while the spider bites. In mygalomorph spiders the poison glands lie within the basal segment of the fangs, but in other spiders they are found further back in the cephalothorax.

Fig. 1.4 (below left): Porrhothele antipodiana, *the tunnelweb spider. Typical of the Mygalomorphae, this spider lifts itself up, spreads its paraxial chelicerae apart and prepares to strike downwards. Note the eye arrangement and drop of digestive fluid between the chelicerae.*

Fig. 1.5a (below right): Fang at rest in the groove of the basal segment. Note the cheliceral teeth and poison slit in fang. Greatly magnified.

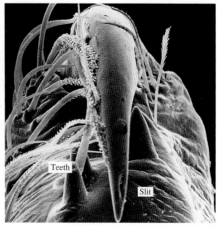

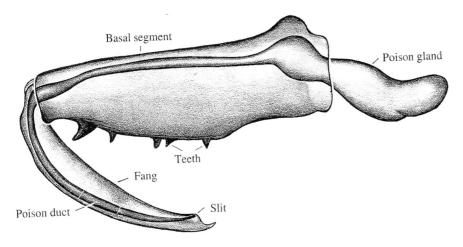
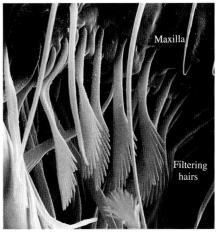

How spiders feed

Spiders use many different methods for catching live prey (as described in the following chapters) before injecting poison. However, they all eat their food in much the same way although some spiders masticate it more thoroughly than others. Prey is first physically pierced or even crushed by the fangs and maxillae. This breaking up assists the action of the digestive enzymes, ensuring that food is well digested and transformed into a broth before the spider sucks it up. A thick border of hairs (Fig. 1.5c) along the inner margins of the maxillae acts as a filter to keep the larger fragments from passing into the mouth. A second, much finer filtering system, called the *palate plate*, is present inside the mouth and this traps indigestible particles which the spider later spits out.

The mouth is a simple opening behind the chelicerae and hidden by the labium. The predigested food passes from the mouth along a tubular *oesophagus* which leads to a strong muscular *sucking stomach* (Fig. 1.6). Behind the stomach the gut may be provided with numerous lobes in both the cephalothorax and the abdomen. These lobes not only act as food storage organs but also produce some of the digestive enzymes which flow from the mouth when the spider is chewing its prey. This means that the digestive system of spiders is better adapted to the ingestion of liquefied food rather than solid tissue. Some spiders suck up the entire liquefied contents of their victim leaving the carcase intact, but others chew the whole catch thoroughly, discarding only the mangled hard parts. Waste products are expelled through the anus, an opening marked by a lobe called the *anal tubercle*.

Fig. 1.5b (above left): Side view of chelicera showing the poison gland and duct leading to the slit above the fang tip. Greatly magnified.

Fig. 1.5c (above right): Maxilla (upper area of picture) lined with modified hairs which act as a filter as the spider feeds (x 1,000).

Fig 1.6 Lateral section through a spider showing the arrangement of internal organs.

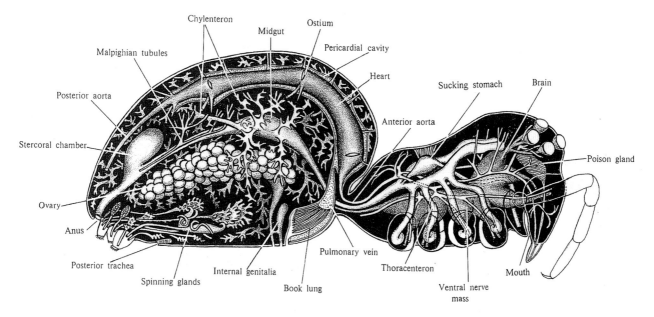

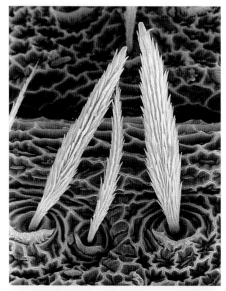

The pedipalps

Spiders have a small pair of leg-like segmented appendages in front, the *pedipalps*, often referred to simply as *palps*. They are not, however, walking legs but have a sensory function (rather like the antennae in insects), a prey-handling role and, more importantly, a reproductive function in the adult male. Unlike the legs, the palps have no metatarsus, which means there are only six segments: the *coxa, trochanter, femur, patella, tibia* and *tarsus*. The pedipalps of the female are not modified for any reproductive purpose and the palpal tarsi often retain the small claw found in juveniles. The mature adult male, however, uses his specially modified palps to convey sperm to the female during mating.

The walking legs

Four pairs of legs are attached to the side of the cephalothorax between the carapace and the sternum. Each leg consists of seven segments which are named from the attachment base: the *coxa, trochanter, femur, patella, tibia, metatarsus,* and *tarsus*. The trochanter is usually the shortest segment and the femur the longest. In some families, such as the Thomisidae, the two front pairs of legs are twisted backwards so that they are directed to the side and what is usually the front side of the leg becomes the dorsal surface. When this happens the spiders are called *laterigrade*. In certain spiders, particularly those such as *Ariadna* and its relatives which inhabit narrow burrows, three of the four pairs of legs are directed forward and are said to be *prograde*.

Hairs and spines

All the legs are clothed with hairs and many have strong spines and bristles which, like the hairs, are movable at the base and probably evolved from them. Generally the hair covering of the legs is finely setose or feathery (Fig.1.7a, b) but it may also be smooth. In some spiders there may be specialised *tenent* (holding or gripping) hairs (Fig. 1.7c) forming a dense bunch on the ventral surface of the tarsi (*scopula*) while somewhat similar hairs are also found either as single hairs or more often as a thick clump beneath the claws (*claw tuft*). These hairs are flattened and the lower surface, which comes into contact with the substrate, has a covering of fine *cilia* (very fine hairs). The purpose of both the scopulae and the claw tufts (see Fig. 1.8c) is to enable the spider to cling to different kinds of surfaces. This is achieved either by the presence of a thin film of moisture or simply by the forces

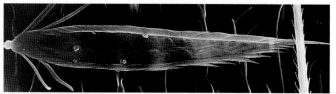

Fig. 1.7 *Specialised hairs. **a** (top): setose. **b** (second from top): feathery. **c** (above left): tenent. **d** (left): bunch of scale hairs (gnaphosids). **e** (above right): a single scale hair. (All greatly magnified.)*

of physical adhesion. The iridescent or metallic hues seen in some spiders are brought about by the presence of flattened scale hairs (Figs 1.7d, e) mostly found on the abdomen but sometimes on the legs. These scale hairs are responsible for the shiny appearance of a number of gnaphosids, for example.

Claws

The primitive number of claws is three, consisting of a pair of upper (*superior*) claws and a single (*inferior*) lower one. The upper claws are generally similar in size but in a few spiders, such as *Celaenia* (Fig. 1.8a), even these may be different. More extreme variations are found in the Gradungulidae, where the first two pairs of legs have one upper claw, the *proclaw*, that is much larger than any of the others (Fig. 1.8b). However, all three claws usually have small teeth on the ventral surface, and in some spiders such as the Orsolobidae (Fig. 1.8c) the upper pair of claws may have a double row of these teeth. Orsolobids are also unusual because the claws are borne on a flexible extension of the tarsus (*onychium*). While the spider is walking, these claws are bent back against the tarsal segment and so press against a pair of specialised hairs (*proprioceptors*) thus providing the spider with sensory information as it moves along.

The loss, during the course of evolution, of the inferior claw leaving just two superior claws, as shown in the Orsolobidae, is common in spiders which have tarsal scopulae or claw tufts. Moreover, some of the three-clawed spiders, such as the orbweb spiders, have specialised spines on the ventral surface of the tarsus just behind the claws. These spines, usually curved with a row of strong denticles along the ventral surface, are known as *false claws* (see Fig. 1.8c). Many theridiid spiders have a row of serrate bristles on the fourth tarsus (*tarsal comb*) which they use to pull out silk and throw at prey, while *cribellate* spiders have a group of metatarsal bristles (*calamistrum*) which similarly scrape silk from the *cribellum* or spigot plate (Fig. 1.8d). Preening combs (Fig. 1.8e) are found on some spiders (e.g., *Huttonia*) because specialised hairs are critical to their survival and must be kept clean and in good condition.

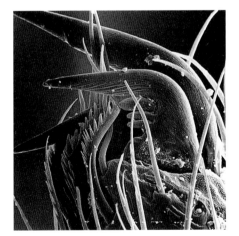

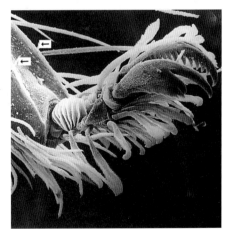

Fig. 1.8 Micrographs. *a* (below left): the three claws of Celaenia, an orb-web spider, with the two upper ones of different size. The serrated false claws lie below; *b* (below centre): tarsus of first leg of a gradungulid spider with extra long proclaw pivoted towards several stout spines. *c* (below right): an orsolobid spider's foot showing two claws, double row of teeth, claw tufts, onychium and tiny proprioceptors (arrowed). *d* (bottom): Calamistrum (row of bristles or hairs) found on the metatarsus of cribellate spiders

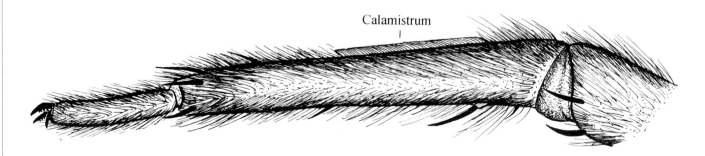

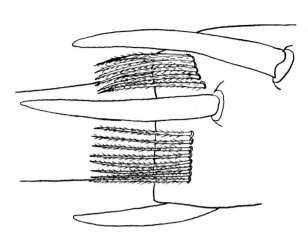

Fig. 1.8e *Preening combs found on the legs of some spiders (e.g.,* Huttonia*). All figures greatly magnified.*

Features that deceive

The shape of the cephalothorax is relatively similar throughout the spider world except for a few modifications such as those affecting the relative sizes and disposition of the eyes; in some spiders the eyegroup is raised well above the carapace level. By contrast the abdomen may be grossly modified in shape and form, while coloration varies greatly between species (Fig. 1.9a). Much of this variation appears to be designed to deceive or deter predators. There are two main ways this can be done. First, a potential predator is discouraged by defensive structures such as spines or perhaps a colour pattern which mimics another animal's active defence mechanism or noxious taste. More commonly the strategy involves camouflage and/or behavioural patterns which reduce the possibility of the spider being seen or recognised as food by a predator (Fig. 1.9b). For instance, many of the foliage inhabiting spiders are green and blend closely with the leaves they inhabit while another spider with grotesque modifications of the abdomen (e.g., *Celaenia*), resembles a bird-dropping when at rest during the day.

Fig. 1.9a *(below left):* Poecilopachys, *the two-spined spider from Australia, exhibits its bright colours and spiky protuberances in defiance of predators.*

Fig. 1.9b *(below right):* Moneta, *a theridiid spider which tricks its enemies by looking just like a bud on a twig.*

Sense organs

Spiders obtain information about the habitat they frequent, their *conspecifics* (members of the same species) and the other animals they meet through their sense organs. They send messages with leg movements, visual badges, chemical emissions and special structures such as stridulatory organs. Although hairs and spines on the legs are innervated and relay information to the spider, other more specialised organs are also present. These structures are often used in the classification of spiders.

The brain (see Fig. 1.6) in spiders acts as the visual centre and when combined with the cheliceral ganglia is known as the *supraoesophageal* ganglion. The ventral nerve mass consists of the fused ganglia of the appendages as well as the abdominal ganglia. From these nerve masses extend neurons serving the sense organs and controlling the musculature.

Trichobothria

Organs such as *trichobothria* consist of a long slender hairs, called *trichemes*, which are set in sockets *(bothria)* (Figs 1.10a–d). In liphistiids and some mygalomorphs, however, the trichemes may be club shaped. Trichobothria are commonly present on the tibia, metatarsus and tarsus of the legs as well as the tibia and tarsus of the palps, but may also occur on the femora of certain orbweb spiders. These organs are used to detect vibration on a substrate such as a web or the ground, and can also pick up airborne vibrations.

Fig. 1.10 *Trichobothria are long slender hairs, easily recognisable in the living spider because they stand out at right angles to the leg. These micrographs show the characteristic features of some different groups.* ***a*** *(top left):* Clubiona. ***b*** *(top right):* Badumna. ***c*** *(bottom left):* Toxopsiella. ***d*** *(bottom right): liphistiids, the most primitive group of spiders, in which the trichemes, or hairs, may be club shaped.*

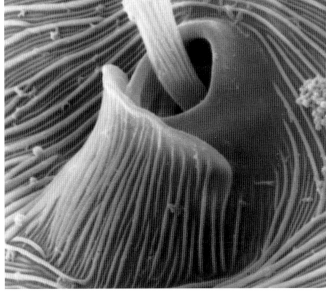

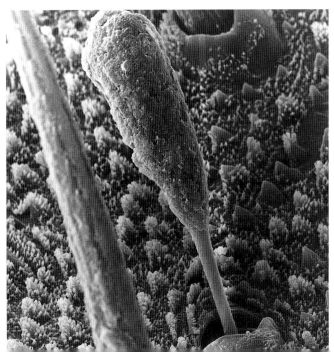

Tarsal organs

Most arachnids have a single specialised sensory organ on the tarsus of the legs and often the palp as well, and this is believed to function as a taste receptor. In spiders it is called the *tarsal organ*. The structure of this organ is characteristic of mainly the higher taxonomic groups, such as the family, and is now commonly recorded at this level. By comparison with the arachnid order Amblypygi, sometimes considered as the sister group of the spiders, the ancestral form of this organ in spiders is a simple saucerlike depression on the bottom of which are usually two or three small sensory nodes. Each of these nodes is served by a bundle of three nerve fibres. The precise message these neurones pass is not clear although it is generally agreed that one function of the tarsal organ is taste or smell detection. This simple saucer-like form (Fig. 1.11a), virtually identical with the Amblypygi organ, is preserved in some of the primitive gradungulid spiders of Australia and New Zealand. It also occurs in a slightly modified form in trapdoor spiders and some of the *haplogyne* spiders (e.g., Segestriidae, Dysderidae). (The terms haplogyne and entelegyne are explained in the section on reproductive organs below.)

This supposedly early form of tarsal organ has developed in a number of ways. For example, the sensory nodes have merged to form a central spine as seen in the Mecysmaucheniidae and some of the Gradungulidae (Fig. 1.11b) or the whole organ has been raised with various lobes surrounding the sensory nodes as found in the Orsolobidae (Fig. 1.11c). A further development, often considered as the most highly specialised, was the elevation of the surrounding integument above the sensory lobes to form keyhole-shaped (Figs 1.11d, e) or oval openings (Fig. 1.11f) above the sensory nodes. This stage is characteristic of all entelegyne families as well as a few haplogyne families. The enclosed forms are called *capsulate tarsal organs*, while the exposed forms are known as *non-capsulate tarsal organs*.

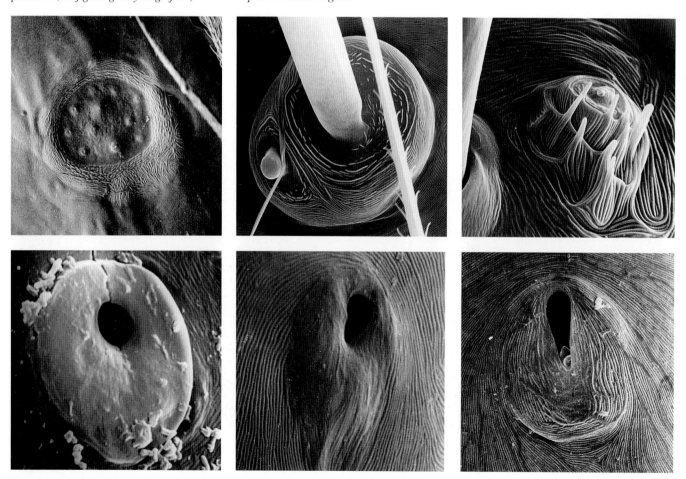

Fig. 1.11 *Micrographs of exposed (a–c) and capsulate (d–f) tarsal organs, (clockwise from top left):* **a** *in* Pianoa isolata *the tarsal organ is extremely small;* **b** *sensory nodes have come together as a central spine (*Gradungula*);* **c** *a raised structure surrounds the sensory nodes (*Subantarctia*); Keyhole-shaped opening found in* **d** Toxopsiella; **e** Neolana; *and oval in* **f** Forsterella. *Note that in these latter forms the sensory nodes are both concealed and protected. (All figures greatly magnified.)*

Lyriform organs

Simple *slits* on various parts of the legs may occur singly or in groups, particularly near the joints. They are supplied with nerves and apparently measure the torsion of the limb and so, by relaying relevant information to the musculature, play an essential part in regulating the spider's posture and movement. Groups of these organs are often referred to as *lyriform organs* (Fig 1.12).

Sound-producing organs

Stridulatory structures which produce sounds are present in a number of groups of spiders. As they are usually found in males they probably come into action during courting. The most common types are just simple '*peg and file*' devices generally associated with the chelicerae and the basal segments of the palp. The pegs, most often located on the inner surface of the palp, are commonly just modified bristles while the file on the opposing surface of the chelicera consists of a series of ridges. More elaborate structures are found on the carapace and abdomen in some New Zealand spiders and are activated by rubbing these two parts of the body together. For example, *Cambridgea* males have a spur on the upper surface of the pedicel which rubs against a series of hard ridges on the front of the abdomen (Fig. 1.13a). The number of ridges differs in each *Cambridgea* species and presumably the vibrations or sound produced by each is species-specific. This means that the signals generated are recognised by potential mates. In the related genus *Nanocambridgea* much the same system is present but the stridulating structures are on the ventral rather than the dorsal surface. *Steatoda*, a theridiid genus, has a similarly activated mechanism but here the ridges are on the carapace and the scraper is on the opposing part of the abdomen (Fig. 1.13b).

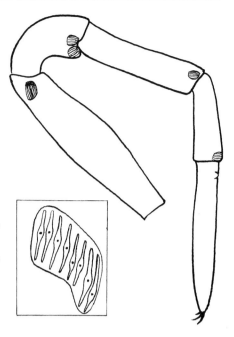

Fig. 1.12 *Leg of a spider showing the common positions of lyriform organs each of which (**inset**) is composed of a number of slit organs.*

Eyes

The eyes are simple *ocelli* (sensory cells) which have a single lens system. There are never more than eight eyes, so this is believed to be the primitive number. They may be arranged in two or three rows near the front of the carapace and the area encompassed by all the eyes is known as the ocular region. The eyes in the front row are described as anterior and in the back row as posterior; the central eyes are called median and side eyes lateral. (For the standard terminology and eye formulae see Figs 1.14a, b.) However, as their arrangement and relative sizes vary greatly this is often a good way to separate species as well as the larger groups of spiders (Figs 1.14c–h). Eye patterns shown here illustrate this variation. More examples are given in the following chapters, or can be found in Forster (1967). Although the range of sizes and positions suggests that eyes are important to spiders, relatively few studies

Fig. 1.13 *Stridulating structures in male spiders. **a** (below left): The* Cambridgea *male has a pedicel spur which scrapes against abdominal ridges, no doubt producing appropriate sounds as he approaches a mate on her web. **b** (below right): The* Steatoda *male has ridges on the carapace and a pick on the abdomen.*

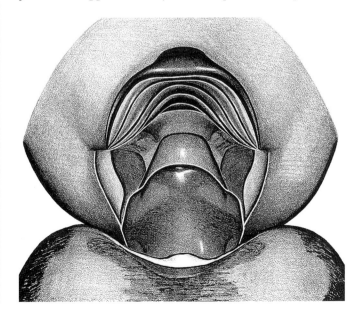
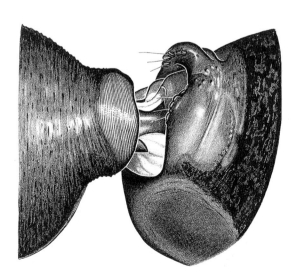

investigating their usefulness have been carried out.

In some groups of New Zealand spiders one pair of eyes is missing and in all of our six-eyed spiders it is the middle front pair (AME) which has disappeared (Figs 1.14i, j). In some overseas spiders there may be a further reduction in the number of eyes, and species have been recorded with four, two and even a single eye. Only jumping spiders have good eyesight. Others such as wolf spiders can detect and respond to movement, but most spiders use one or more pairs of eyes to distinguish light from dark and to measure day/night lengths. So, for example, when the nights are much longer than the days, the spider becomes less active. However, some spiders also make use of polarised light from the sky to find their way about.

The three main organ systems associated with the abdomen are the reproductive organs, the respiratory systems, and the spinnerets and silk glands.

Reproductive organs

Almost all spiders reproduce sexually. Male spiders have developed two intromittent organs of varying complexity associated with the (pedi)palps while females possess one of two quite different genitalic systems. Courtship behaviour is a normal prelude to copulation during which sperm is transferred to the female by the male palps. Parthenogenetic reproduction has been reported occasionally in a few species, (e.g., *Dysdera hungarica*) but there have been few detailed studies.

The male palpal organ

By the time a male spider becomes mature a number of changes to the palps have taken place. Each palpal tarsus is enlarged and has become concave or hollow on the undersurface at which stage it is usually referred to as the *cymbium*. A structure known as the *palpal organ* has developed and this is attached to and

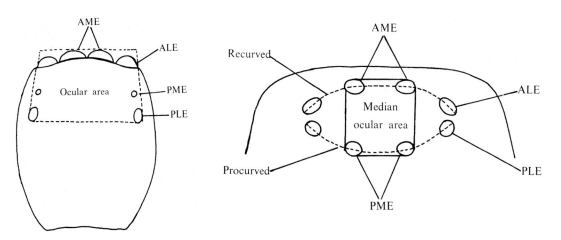

Fig. 1.14a–b Names of eyes and their variation in size and arrangement for eight spider groups: **a** (above left): carapace of a salticid spider showing the ocular area. An eye formula is used to number eyes in each row, e.g., 4.2:2. means four in the front row, two in the middle row and two in the back row. **b** (above right): eyes of a spider, named in pairs: AME, anterior median eyes; ALE, anterior-lateral eyes; PME, posterior-median eyes; PLE, posterior-lateral eyes. The rows of eyes may be curved and, depending on whether this curve is towards or away from the front of the carapace, are said to be recurved or procurved.

Fig. 1.14c–h (facing page, from top left): Eight-eyed spiders. **c** Gnaphosidae (e.g., Hypodrassodes), eyes in two rows; **d** Hexathelidae (e.g., Porrhothele), eyes grouped together; **e** Araneidae (e.g., Eriophora), note four eyes at carapace margin; **f** Cycloctenidae (e.g., Cycloctenus), note size increase in rear eyes; **g** Lycosidae (Lycosa), two rows with four small eyes and four larger ones; **h** Salticidae (e.g., Trite), prominent eyes and good vision. **i, j** Six-eyed spiders; **i** Periegopidae (e.g., Periegops) small eyes, in well-spaced pairs; **j** (centre) Dysderidae (e.g., Dysdera), medium-sized eyes grouped together. For more examples see Forster 1967 and later chapters.

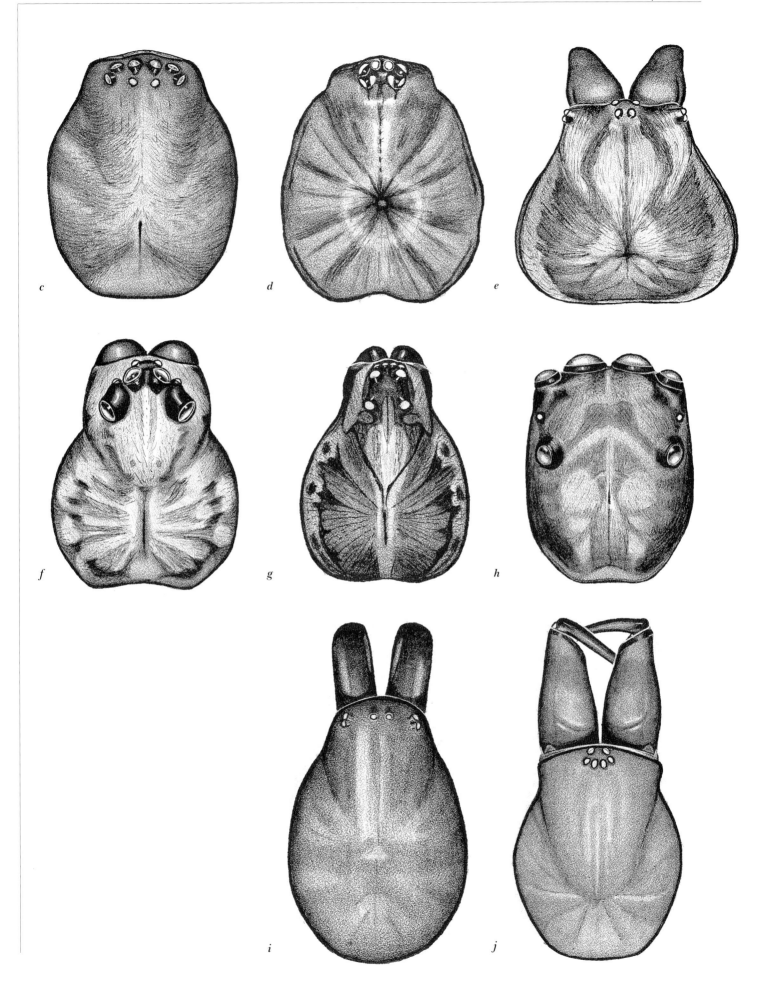

fits within the concave surface of the cymbium (Fig. 1.15a). The simplest form of palpal organ consists merely of an embolus containing a duct which leads to a coiled spermophor inside a bulbous sac (Fig. 1.15b). The coiled spermaphor is a common feature in most palpal organs and is surrounded by secretory glands. These glands become active during mating and probably play a role in the emission of sperm. Although palpal organs vary greatly in complexity (Fig. 1.15c), it seems that the simplest structures are not necessarily the most primitive.

The basic shape of most palpal organs consists of three separate elements: (i) a slender process – the *embolus* – which carries a duct leading from an internal reservoir, the *spermaphor*, for storing the sperm; (ii) a structure associated with the embolus – the *conductor* – which provides mechanical support for the embolus; and (iii) a hinged process – the *median apophysis* – which appears to assist in the positioning of the palpal organ during copulation. These structures may vary in different groups, perhaps by additions or modifications or alternatively by the loss of the conductor and/or the median apophysis. Furthermore, they may become divided into sections by inflatable membraneous regions called *haematodochae*, which are expanded during mating and serve to position the various parts of the palp in ways that enable it to be linked to the female *epigynum*.

Indirect fertilisation

In all spiders, males fertilise females by an indirect method. Before mating the male must first charge his palps with sperm from a gonopore on the underside

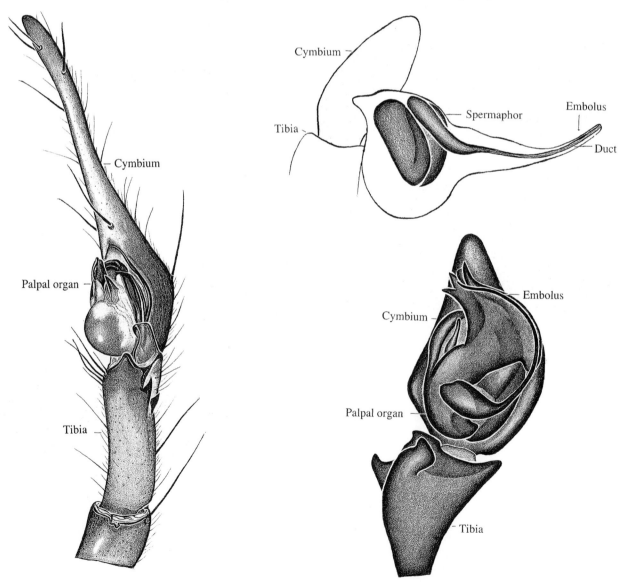

Fig 1.15 Male palpal organs. ***a*** *(below left): cymbium (the pre-adult tarsus) of the male pedipalp showing development of the palpal organ within the cymbial cavity (e.g., Cambridgea).* ***b*** *(below right): simple palp with coiled spermaphor and duct inside a pear-shaped sac (e.g. Wiltonia sp.).* ***c*** *(bottom): palps show considerable complexity in some groups of spiders. For example, in Orepukia it has become very elaborate and here almost covers the cymbium.*

of his abdomen. Here is how this is done. Shortly after reaching maturity, the male spider constructs a sperm web using the main set of spinnerets and spigots. Then, with a special set of silk-producing *epiandrous* glands situated behind the epigastric furrow and ducted to spigots (Fig. 1.16a), the male adds a silken pad to the sperm web. Upon this silken substrate he deposits a drop of sperm which arises from the testes and exits through the gonopore. The tip of each embolus is dipped into this globule of sperm (Fig. 1.16b) and then, rather like using a pipette, it is drawn into the embolus duct and subsequently stored in the internal spermaphor of the palp. With the two palps charged with sperm, the spider is now ready to search for a suitable female.

It is not difficult to understand why the first observers who noticed this unusual method of sperm transfer concluded that there must be some internal connection between the testes in the abdomen and the palpal organs but the real story has been known since the nineteenth century. There are some variations to this general pattern, however. For example, using its third pair of legs, the daddy-long-legs spider (*Pholcus phalangioides*) pulls a single thread across the genital opening, takes the exuded sperm globules from this thread with his chelicerae and then charges his palpal organs. Although for some spiders with greatly elongated palps it would be physically possible to charge the palps directly from the gonopore, none of them do so.

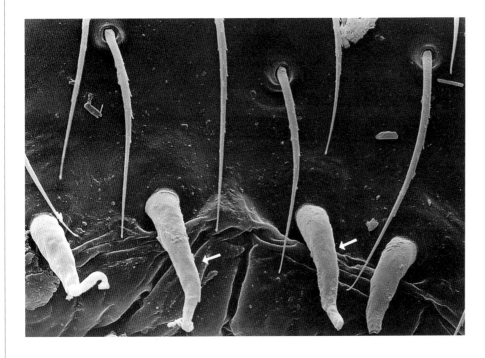

Fig 1.16a *(above left): spigots (arrowed) of epiandrous glands in male spider (greatly magnified)* ***b*** *(above right): male spider using his palp to draw up a droplet of sperm from beneath the sperm web.*

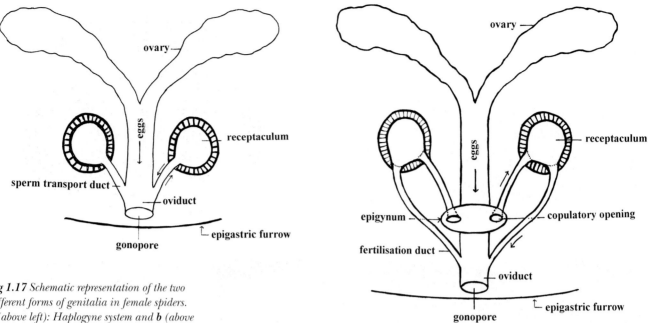

Fig 1.17 *Schematic representation of the two different forms of genitalia in female spiders.* ***a*** *(above left): Haplogyne system and* ***b*** *(above right): Entelegyne system. (Details in text.)*

Female genitalia

There are two distinct reproductive systems in females. The simplest of these is the *haplogyne* system and the more complex one is known as the *entelegyne* system (Figs 1.17a, b). There is one fundamental difference between these two systems. In haplogyne spiders, the deposition of sperm by the male and the eventual egg-laying involve the same opening, the female gonopore. In entelegynes, the development of an *epigynum* has led to one or two openings in this structure, through which sperm are deposited directly into the *spermatheca* (receptaculum). However, eggs are still laid from the gonopore; hence, in this system, insemination and oviposition involve different openings.

Note that the gonopore, in both haplogynes and entelegynes, is the opening to the oviduct. It is situated in the centre of the epigastric furrow on the ventral surface of the abdomen. This furrow marks the division between the second and third abdominal segments. The second segment extends from the pedicel (which is the first segment) to the furrow, while the third segment extends back to the spinnerets.

Haplogyne system

In haplogyne spiders, sperm are deposited by the male directly through the gonopore into the oviduct. In some spiders, sperm may remain here until the eggs pass down the oviduct where they are fertilised. In other spiders, sperm may be transported via narrow ducts to various blind receptacula for storage. In all haplogyne spiders, sperm have to return through the same ducts to fertilise the eggs as they are laid.

Entelegyne system

Entelegyne spiders have an area in front of the epigastric furrow (see Fig.1.2) which has been modified to form a structure known as the *epigynum* (plural *epigyna*). This may be just a simple but distinct lobe or mound, although more often there is also a plate ornamented with ridges, or grooves, or spines. It may have one or two openings through which the two male emboli are inserted separately (Fig. 1.18a). Once inserted, each embolus is forced along a copulatory duct to a receptaculum where sperm are deposited and stored until the eggs are ready for fertilisation (Fig. 1.18b). An additional duct leading from this receptaculum to the oviduct is present. Sperm travel along this fertilisation duct to fertilise the eggs in the oviduct as they are being laid. This

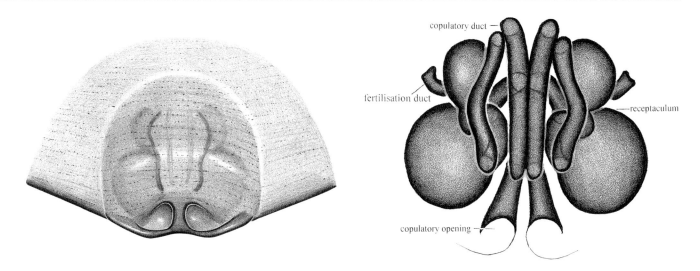

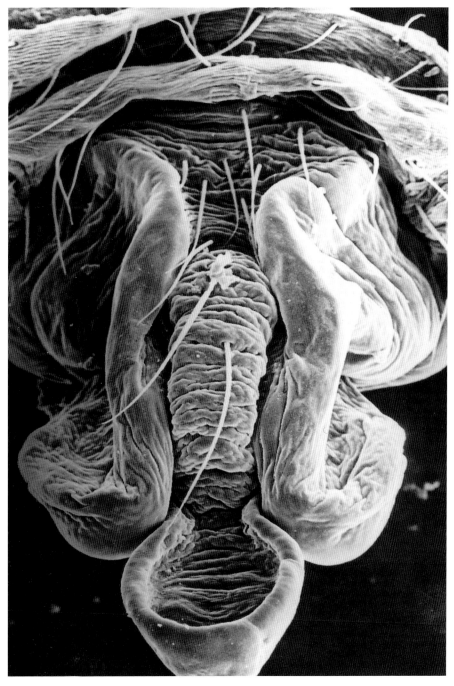

Fig 1.18a *(above left): external view of epigynum of* Clubiona. **b** *(above right): internal structure of the reproductive system in* Clubiona *showing ducts and receptacula.* **c** *(left): micrographs of external epigyna of* Cryptaranea subalpina *showing elaborations which aid in the positioning and securing of the palpal components. (Greatly magnified.)*

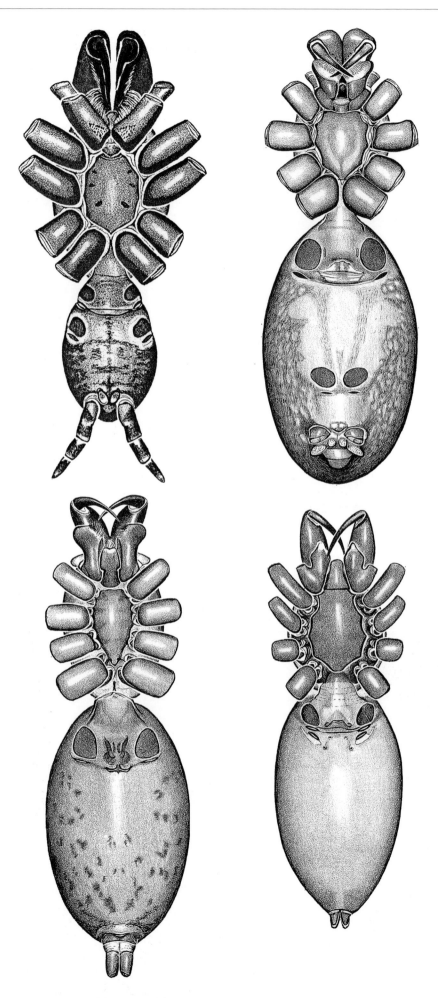

Fig 1.19 *Ventral views of four spiders showing external appearance of respiratory organs.*
a *(top left):* Porrhothele antipodiana *(Hexathelidae), four book lungs.* ***b*** *(top right):* Gradungula sorenseni *(Gradungulidae), four book lungs.* ***c*** *(bottom left):* Clubiona *sp. (Clubionidae), one pair of booklungs, single spiracle near spinnerets leading to tracheae.*
d *(bottom right):* Dysdera crocota *(Dysderidae), one pair of booklungs, pair of spiracles leading to tracheae behind epigastric groove. (Note also paraxial, semi-diaxial and diaxial chelicerae, size and position of spinnerets.)*

second duct is another distinguishing feature of the entelegyne system.

There are numerous modifications of the epigynum, each characteristic of a particular species of spider (Fig. 1.18c). These modifications invariably have some significance with respect to processes on the male palp. When the male and female organs are brought together, the embolus is guided by these processes into the external opening of the epigynum.

Respiratory systems

There are a number of differing respiratory systems in spiders, and these may consist of one or two pairs of booklungs and/or tracheae of various origin, design, and extent. The booklungs are visible as pale patches on the ventral surface of the abdomen, while small, barely visible, slits mark the openings of the tracheae (Fig. 1.19 a–d).

Booklungs

A booklung consists of an *atrium* or hollow pouch from which extends a series of internal leaves or elongations of the integument which lie one above the other (Fig. 1.20) like a book. The atrium opens externally by a *spiracle* or slit. The blood circulatory system surrounds the leaves of the booklung so that they are bathed in blood, allowing oxygen to pass from moist air inside the leaves through the integument and into the blood. This process also permits the exchange of carbon dioxide from the blood into the leaves and finally out through the spiracles. The inflow of air and release of carbon dioxide from the atrium are controlled by the opening and closing of the spiracles.

In the earliest spiders, the respiratory organs consisted of two pairs of booklungs with the anterior pair opening onto the ventral surface of the second abdominal segment to the front of the epigastric furrow, and the posterior pair opening somewhere in the third segment behind the epigastric furrow. Among present-day living spiders this four-booklung arrangement is relatively rare. It is found only in the liphistiids, the trapdoor spiders and their relatives. In addition, three small groups of True Spiders (e.g., Gradungulidae, which is found in New Zealand and Australia, see chapter 5) also retain the original two pairs of booklungs.

Internally, the atria of the posterior booklungs are always connected by a transverse duct and, in most spiders, anterior booklungs are also linked in this way. However, only the posterior transverse duct possesses *apodemes* or muscle attachments, which are merely hollow extensions of the duct and similarly lined with chitin.

Fig 1.20 Cross-section through a booklung shows how leaves are arranged in the atrium. Air flows between the leaves in one direction while blood flows around them in the opposite direction.

Tracheae

In most of the True Spiders or Araneomorphae the posterior booklungs have been replaced by tracheae – slender chitinous tubes extending into the abdomen and sometimes into the cephalothorax (prosoma). Like booklungs, they open to the exterior via *stigmata* (spiracles). Tracheae are generally thought to be more effective in delivering oxygen to the vicinity of various tissues and organs than the open blood vascular system, although this latter system is also found in all spiders.

In the course of millions of years, replacement of the posterior booklungs by tracheae (Figs 1.21a–h) probably took place in a number of ways. The leaves of these booklungs gradually, or even quite suddenly, became tubular thus resulting in bundles of tracheae as shown in some of our six-eyed spiders, (e.g., Dysderidae (Fig. 1.21a). Alternatively, in some Filistatidae (Fig. 1.21b) for example, the leaves of the booklung disappeared entirely; the spiracles merged and now open into a simple atrium or empty pouch with apodemal extensions. Sometimes slender apodemal tubes grew forwards from the atrium to function as tracheae, but they did not extend beyond the abdomen. This situation is found in a few spider groups, e.g., Lycosidae (Fig. 1.21c). In Cyatholipidae (Fig. 1.21d), however, the posterior booklungs have developed into tracheae which extend into the cephalothorax, while the two apodemes in the middle of the duct are unchanged.

Nevertheless, those complex branched tracheal systems that are limited to the posterior abdomen of many spiders probably did not develop from booklungs, but from apodemes which extended anteriorly and became multi-branched systems. Of course, as we noted above, sometimes single tracheal tubes could also arise from these apodemes. Hence, if the atria of the now-lost posterior booklungs developed simple tubes, the end result could be either four simple tracheal tubes such as found in the Araneidae (Fig. 1.21e) or a

Fig 1.21 *Different forms of tracheae found in spiders.* **a** *Dysderidae.* **b** *Filistatidae.* **c** *Lycosidae.* **d** *Cyatholipidae.* **e** *Araneidae.* **f** *Desidae.* **g** *Anapidae.* **h** *Sicariidae. (See text for details.)*

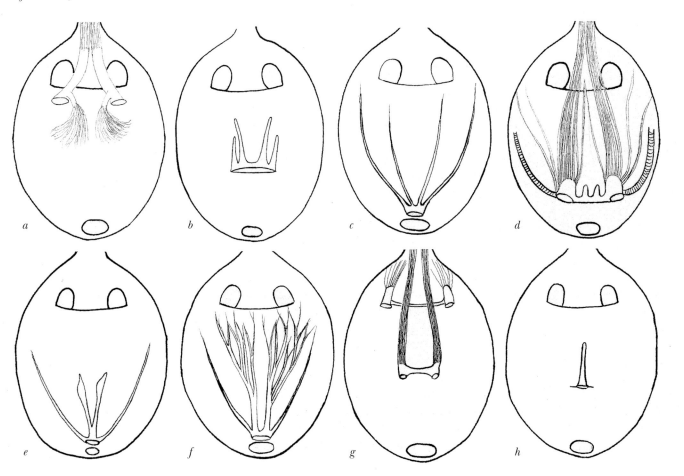

system in which the outer pair is simple and the inner apodemal pair is branched, e.g. Desidae (Fig. 1.21f).

In a few groups of spiders, the anterior pair of booklungs have also become tubular. In these groups, because there were no apodemes associated with the connecting duct, tracheation has been limited to the booklungs. Such spiders are called *apneumone* spiders and, in New Zealand, one example is the family of midget spiders known as Anapidae (Fig. 1.21g). Posterior tracheal systems with four simple tracheal tubes are always restricted to the abdomen. If the middle ones are branched, they may pass through the pedicel into the cephalothorax. But if an anterior tracheal system extends into the cephalothorax, the posterior system does not; and if a posterior system does so, then the anterior system does not. Moreover, in some spiders, tracheae passing into the cephalothorax may even extend into the appendages. In Sicariidae (Fig. 1.21h), however, the posterior spiracles have merged, resulting in a single opening with a short trunk formed from the two apodemes. In a few spiders, (e.g., Pholcidae) there are no posterior respiratory organs at all.

Evolution at work

Although doubts are sometimes expressed about the likelihood of such changes from one form of respiratory system (booklungs) to another (tracheae), studies show that the booklungs of some living spiders have fewer, but greatly elongated leaves. This situation may represent an intermediate stage between booklungs and tracheae, presupposing a gradual change. However, modern thinking favours an evolutionary process called 'punctuated equilibrium', meaning a big development or change that takes place suddenly but which then remains stable for a long time. Earlier thinking favoured 'gradualism' in which evolutionary changes took place consistently but almost imperceptibly. Despite such opposing theories, it is widely accepted today that both kinds of evolutionary change probably take place. Processes such as these, and others not mentioned here, may have been responsible for the different kinds of respiratory systems now known to occur in spiders.

It is very likely that these evolutionary pressures were concerned as much with the need to conserve moisture within the body as with more efficient oxygen uptake. As spiders became aerial and began to build web snares above the ground to trap their prey, they would not only need to become lighter, more agile creatures, but would also require a more efficient supply of oxygen to busy organs. Water loss from the very large surfaces thus exposed to air by the booklungs would be a handicap and a more rapid removal of carbon dioxide from the body would also be needed. Hence, natural selection would favour systems that minimised the disadvantages of booklungs and provided the improvements required. For many spiders, tracheation was the solution.

The spinnerets and silk glands

All spiders produce silk from secretory glands in the abdomen and exude it from spinnerets. Silk is used by spiders in many different ways (Figs 1.22a–e). For example, as spiders move about, a silk dragline stretches out behind them. Nests and retreats are constructed from silk, and eggs are bound and wrapped in it. Many spiders fashion silk webs to catch their prey, and these range from simple traplines to elaborate structures. Moreover, silk is a protein and some spiders regularly eat their webs to recycle the protein components.

Spinnerets are found just in front of the anal tubercle which, in most but not all spiders, marks the posterior limit of the abdomen. The primitive number of spinnerets is eight. Only six of these are found in living spiders, although vestiges of the missing pair are still seen in some groups. It is generally agreed that the origin of the eight spinnerets arises from an ancestral 'spider' at which time two pairs of two-branched appendages were associated with the fourth and fifth abdominal segments. Today, the spinnerets lost in most living spiders are the anterior median (AM) pair. This leaves the anterior lateral (AL), posterior lateral (PL) and posterior median (PM) pairs as the six spinnerets normally found (Fig. 1.23a). However, a few New Zealand spiders, including some Mygalomorphae and Mecysmaucheniidae, have fewer than six spinnerets although in the latter family, which has only one (AL) pair of spinnerets, some of the spigots of the four missing spinnerets remain.

Changes over time

Early in the evolution of true spiders, the anterior median spinnerets were apparently modified to form a flat plate or *cribellum* (see Fig. 1.23a) which bears many distinctive lobulate spigots (Fig. 1.23b, c). In some of these cribellate groups the original paired spinnerets are indicated by the separation of the spigots into two distinct areas, while in others there is only a single field of spigots. With the development of the cribellum, a group of serrate bristles also formed on the metatarsus of the fourth pair of legs. This brush of bristles or *calamistrum* (see Fig. 1.8d) is used to tease out the cribellar silk. Until recently, it was thought that the cribellate type of spigots were found only on the cribellum but it is now known that similar spigots are found in some of the cribellate spiders (e.g., Neolanidae) on either the PL and PM spinnerets or both. These are called *paracribellar spigots* but the exact function of the silk produced by them is not yet clearly understood. Furthermore, in many of the True Spiders, the spigots of the cribellum have been lost, the only vestige remaining being a flat plate or lobe called a *colulus* (Fig. 1.23d).

Each spinneret is studded with many small nipples known as *spigots* (Figs 1.23e-h). Ducts run from each silk gland to a spigot on a particular spinneret. From these glands a viscous fluid flows and this is expressed from the tip of the spigot. When the spider attaches a droplet to the substrate and pulls on it, it is transformed by the realignment of the molecules into a silken thread. For

Fig 1.22 Ways in which spiders use silk. **a** *(top left):* Dolomedes minor *makes a silk tentlike nursery on the top of low shrubs. Inside, the eggsac and eventually the newly hatched spiderlings are protected.* **b** *(top centre): from its hiding place,* Badumna longinqua *keeps two or three legs on its web, dashing out quickly if it feels an insect struggling in its snare.* **c** *top right,* Argiope protensa *monitors its orbweb snare from the central hub. Note the spinnerets centred on the underside of its abdomen.* **d** *(bottom left):* Cyclosa *makes a prominent* stabilimentum *in the centre of its orb-web. This thick web structure is believed to strengthen the web and to act as a warning device to birds which might otherwise fly through the web.* **e** *(bottom right):* Clubiona sp. *seen inside her silk cocoon with newly hatched spiderlings. Similar cocoons are also used by many spiders during moulting.*

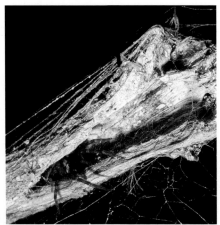

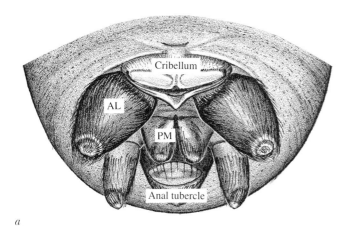
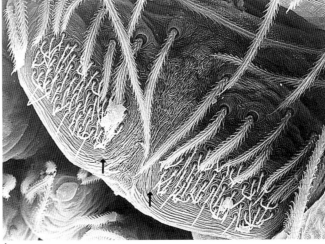
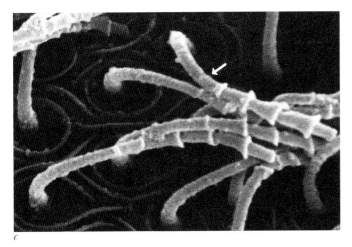
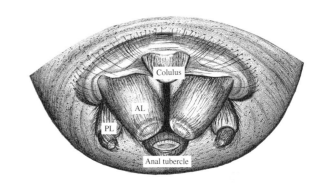
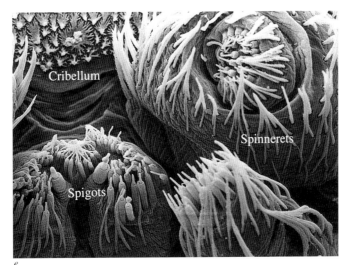
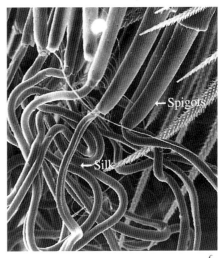

Fig 1.23 (from top left). **a** *Arrangement of three pairs of AL (anterior-lateral), PL (posterior-lateral) and PM (posterior-median) spinnerets in* Badumna longinqua. *The cribellum, with two spinning fields, is shown here above the spinnerets.* **b** *Micrograph of cribellum with spigots.* **c** *Highly magnified view of spigots showing lobulated nature.* **d** *Three pairs of spinnerets (PMs hidden by ALs) in* Cambridgea antipodiana *with non-functional lobe or colulus centred above.* **e** *AL, PL and PM spinnerets with spigots and section of cribellum in top left corner* (Notomatachia *sp.*). **f** *Distal portion of pyriform spigots showing silk being drawn from the spigots (*Hypodrassodes *sp.*).

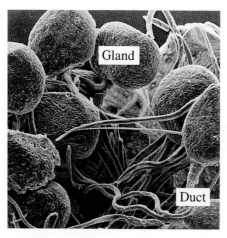
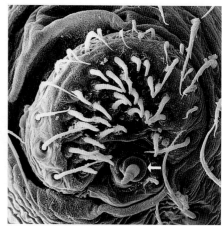

Fig 1.23g (right): *internal pyriform glands of* Myro *sp. (dissected out).* **h** *(far right) close-up of AL spinnerets of* Mimetus *sp. showing numerous pyriform spigots and a single large ampullate spigot in the lower middle area.*

example, silk drawn out in this way from the ampullate glands via the AL spinnerets forms the spider's dragline.

Many glands

A single spider may have hundreds of glands, each opening from a separate spigot. Up to nine different kinds of silk glands have been identified (Figs 1.24a–g), each producing silk with different properties for specific purposes.

Aciniform glands (Fig. 1.24a): Each aciniform gland is spherical with a short duct but numbers of them are usually aggregated in berry-like clumps. Their silk is exuded from spigots on the PM and PL spinnerets so as to produce a swathing band. These same glands are also used to produce silk for the outer layers of the eggsac as well as constructing the male sperm web.

Pyriform glands (Fig. 1.24b): Most are pear-shaped glands which may be also present in clumps but each gland is longer and narrower than an Aciniform gland. Linked to the AL spinnerets, their silk is used for attachment disks which spiders press on to the substrate every few centimetres as they move along. If they drop suddenly, their dragline is anchored by these disks.

Major Ampullate glands (Fig. 1.24c): These large glands, never numerous, are shaped like a cylindrical tube with the middle part enlarged. It is common for only one pair to be associated with the AL spinnerets of True Spiders and even this pair may be reduced to a single gland with a functionless nubbin representing the second. This gland produces dragline silk and generally provides the dry silk for making web frameworks. However, some of the more primitive groups such as the Gradungulidae, may have a large number of these glands, although the reason for this is not known.

Minor Ampullate glands (Fig. 1.24d): These smaller glands often differ in shape and secretory function from the major ampullate glands. They open from the AL spinnerets.

Cylindrical or **Tubular** glands (Fig. 1.24e): These long, cylindrical-shaped glands are more or less even in width and are sometimes convoluted. Found only in female spiders but missing in the Dysderidae and Salticidae, they are used to make the inner cocoon for the eggsac.

Aggregate glands (Fig. 1.24f): These are irregularly branched, lobed glands which bear cellular knobs on the duct. Their spigots are found on the PL spinnerets and are associated with the production of sticky silk for snares.

Flagelliform or **Lobed** glands (Fig. 1.24g): These glands have irregular lobes similar to the aggregate glands but without the knobs on the duct. They are associated with the PL spinnerets and are used by orb-weavers for making their sticky spiral webs. In the Theridiidae they produce the liquid swathing silk which these spiders fling over their prey with the aid of bristle combs on the fourth tarsi.

Many spiders also have a cribellum which produces a special kind of silk.

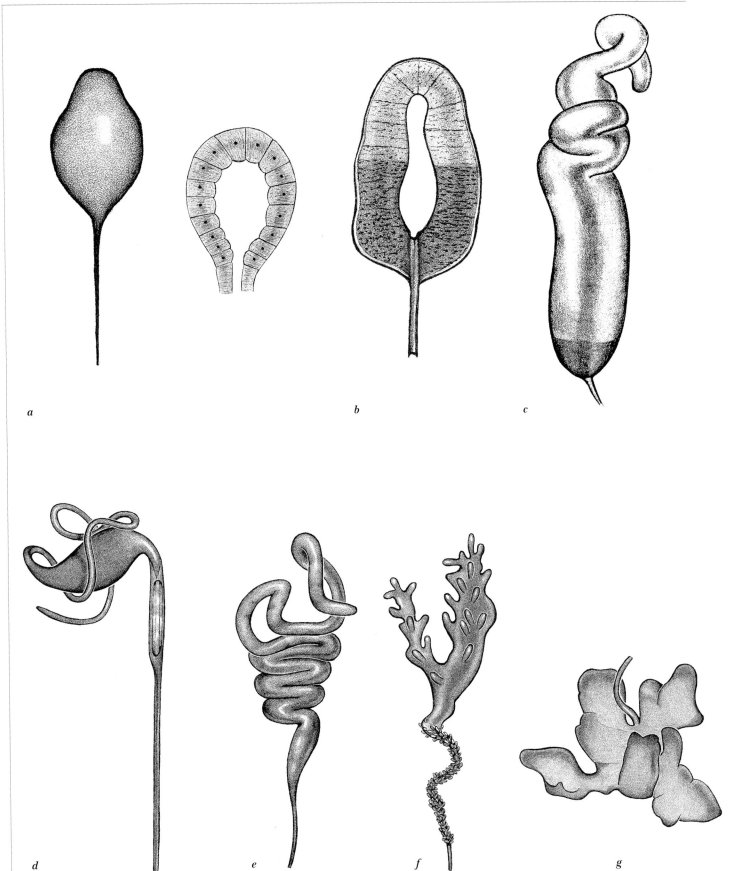

Fig 1.24 *Seven of the nine silk glands found in spiders. No spider possesses all nine but many have four or sometimes even five of them.* **a** *Aciniform gland.* **Inset**: *Section through aciniform gland.* **b** *Pyriform gland.* **c** *Major ampullate gland.* **d** *Minor ampullate gland.* **e** *Cylindrical or tubular gland.* **f** *Aggregate gland.* **g** *Flagelliform or lobed gland.*

Cribellar glands (not illustrated): These are small spherical glands which open through the spigots of the cribellum. They produce fine silk which is combed out from the spigots by the calamistrum and wrapped around a stronger axial thread emanating from posterior spinnerets thus forming the hackled entangling band so characteristic of the cribellate web.

Paracribellar glands (not illustrated): These small glands, varying in number, are near the cribellar ones and have similar secretory products. Spigots from these glands may open from the PM and PL spinnerets.

Silk thread formation

Slightly different silk exuded from spigots forms an extremely fine thread and it is only by the aggregation of very many fine threads from different glands that a suitably strong and versatile line is constructed (Fig. 1.25a, b). This aggregation takes place as silk is pulled out from hundreds of spigots so that the spider is able to spin its interwoven thread as it moves about.

Many spiders are described as producing 'sticky' silk, of which there are two kinds. In the first type spiders with cribellar glands as described above produce dry silk which consists of a double-stranded supporting filament surrounded by a tangled network of very fine silk fibrils. When this thread touches a surface with many tiny irregularities, the two 'stick' together. This stickiness works very much like Velcro, a tradename for two nylon tapes, one having tiny loops and the other tiny hooks which interlock and hence adhere to each other. Spiders which spin this sticky (retentive) silk are said to be *cribellate*. Spiders which do not produce this kind of cribellar silk are said to be *ecribellate*. The second type of sticky silk is produced by spiders with flagelliform glands. It is wet and viscous and very much like the sort of glues with which we are all familiar. Thus, they are 'sticky' in the same way.

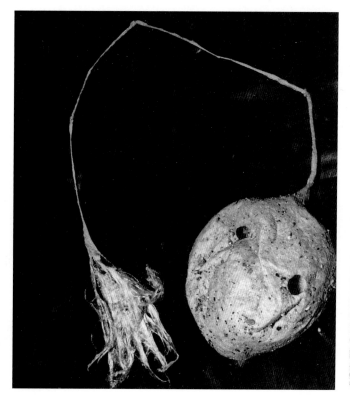
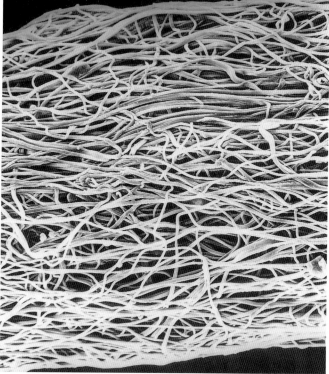

Fig 1.25a *(below left): eggsac of* Pianoa isolata *with long silk line from which it hangs.* **b** *(below right): micrograph of portion of this line showing the aggregation of many fine threads which make up the silk line.*

CHAPTER TWO
THE LIFE OF A SPIDER

The life cycle of a spider is quite unlike that of most insects. To begin with, fully formed but tiny spiders, called spiderlings, emerge from eggs held together by the binding silk of the eggsac. As they grow spiders shed their outer skin or *exoskeleton* some four to eight times. This process, called *moulting*, takes place every ten days or so, depending on temperature and food intake, and maturity is reached in about two to three months. The only change during this time, apart from size, is the development of their respective genitalia as described in chapter 1.

In temperate countries, such as New Zealand, mating and egg-laying takes place in the spring (August to September) or early summer (October to November) and may continue over the summer months. After mating, males and females go their separate ways. He will usually look for another mate; she will catch food – a great deal of it – to nourish her developing eggs. When the time comes, usually in two to three weeks, the female hunting spider builds herself a silken retreat in a hidden spot and lays her eggs there. The web spider will hang her eggsacs in or near the web, often camouflaging them in an attempt to deceive the ever-vigilant predators.

Much of a spider's behaviour is governed by *instinct* – that is, its genes not only determine its morphological and physiological make-up but also how it carries out its everyday tasks and responds to the environment. However, in spiders as in humans, 'practice makes perfect', and so we find that in certain instances spiders learn to do things better. For example, young jumping spiders seldom catch prey after their first stalk-and-jump but after three or four attempts they become efficient predators.

From egg to spider

The eggsac
The hazards spiderlings face during growth to maturity are such that to ensure their line persists most spiders need to produce hundreds or even thousands of eggs each season. However, some small spiders such as the minute Symphytognathidae and Anapidae may limit the number of eggs they lay to as few as three or four in each eggsac. Some spiders, such as the daddy-long-legs (*Pholcus phalangioides*), bind their eggs together with a few strands of silk and carry them below the body where they are held by the chelicerae and palps (Fig. 2.1). Most spiders, however, construct a distinctive eggsac which encloses the egg mass within a protective cover of one or two layers of special silk. While this cover forms a partially effective barrier against predation it also helps to maintain a constant level of humidity and provides protection from fungus infections. Even so, some wasps are able to pierce the silken cover and deposit their eggs so that when the grubs hatch they feed on the spider's eggs. Although an eggsac's shape and colour are often characteristic for a particular species of spider, features found in families such as wolf spiders tend to be much the same. In other families, such as the orbweb groups for example, eggsacs may vary considerably between genera (Fig. 2.2a, cf. Fig. 2.3).

Fig. 2.1 Lightly bound together with silk, the eggs of Pholcus phalangioides *(daddy-long-legs spiders) are carried by the mother with her chelicerae until they hatch.*

Hatching from the egg

Unlike some insects, spiders do not overwinter in the egg stage. Even in those groups which live in protected habitats and are active in winter, the normal annual life cycle is maintained. Once the spiderling has completed its development within the egg it must first break out from the thin, but strong, membrane in which it is enclosed. To assist with this, the first of its continual struggles for survival, it has a small transitory spine on the base of each palp. The new pre-nymphal spiderling is completely helpless (Fig. 2.2b), unable to feed or even spin silk and is dependent on the remnant of egg yolk within its body. For some days it will remain secure within the silken cover which earlier protected the eggs. At least one, and more commonly, two moults will be undergone before spiderlings are sufficiently developed to lead independent lives. So a week and often a considerably longer time will pass before they actually leave the eggsac.

Even the emergence from the eggsac is not necessarily a simple procedure. In some groups, a spiderling cuts a hole in the eggsac – surprisingly, usually only

Fig 2.2a (below left): After hiding her eggs within a tangle of silk alongside her orbweb, Eriophora pustulosa *remains on guard.*
b (below right): After 3 weeks or so the spiderlings hatch out and shortly afterwards moult. Here, the eggsac has been opened to show these pre-nymphal spiderlings and their shed skins.

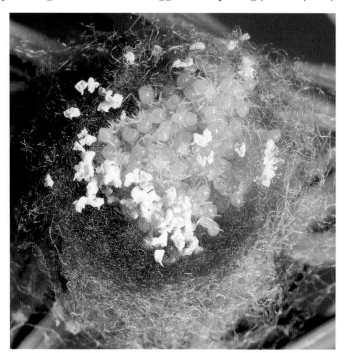

one hole is made, all spiders scrambling out through it one after the other. The process by which one spiderling is 'delegated' to cut this hole is a mystery. In spiders that have loosely woven eggsacs it seems that movement of the spiderlings alone is sufficient to break open the silken cover and provide an exit. In a number of groups the mother spider assists in opening the sac when the spiderlings are ready to emerge. Where the outer cover of silk is hard and parchment-like, as in the gradungulid *Pianoa isolata* (see Fig. 1.25a) and the orbweb spider *Celaenia* sp.(Fig. 2.3), another explanation is called for as spiderlings always escape through small holes in the suspended eggsac. It is possible that a fluid, perhaps released during those pre-emergence moults, accumulates and softens or even dissolves the silk, thus enabling the spiderlings to burst through.

Maternal behaviour

Eggsacs packed with nutrient-rich eggs make tasty morsels for hungry predators, including other spiders, so it is not surprising to find that many species adopt measures to safeguard them. These measures range from cryptic coloration of the eggsac, various disguises designed to conceal it (see Fig. 2.3), hard-to-find hiding places, and – most effectively – the physical presence of the female. Even after spiderlings have left the eggsac there are instances where they are still protected by the female spider. For example, baby wolf spiderlings clamber up and cling onto their mother's back for a week or so (Fig. 2.4), a habit which enhances their chances of survival. The young of many trapdoor spiders remain in the burrow with the female for considerable periods before setting out on a life of their own. It is very likely this also gives them the opportunity to share prey captured by the female.

In his 1958 book on spiders, W.S. Bristowe records a truly advanced form of parental care for one of the European cobweb spiders, *Theridion sisyphium*, where the female actually feeds the young with fluid from her mouth. Mouth feeding continues for several days and then the young begin sharing her prey. More recently, a group of French arachnologists found out how food is transferred from the female spider to her young in a sub-social species, *Coelotes terrestris*. The mother produces tiny trophic eggs which are immediately consumed by the recently hatched spiderlings. Similar behaviour has also been observed in a few other spider species as well as in some insects. Although it is possible that some New Zealand species behave in the same way, we have never seen this happen.

Ballooning

Once spiderlings have passed successfully through their very early moults and can live on their own, they disperse rapidly. While the more sedentary ones settle close to their birthplace, many of the web spiders take advantage of the wind to carry them far and wide. When these spiderlings 'come of age' and are ready to leave home, they clamber to the tip of a twig exposed to the breezes. Standing on tiptoe they let out a tuft of silk from their spinnerets until the pull

Fig 2.3 (below left): The eggs of Celaenia *are sealed within their covers of parchment-like silk, with one eggsac showing a hole through which the spiderlings have escaped. With their dark amber colouration and knobbly exteriors, they are easily mistaken for seed capsules.* Celaenia, *seen here hanging from its eggsacs, is often referred to as the 'bird-dropping' spider.*

Fig 2.4 (below centre): Baby wolf spiderlings perch precariously on their mother's back. Unless they clamber back quickly, those that fall off are left behind.

Fig. 2.5 (below right): Standing on tip-toes and raising its abdomen, a spiderling lets out silk as it prepares to balloon.

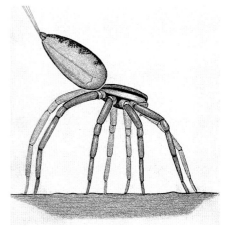

of the wind, or a thermal updraft, lifts them up and carries them through the air (Fig. 2.5). Other techniques involve dropping on a dragline, or allowing a web loop to be drawn out by the wind until it breaks and the spiderling is carried aloft. The term *ballooning* describes this habit and the tangled threads they float away on are called *gossamer*. In some countries in certain seasons, this gossamer comes to rest in such vast quantities that it is like a fall of snow. Such an event presupposes that large numbers of spiders have aggregated and eventually balloon together when particular climatic conditions trigger this aeronautical behaviour.

Gossamer flights

In many overseas countries gossamer showers are expected every year, but only rarely is a significant one reported in New Zealand. Most records are from the North Island, but we were lucky enough to witness one while we were driving near Dunedin one sunny morning in early summer, when a light wind was blowing out to sea. The site was near Brighton, just above the beach where the ground was covered in long grass. We came to a halt when we noticed that the grass was coated with a thick white layer which we suspected might be gossamer. To our surprise every clump of grass had two or three spiders clinging to the tips of the blades, each with a silken thread attached to the spinnerets and blowing up in the breeze. Every now and then as we watched, one let go from the blade of grass and floated gently away.

In thirty-odd years of observing spiders this was the one and only time we have seen such a sight. We collected some of the spiders to identify them. As noted on other occasions in the North Island, most were adult 'money spiders', those tiny black spiders which live mainly in grassy areas. All the money spiders found in New Zealand are introduced species (mainly from Europe) and, indeed, make up most of spiders reported in the gossamer showers overseas. The really interesting discovery, however, was that in this particular event some of these gossamer spiders were not introduced money spiders but immature native species of the mynoglenine group of linyphiid spiders (Fig. 2.6). Hence this was the first record indicating that our native spiders were involved in this kind of mass phenomenon. Although we have always suspected that many native spiderlings do move about by ballooning, this spectacular mass behaviour is certainly not typical. It seems more likely that those mynoglenine spiders which also live in grassy habitats just happened to get caught up with money spiders in this aerial uprising. Although ballooning is a common method by which spiders move from place to place, it has never been as prevalent in New Zealand as it is in other countries. Perhaps ballooning was not a very profitable venture when the country was covered with dense forest; also in such a narrow ocean-bound land, spiders blown out to sea would meet a watery death. Undoubtedly this happened to many of the spiders we watched at Brighton. Hence, selection pressures would favour those native spiders which did not participate in mass ballooning.

Fig 2.6 *Trailing their gossamer threads behind them, these tiny black spiders depend on fair winds and thermal updrafts to waft them aloft.*

Distribution

The present countrywide distribution of several orbweb species, including the common garden spider, *Eriophora pustulosa,* can only be accounted for if it is accepted that they have 'ballooned' from Australia in recent times. Of course, some of these migrants have not found conditions in New Zealand to their liking and hence do not survive. But they continue to turn up from time to time, after travelling across the Tasman. A noticeable feature of their arrival here is that they are frequently reported after bush fires in Australia when it seems they are swept skywards by the huge updrafts these fires generate. Nevertheless, *Eriophora pustulosa* has apparently found New Zealand more hospitable than its homeland, for it has become one of our most 'successful' introductions. It is now found all over the country except in virgin forest although it often appears around bush huts many miles from the road. Furthermore, this spider has managed to cross even wider stretches of ocean being now well established in most of the outlying islands including the subantarctic islands.

Being rather sedentary, trapdoor spiders often settle close to their birthplace so it is common to find colonies containing family groups of different ages separated from other groups with a similar age range (Fig. 2.7). These spiders often live in the same burrow all their lives, which may span ten years or more. Only the male spider, after reaching maturity, seeks new pastures in his quest for a mate. Many other web spiders rely on ballooning to disperse and in early summer the young spiders may be found at the tip of a twig exposed to the breezes. A glint of sunlight on the threads highlights them as they float past.

Moulting

Like all arthropods, spiders are enclosed in a firm outer skeleton and so, apart from the elasticity of the abdomen, they are not able simply to become larger as they get older. To grow, they must shed this hard outer skin at intervals and then expand rapidly before the new exposed skin has hardened. This process of shedding the skin is known as the moult or *ecdysis* and the period between ecdyses is known as the *instar*. It is characteristic of spiders that there is not a very great change in appearance between instars except for a sudden increase in the size of the cephalothorax and appendages. The method of moulting is much the same in all spiders. A few days before a moult takes place, the spider ceases to feed and becomes sluggish. At this time, some of the tunnel dwellers seal up the opening of their hole, many of the web dwellers suspend themselves up-side down from a web with the legs grouped together, while others may lie on their back or on their side.

Fig 2.7 *Five trapdoor spider tunnels of different-sized juvenile* Misgolas vellosa *(Idiopidae) in close proximity. Note the opened tunnel (lower left), a young spider lurking within and the larger, partly opened tunnel at the upper right, as well as scattered prey remains.*

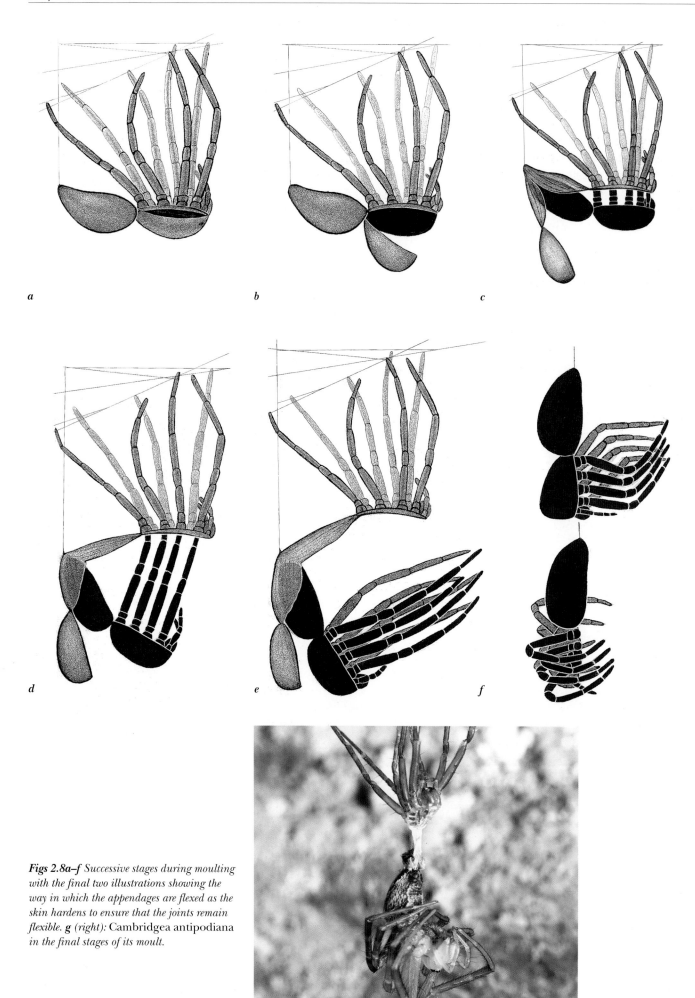

Figs 2.8a–f *Successive stages during moulting with the final two illustrations showing the way in which the appendages are flexed as the skin hardens to ensure that the joints remain flexible.* ***g*** *(right):* Cambridgea antipodiana *in the final stages of its moult.*

Apart from this variability, the methods used by the spider to shed the old skin are remarkably stereotyped (Figs 2.8a–g). First the skin around the carapace above the legs splits (a) and the carapace lifts off like a lid but remains attached to the pedicel (b). The skin of the abdomen then splits along the sides (c) and as the abdomen comes free, the legs, palps and chelicerae are drawn out in a series of rhythmical movements (d), rather like the fingers from a glove, so that eventually the spider is hanging from the tip of its abdomen (e). At this stage the spider is soft and defenceless and must remain in this position until the skin has hardened. However, it is not entirely inactive during this time because the limbs must be repeatedly flexed (f) or they will become stiff and distorted, and the spider will not be able to carry out its usual activities. In Fig. 2.8g we see *Cambridgea antipodiana*, having pulled itself clear of its old skin, hanging downwards and flexing its legs as it waits for the new skin to harden.

Developmental stages

The number of instars a spider undergoes to reach maturity is related to its ultimate size, not only with respect to the species but also between male and female of the same species especially where the male is much smaller than his mate. However, in general, four to eight moults occur. In some species, the male is able to take advantage of his earlier maturity, since he may be found closeted with the female in her retreat as she awaits her final moult. Often he is able to copulate with her while she is still soft and helpless. Amongst New Zealand spiders this is commonly seen in clubionids and some gnaphosids and is a common practice with the large jumping spider *Trite planiceps*. It is only after the final moult that the secondary sexual characters are visible. The epigynum and internal genitalia in the female and the palpal organ of the male are fully developed while the bright colours and ornamentation so often a characteristic of the adult spider are revealed (Fig. 2.9).

Regeneration of limbs

At any stage of its life, a spider may lose a leg – perhaps in a struggle with another spider, or in trying to escape from a predator, or even simply getting it caught in some way. Even a slight pull on the leg may result in it being discarded but the break is quickly sealed off by the constriction of muscles in the region and coagulation of blood over the gap so that there is little loss of body fluids. Under most circumstances the break occurs at the junction of the coxa and trochanter, a weak joint designed to give way easily thus allowing the spider a rapid escape if necessary. If the leg is ruptured other than at the weak joint, the spider will detach its own leg (*autotomy*) by tugging at it or even by anchoring it with silk and pulling it away (Fig. 2.10). Undoubtedly this ability to lose a leg without any immediate inconvenience is an excellent adaptation for survival when attacked by a predator, but this is by no means the complete story because all appendages, including the chelicerae and spinnerets, are capable of regeneration.

Much of our knowledge of limb regeneration in spiders comes from research by the eminent French araneologist Pièrre Bonnet. In a series of experiments, Bonnet even went so far as to demonstrate that after he had removed all of a spider's limbs they would all regenerate. However, this would obviously be of little use in nature because Bonnet had to feed the spider carefully by hand as it rested in a soft bed of cottonwool. More recently, scientists have been studying this phenomenon in some insects to try and understand the processes involved in remaking a particular part of the body.

This is what happens in spiders. Within the remaining coxal cavity of the lost leg, the new limb takes shape. Here it grows, coiled up in this tiny coxal segment until the next ecdysis. When the old skin is shed this new leg is unfolded and stretched out. For this to happen, however, the limb must be lost within the first part of the instar, and even so will not be as long as the original leg. At least three moults are required before the appendage reaches full size. The regen-

Fig 2.9 (top): Several species of jumping spider (Salticidae) live at high altitudes on the Remarkables, a mountain range in Central Otago, New Zealand. This undescribed female spider is inconspicuous from above as it hunts for prey amongst the lichen-covered rocks. However, on the ground she is easily recognisable to an all-black male by the luxuriant growth of orange hairs below her eyes. This feature only appears at maturity.

Fig 2.10 (above): This female Subantarctia trina (Orsolobidae) has lost two left front legs and a right front leg and is seen here sucking the fluids from a front leg. Retrieving nutritious fluids after such mishaps is common practice in spiders, more particularly when they have automised their own limbs.

eration of the unmodified female pedipalp presents no difficulties but, because of its complexity, the male palp requires at least three moults to become fully restored and functional. Limbs lost by adult spiders cannot usually be regenerated. However, in some spiders such as mygalomorphs or filistatids, mature females sometimes moult again, although it is not known whether this allows for the regeneration of a limb.

Silk and its uses

Silk is the substance which ramifies throughout the whole of a spider's life. No other group of animals has exploited the possibilities of this versatile substance to the same extent. Although some insect larvae, for example, use silk to make cocoons for pupation this ability is usually restricted to a single stage in their lifespan. As we saw in chapter 1, spiders produce up to nine different kinds of silk for a multiplicity of purposes. In addition, silk has three properties – tenacity, elasticity and the potential for recycling – which make it valuable for these various purposes. If we compare the strength of spider silk to a similar thickness of nylon thread, for example, we find that that they would both have to be about eighty kilometres long before breaking under their own weight. With respect to elasticity, though, the dragline is superior to a nylon thread for it can be stretched to an extra one-third of its length compared to about one-fifth for nylon. It has been shown by Colgin and Lewis that in the large orbweb spider, *Nephila clavipes*, the dragline silk has a tensile strength about nine-and-a-half times that of nylon and lycra of a similar denier. Moreover, the silk proteins of the web are conserved by certain orbweb spiders such as *Araneus diadematus* and *Arachnura* species which eat the old web before building another (Fig. 2.11). By the radioactive labelling of old webs, Peakall discovered that most of this silk was recycled by the spider into a new web within some 30–40 minutes. Other spiders, however, which do not recycle silk, use the same webs for a long time, merely cobbling together the parts that are broken by prey capture, wind and rain or marauding animals.

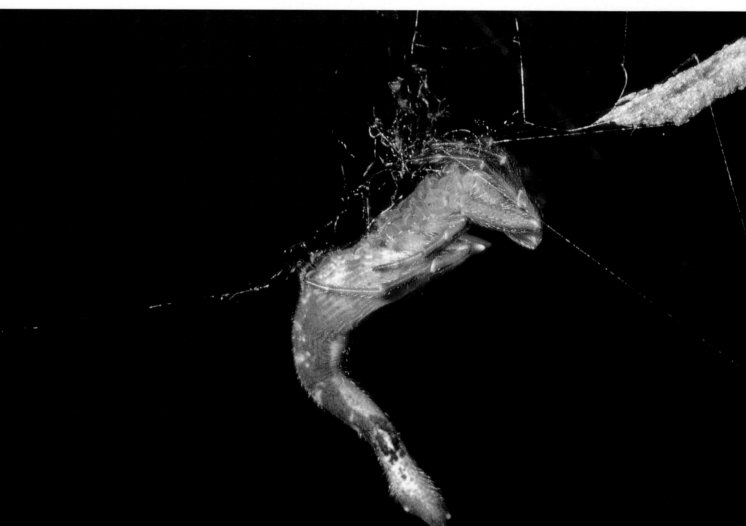

Fig 2.11 Arachnura *sp. (the tailed spider) hangs from a silk line as she rolls up her previous orbweb prior to ingesting it.*

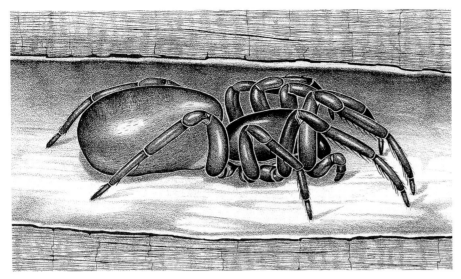

Fig. 2.12a (above left): Holes bored by insects make good homes for spiders such as the Ariadna *species seen here. Thickly lined with silk and 'wired' with radiating triplines, this spider waits inside for prey to make its presence felt.*

Fig 2.12b (above right): Tunnel of Ariadna septemcincta *opening from a tree trunk and showing triplines.*

Dragline versatility

During its first venture into the outside world the spiderling lays down the dragline that will lead it back home, enable it to traverse from twig to twig, and from which it will dangle if it loses its footing. As the spider moves it attaches the dragline at intervals so that there is never a great length of free thread. It can be seen, therefore, that the dragline has three functions, one as a safety line, two as a guide to take the spider back to its retreat and three, to bridge gaps. In later life, it may be used as a means to locate a mate. Perhaps this early habit of relying on a safety line lies at the root of the evolution of the multitude of ingenious snares which these spiders use to catch their prey today. A quick survey of their snares demonstrates how many of these seemingly divergent forms of web might have developed.

It is not difficult to see, from the early use of silk by spiders, how this original reliance on a dragline might have led to the elaboration of snares that we see today. For example, if we consider the webs of six-eyed spiders such as *Gippsicola* and *Ariadna* (Fig. 2.12a) which live in tunnels in twigs, we can see that it is a small evolutionary step from the use of a dragline to the development of a series of triplines (Fig. 2.12b) radiating out from the tunnel. These triplines warn the spider of nearby prey and in which direction it is moving. A further elaboration of these external threads results in the dense wide-mouthed extension constructed by the tunnelweb spider, *Porrhothele*. Looking at their counterparts in True Spiders we can see that a later development might be a sheetweb linked to a tube retreat such as that built by the grey house spider *Badumna longinqua* or the European house spider *Tegenaria domestica*. Once flying insects arrived on the scene, spiders took advantage of this new food supply by developing space webs – that is, snares which could be built amongst foliage or between different substrates and designed to trap a wide variety of insects. The vast range of aerial webs that are used today show us that for spiders, it was clearly a very profitable venture to migrate above the ground.

Snares and prey capture

Many kinds of snares

Aerial snares range from very basic to highly sophisticated. The simplest is a single thread – a trapline used by minute cobweb spiders of the genus *Phoroncidia*, common in forest and tussock grassland and occasionally seen in the garden. This trapline attracts insects which land on the web and are captured by the spider. Most theridiids, generally known as cobweb spiders, construct a small tangle web in which some of the threads are sticky. When a beetle, for example, is restrained by the web, the spider is alerted to its presence, and rushes to secure it further by throwing sticky silk over it. The native *Cambridgea* makes a large sheetweb above which is a maze of threads designed to intercept flying insects. Below the sheet the spider waits to seize whatever is knocked down. Orbwebs are familiar to most people, easily recognised by the characteristic sticky spiral structure, its silk often sprinkled with dewdrops. These webs come in many designs and sizes – some having widely spaced spirals, others with threads close together – but all effectively intercept particular kinds of flying insects. Once prey is entangled in a snare further steps may have to be taken before a meal is assured; hence we find that many spiders have developed ingenious methods to ensure that prey does not escape. Whereas theridiid spiders fling sticky swathing silk over their victims, many orbweb spiders wrap their prey in silk (Fig. 2.13) before actually biting it. Even more resourceful techniques may be employed. For example, in the orbweb family Theridiosomatidae, spiders hold their webs under tension by pulling on a thread from the centre, then releasing it when an insect flies into the web, thus causing it to roll up. Other spiders vibrate the web vigorously after contact to ensure that the prey is well entangled.

Web reduction

The evolutionary path to more efficient prey capture does not always follow from bigger and better silken snares. This is shown dramatically by some of the orbweb spiders which have abandoned formal snares for other rather astonishing techniques. For example, an Australian orbweb spider, *Dichrostichus* sp., does not construct a snare at all but instead pulls out and prepares a short length of silk with a sticky blob on the end which it holds by the second pair of legs. When a moth approaches, the thread is twirled around like a bolas (hence its common name) until it hits the moth and snares it by the blob. The catch is then hauled up like a fish at the end of a line. A similar spider has been found in America and South Africa, but not in New Zealand. However, the more distantly related orbweb spider, *Celaenia*, which manages to capture its prey without any use of silk, is quite common in New Zealand and Australia. In this genus, which contains several species (see Fig. 2.3), there are a number of glands on the front pair of legs and these release a fluid apparently mimicking an odour exuded by female moths to attract a male. The spider hangs down from a little silken platform at night, and as male moths approach it grabs them with the front two pairs of legs (Fig. 2.14). At first glance this may seem a precarious way to 'earn a living' but the sophistication of attracting potential prey by mimicking its mate's scent is undoubtedly very successful.

Fig 2.13 (below left): Eriophora pustulosa *holds her silk-wrapped catch as she prepares to suck up this predigested meal.*

Fig 2.14 (below right): Attracted by a scent emitted from Celaenia, *this large male moth has ventured too close and has been seized by the spider.*

Different tactics

Many hunting spiders which do not construct snares do use silk in some other way to capture their prey. The rapidly moving gnaphosids use a strong band of silk which streams out from a large number of spigots as they attack their victim and pin it down as they close in for the kill. The strange spitting spider, *Scytodes*, shoots out a stream of sticky fluid from its fangs which glues its prey to the ground. Other hunting spiders catch their prey 'on the hoof' and have abandoned the use of silken snares altogether. Many, such as the gradungulid *Pianoa isolata*, simply grab edible items as they cross its path; others, such as wolf spiders, alerted by vibration and/or vision, turn and seize prey from 4 to 6 cm away. Perhaps the most sophisticated predators are the jumping spiders, whose sharp eyesight enables them to catch their prey in daylight, and to hunt their victims in the manner of a cat catching a mouse.

Courtship and mating

Of all the extraordinary behaviour patterns which make spiders so fascinating, those associated with mating are certainly the most unusual. As we explained in chapter 1, the male accomplishes the transfer of the sperm package to the female with special structures on his palps. This procedure is usually referred to as the male 'applying' his palps. Before he can do this, however, both spiders will engage in a series of species-specific actions.

Messages and meanings

The coming together of two carnivorous creatures can be hazardous at times and so it would be with male and female spiders if some kinds of recognition signals were not used (Fig. 2.15a). It is always the male who searches out the female and the urge to mate is presumed to overcome his predatory instincts for food. Nevertheless, he himself would be in constant danger if he could not in some way persuade his prospective mate that the time was opportune for mating rather than eating. Male web spiders transmit their messages with visual or chemical signals, by plucking at silk threads, by vibrating the whole body, or even using a special stridulatory structure to send information that the female will receive and recognise. His approach is always cautious until he is near enough to touch her legs or body. Mating may take place on the spot or the male may spin a special mating thread onto which he coaxes the female. Whatever the circumstances, both spiders take up the mating position characteristic of that species (Fig. 2.15b). Once the preliminary approaches have been accomplished, the female remains passive as the mating proceeds and, after its completion, the male is usually able to retreat safely.

While many of the poorer-sighted hunters such as the clubionids and thomisids still rely to a limited extent on vibration, most recognition is achieved through touch and taste and smell. As we examine courtship in the lycosids and pisaurids, and then in the longsighted salticids, it becomes evident that recognition is increasingly dependent on sight. Among those spiders with keener eyesight, visual recognition plays a greater part in the preliminary courtship, and so jumping spiders have much more conspicuous rituals consisting of adornments, postures, body movements and leg actions. The pioneers studying this more elaborate courtship concluded that it would enable the female to choose the most vigorous and ornate performer as her mate. By selection, this would result in stronger stock and the development of even more complex displays. For many years these suggestions were not greatly favoured as studies tended to show that the real functions are recognition and suppression of the feeding instinct followed by stimulation of the female. These processes ensure that her reactions during mating are coordinated with the male's endeavours. Today, as so often happens, we are beginning to realise that the early suggestions are also valid – that natural and sexual selection work in tandem to produce the fittest individuals.

Fig 2.15a *(top): A small Australian redback spider male (L.* hasselti*) carefully advances towards his would-be mate. His courtship is a lengthy affair as he plucks and twangs and drums on the silk mating threads in which she hangs. His long legs probe the web cautiously as he gets closer.*

Fig 2.15b *(above): Here,* Aotearoa magna *(*Mecysmaucheniidae*) spiders mate in the head-to-head stance shown in Fig 2.16. (Photograph by Frances Murphy)*

Mating stances

Although the end result leads to copulation for all spiders, the stance which spiders adopt during the actual insemination varies considerably in different groups (Figs 2.16a–f). It is convenient to separate the various stances into six types although numerous variants are known:

Stance (a) The two sexes face in opposite directions with the dorsal surface of the carapace of the male against the ventral surface of the female. Web builders such as the linyphiids which use this stance are usually both suspended upside-down with the male above. In those ground dwelling spiders which also use this stance, the male stands with his body at an angle to the female.

Stance (b) The male and female are venter to venter facing each other, often with one or both partly upright. The male may insert one palp, or both palps at the same time, into the female genital opening. This stance is used by *Scytodes* and is also typical of mygalomorphs and many families of six-eyed spiders.

Stance (c) In this position the male and female are facing in opposite directions after the male has climbed on top of the female to reach down with his palps beneath her abdomen. This is the mating posture commonly found in ground-hunters such as the wolf and jumping spiders.

Stance (d) In Thomisidae, illustrated by *Xysticus*, the two sexes are venter to venter and facing in the same direction. In some species, such as *Diaea*, the female is tied to the substrate with silk before mating. In other crab spiders, the male creeps over the body of the female and forces himself beneath her.

Stance (e) The two sexes take up a position venter to venter but facing in opposite directions. This stance, which allows mating to take place in confined spaces, has been recorded for the clubionid genus *Cheiracanthium*.

Stance (f) In this position, adopted by most of the web-building araneids and theridiids, the male is above the female but they are venter to venter and facing in the same direction. In these families, males are often very much smaller than females.

Coupling

As far as we know all entelegyne spiders apply one palp at a time, alternately, during mating, while it is characteristic of most haplogyne families for both

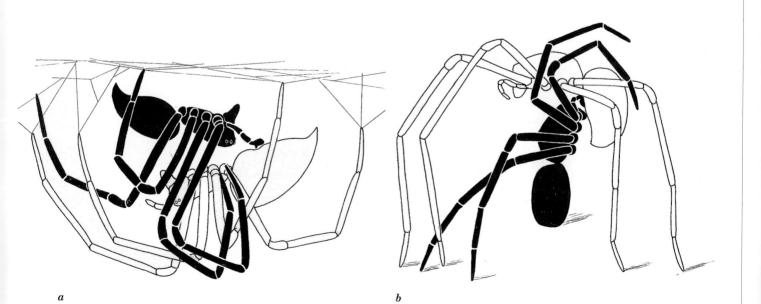

a   b

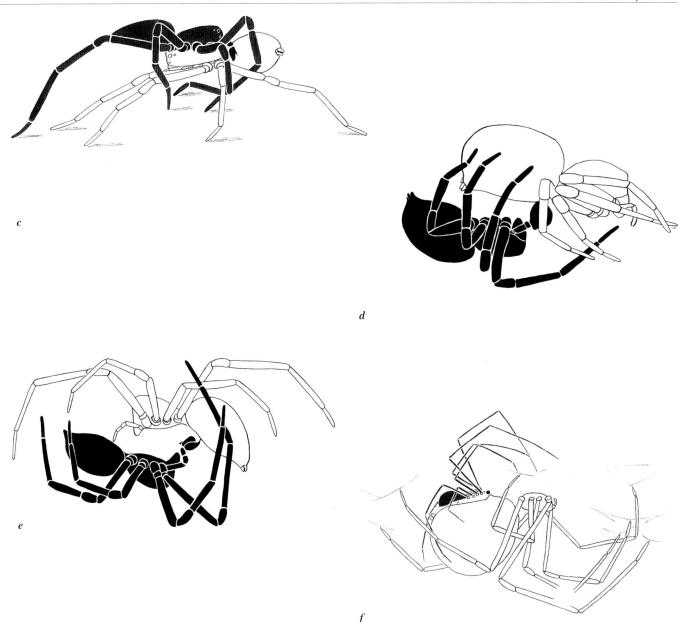

Fig 2.16 (left and above): *Characteristic mating stances used during the transfer of sperm from males (shown in black) to females.* **a** *A position adopted by the Linyphiidae (e.g.,* Ostearius melanopygius*) in which spiders are suspended from silk and face in the opposite direction.*
b *The stance seen here in* Scytodes thoracica *(Scytodidae) is also typical of mygalomorphs and many of the six-eyed spiders.* **c** *Hunting spiders such as Agelenidae, some Clubionidae, Drassidae and Salticidae, mate in the position shown here by* Lycosa rabida *(Lycosidae).* **d** *In the Thomisidae (e.g.,* Xysticus*), spiders are venter to venter and face in the same direction.*
e *Recorded for* Cheiracanthium *and most other Clubionidae, the female venter and male dorsum are together and spiders face in opposite directions.* **f** *Found in most of the Araneidae and Theridiidae, this position is appropriate for web-based matings. The stance shown here is typical of* Latrodectus katipo.

palps to be applied simultaneously. Until recently it was thought this was a definitive behavioural character which separated these two groups of spiders. However, like many of our early conclusions, recent discoveries show that this distinction is not strictly correct. We now know that spiders of primitive haplogyne families such as the Hypochilidae and Mecysmaucheniidae apply each palp separately as do other haplogyne spiders belonging to predominantly entelegyne families. Nevertheless as a general rule this differentiation is useful.

In some spiders the actual copulation may be over in a few seconds while in others it may last for an hour or more during which the embolus of the palpal organ is repeatedly inserted. In his excellent book *American Spiders* (1979), Willis Gertsch describes the action of the palp as follows:

The palpus may be scraped across the epigynum until a spur on the tibia, on the tarsus, or on the bulb itself, becomes fixed in a particular groove. Once firmly anchored in this starting point, the palpus swings to assume a position that, with the aid of ridges, grooves, and other processes on the epigynum corresponding to its own outlines makes it possible to guide the embolus to exactly the right point to enter the orifice. At this stage, the bulb of the palpus is still largely in its resting position, lying folded in the cup of the tarsus and preliminary contacts serve to hold it firmly in place. Most spiders have at the base of the bulb various thin pouches or haematodocha that swell up with the influx of blood until they attain enormous size. This distension causes the entire bulb to turn on its axis, which action forces the embolus into the appropriate opening.

Once sperm have been deposited within either the seminal receptacles or the oviduct of the female, they may remain viable for a long time. It is common for spiders in captivity to keep producing fertile eggs over the entire summer period and we have had female *Steatoda* which were still producing fertile eggs after two years in captivity.

To conclude this brief survey of the life of a spider, we must accept that the most important task of the male has been achieved with the fertilisation of the female and after that he will very shortly die. The female's main responsibility is just beginning because she must nourish the eggs and fashion the eggsacs which will perpetuate the species. At least one pair from the hundreds or even thousands of her progeny must eventually reach maturity to ensure that the population is maintained.

CHAPTER THREE
SPIDER RELATIVES

Spiders and their relatives belong to the large class of animals known as the Arachnida, which have four pairs of legs and a body divided into two portions – the cephalothorax and the abdomen. Insects, on the other hand, have bodies divided into three parts, and only three pairs of legs, and such distinctive features are easily apparent to the naked eye. A closer look at an arachnid with a magnifying glass will also reveal that the eyes, when present, are always simple and lack the multifaceted lenses that characterise the two main insect eyes. Moreover, unlike many of the insects, arachnid mouthparts are never complex.

Sixteen different groups or orders of Arachnida are generally recognised but six are known only as fossils. Some of these date back at least 200 million years to the Carboniferous Period which means that arachnids are among the oldest arthropods. These orders evolved over many millions of years at the same time as the insects which constitute the bulk of their prey. Of ten living orders only four, apart from spiders (Araneae), are found in New Zealand. These are the Acari or mites; Chelonethi or false scorpions; Opiliones or harvestmen and Palpigradi. Most of the other orders, such as scorpions (Fig.3.1), whip scorpions (Fig.3.2), tail-less whip scorpions (Fig. 3.3) and Ricinulei (Fig.3.4) which are not found in New Zealand, are more at home in tropical climates. Despite the occasional introduction of some of these exotic creatures including spiders such as the banana spider and various tarantulas, very few of them, except those spiders and mites from more temperate countries, are able to become established.

*Fig 3.1 (top left): **Scorpions (Scorpiones)**. Scorpions, well-known denizens of hot and arid deserts, are actually found in a wide range of habitats and in most tropical and semitropical countries. They occur in Australia but not in New Zealand.*

*Fig 3.2 (top right): **Whip scorpions (Uropygi)**. These creatures are often called vinegaroons because they squirt acetic acid from the long flexible tail used for defence and prey capture. They feed largely on insects. Eggs are carried in a sac under the abdomen, and after the young hatch they ride on the mother's back for a while.*

*Fig 3.3 (bottom left): **Tail-less whip scorpions (Amblypygi)**. As in vinegaroons, eggs and young are carried by the female. They do not occur in New Zealand, being found only in warmer climates. Neither of these two whip-scorpion groups have a poisonous sting but subdue their prey with strongly spined pedipalps.*

*Fig 3.4 (bottom right): **Ricinuleids (Ricinulei)**. These hard-bodied, tick-like arachnids were once little known, but in recent years the extensive exploration of caves and forest litter in tropical America has led to an increase in the number of species and revealed some aspects of their life. Apparently only a single egg is carried by the female below the front plate and held there by the palps. The species illustrated lives in a cave in Mexico.*

***Fig 3.5** (clockwise from top left):* **Mites (Acari).** *Mites are perhaps the most abundant and widespread of arachnids. New Zealand is home to some thousands of native species yet relatively few are named. The species illustrated here give some idea of their diversity.*

a *This photograph shows a tick on the feathers of a red-billed gull. Ticks are parasites of vertebrate animals.*

b *Fast moving eupodid mites are commonly found in decaying wood on the forest floor. They prey on other small invertebrates living in this habitat.*

c d e *Three strikingly different trombidiid mites common in forest leaf litter.*

f *Collection of mites from leaf litter.*

Harvestmen, mites and false scorpions are abundant in New Zealand and, together with spiders, are represented by some thousands of species. Moreover, various minute, blind, litter- and soil-dwelling arachnids belonging to the Palpigradi have been recorded from widely separate localities and, although found only rarely, it appears this group is well established in this country. One species is new but there is still some doubt as to whether the others are native or were introduced with other soil animals from Europe many years ago.

Mites: Order Acari

Mites by the million

Mites are ubiquitous and have been found in practically all habitats where animals are found. While the majority are land dwellers, some are found living in streams and lakes and many live in the sea. Even the frozen wastes of Antarctica have their fauna and in New Zealand mites are probably the most plentiful group of animals, both in numbers of species and numbers of individuals (Figs 3.5a–e).

Understandably, the best known are those which directly affect humans – the ticks, chiggers, red spider mites, cheese mites and gall mites – some of which transmit disease while others attack crops and produce. However, by far the most abundant are those free-living mites which swarm in the upper layers of the soil performing an essential role in the breakdown of organic material. Like spiders, mites are fluid feeders and liquefy their food with enzymes. However, they vary considerably in their feeding habits, a few being generalists but most being specialists. Among them are carnivores, herbivores, omnivores, fungivores, saprophages and gramnovores. A mere handful of leaf litter from the forest floor supports many hundreds of individuals and a bewildering variety of forms (Fig. 3.5f).

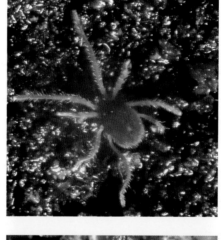

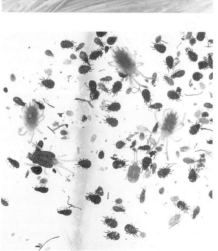

Body features

As a group, the mites are characterised by having the mouthparts, chelicerae and pedipalps more or less distinctly separated from the rest of the body as a false head or *gnathosoma*. Because the remaining parts of the body are fused into one piece, there is no pedicel as in spiders. The abdomen is smooth and not divided into segments by regular transverse grooves, hence they may be distinguished quickly from the mite-like harvestmen of the suborder Cyphophthalmi. Note that many mites do have a number of separate hard plates on their bodies but this is not true segmentation.

Separating mites into suborders depends mainly on the number and position of openings in the respiratory tracheae. On this basis six suborders are accepted and most are found in New Zealand. As expected with such a large and diverse group, the life history is varied, but in general eggs hatch into six-legged larvae which are similar to the adult except that the fourth pair of legs has not yet developed. From this larval stage they pass through a series of eight-legged nymphal forms until they reach maturity. While many of the nymphal forms are very similar to the adult, differing mainly in the lack of a functional reproductive system and associated structures, the beetle mites which dominate the soil fauna may be markedly different during the nymphal stages.

False Scorpions: Order Chelonethi

What are false scorpions?

False scorpions, as their name suggests, seem at first glance to be baby scorpions which have not yet grown a stinging tail (Fig. 3.6a). Often these fascinating little animals are sent to us in the mistaken belief that they are baby scorpions but surprisingly, despite their appearance, they are not even closely related to true scorpions. These creatures are never very large, the body length of our largest species being only about 5–6 mm. The small and inconspicuous chelicerae have a fixed as well as a moveable finger which form nippers in the same way as the palps. The palps, however, are relatively large and usually much longer than the body of the animal. Eyes play only a small part in the life of these animals; indeed, many are blind, but some have one or two pairs of simple eyes near the margins of the carapace.

As well as being small, false scorpions are retiring and so, although common, they are rarely seen. Often they can be found on foliage or under stones or by

Fig 3.6a–c (left and following pages): **False Scorpions (Chelonethi).**

a *A false scorpion with large palps and conspicuous nippers. Its body length is about 5 mm. Note that it does not have a tail, hence the name 'false' scorpion.*

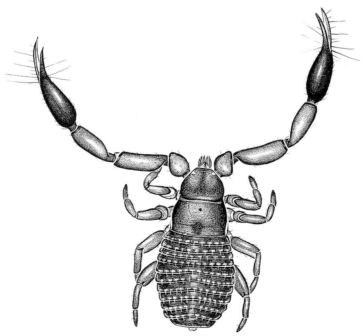

Fig 3.6b *Two silken retreats, on the underside of a rock. One of these reveals a female* Opsochernes carbophilus *with her eggs held in the membraneous sac beneath her abdomen. The other retreat is empty.*

peeling away the loose bark of trees, but the surest way to find some is to sift leaf litter from the forest floor into a flat white dish and wait for them to move. However, they are not restricted to the forest and many species are found in tussock grassland and subalpine bush or even under stones along the seashore (Fig. 3.6b). Their prey is always caught alive and, although they undoubtedly eat many of the smaller forms of life which abound where they live, they seem to prefer those minute, wingless insects – the springtails. Hunting methods are much the same in all species and although, as we mention later, they are capable of producing silk, this is not used in the capture of prey.

Catching prey

If you watch a false scorpion, you will see that it moves along slowly with its relatively large pedipalps probing ahead. As soon as something edible touches the fine hairs at the tips (see Fig. 3.6a) it is seized by the nippers. If, however, the prey is too big, the pedipalps are rapidly withdrawn and the false scorpion retreats backwards with a peculiar scuttling movement that is so characteristic of them. However, once a suitable insect is grasped, it is rapidly killed by venom oozing from one or both nippers which then pass it to the chelicerae. Here it is torn up and, simultaneously, digestive fluids flow from special glands in the cheliceral fingers and liquefy the prey's tissues so that it can be sucked up through the mouth into the stomach.

The use of silk

Other than capturing prey, the chelicerae have another function – they produce silk. Amongst arachnids, only spiders, mites and false scorpions produce silk and this plays an essential part in the life of each. In false scorpions silk comes from glands which open from rows of variously shaped teeth forming a peculiar comb-like structure called the serrula, attached to both fingers. Unlike spiders, which use silk in all phases of their lives, false scorpions use it sparingly for a few limited purposes. Principally it is used to fashion a retreat within which the animal may safely pass through those defenceless stages of its life – during moults, laying of eggs, the early stages of development of the young, and for hibernation.

Construction of domes

Retreats are dome-shaped and, because they are usually covered with small pieces of nearby debris, they blend with the background and are difficult to see. To make a retreat the false scorpion uses a method best compared with traditional igloo construction. First, a collection of fragments is gathered in the chelicerae and laid out in a circle. Then a layer of silk is spun over them to bind them together. As more debris and silk are added the wall is built higher, but the diameter is reduced so that, as the dome begins to take shape, the opening gets progressively smaller and smaller until it is finally closed at the top with the false scorpion securely sealed within.

We discovered that domes made by a species living under stones in Central Otago (South Island) were lined with a clean white sheet of silk despite the outer covering being made of small fragments of earth and debris. There were up to twelve of these little domes dotting the underside of every stone we examined. Others found on the underside of dead leaves lying on the forest floor were covered with fragments of leaf-mould but the ones we photographed (see Fig. 3.6b) from under stones at the beach were constructed with silk alone. Domes containing pregnant females are built from early summer (November) onwards and the female does not leave her dome until the young nymphs have moulted once or twice. At other times of the year, domes house only single immature animals undergoing a moult, although during the winter months some may contain hibernating adults.

Perhaps before we describe how the female guards and nourishes her young we should outline some of the bizarre events which take place earlier, during the courting period, and the method by which she is fertilised. Only fragmentary information is available on the behaviour of our native species, but what little we have seen shows that the extraordinarily detailed observations of these animals in other countries – particularly those by Max Vachon, formerly of the National Museum of Natural History in Paris – apply in greater part to our own fauna.

Courtship, mating and egg development

As in spiders, courtship is initiated by the male who, as he slowly approaches the female, vibrates his body and moves his pedipalps backwards, forwards and upwards in a series of regular postures. At the same time some species extrude a pair of long horns from underneath the abdomen, and these extend forward like fingers. During the male's activity, the female, if receptive, remains passive. When the male is close to her, he first lowers his body to the ground and then, as he lifts it up, a slender rod of viscous material is drawn from the reproductive opening under the abdomen. This quickly hardens into an erect rod firmly fixed to the ground and on its tip the male now deposits a globule of seminal fluid containing sperm. This is known as a *spermatophore*. The male must then ensure that the sperm globule is deposited within the egg duct of the female.

To achieve this, various methods are used, each characteristic of the species involved. In some species, the male, upon backing away from the spermatophore, vibrates his pedipalps, a signal that induces the female to advance slowly until she is straddling the rod. Moving towards her, the male grasps the base of her pedipalps with his nippers and, shaking her vigorously, causes the tip of the rod to be inserted in her genital opening. The sperm package is thereby deposited in the female's oviduct, there to await the passage of eggs. In other species the male moves directly towards the female after depositing the spermatophore. Then, grasping her pedipalps, he draws her forwards until the rod is inserted. Other species may short-circuit the process by grasping the female's pedipalps during the initial courtship as well as during impregnation.

As her eggs begin to develop the female constructs a retreat and seals herself inside. Before the eggs are actually laid, however, advance preparations are made. From special glands in the egg duct a fluid is excreted and this bulges out from the female's genital opening. This hardens to form a small sac or an incubation chamber into which the eggs will be laid, each fertilised as it passes down the oviduct. Very soon the eggs hatch, but the larval stage shows no resemblance to

Fig 3.6c Thelassochernes pallipes *abandons her retreat while her young are still in the early nymphal stage. She continues her normal hunting activities while the nymphs remain in the sac and feed on the fluid she exudes.*

the adult, being little more than a small white bag with one pair of appendages, the pedipalps, plus a short muscular beak.

But the most astonishing part of this story is yet to be told.

How the nymphs are nourished

At hatching, these helpless larvae represent a much earlier developmental stage than any other arachnid. Moreover, as they have no internal store of yolk to see them through to the next moult, they are completely dependent on the mother for nourishment. This she provides in a most unusual way. Her ovaries break down and form a nourishing fluid which pours out to fill the incubation chamber. Here it is sucked up through the muscular beaks of the larvae. A few days later, almost as if this method is too slow, a sudden muscular contraction of the mother's abdomen forces a vast amount of this food into the larvae so that in a few seconds their bodies swell to as much as three times their previous size. The incubation chamber also expands with the result that the female's abdomen, flattened by the transfer of food to the young, is pushed high in the air (see Fig. 3.6b). Over the next few weeks this enormous meal will be absorbed and the typical form of the false scorpion will gradually develop. However, after the nymphs have moulted once or twice the female, still carrying them beneath her abdomen (Fig. 3.6c), leaves her silken dome and resumes her normal prey-catching activities. When the growing nymphs reach a stage recognisable as false scorpions they burst out of the incubation sac and fend for themselves.

Harvestmen: Order Opiliones

Strange creatures

Of all the arachnids, harvestmen epitomise the unusual and grotesque, but they are in fact the most harmless of creatures and one of the few arachnids which make no use of poison. Curiously enough, although native harvestmen are numerous and widely spread, the first one people normally encounter in this country is the common brown European harvestman *Phalangium opilio* (Fig. 3.7a). This harvestman is seen running about in the garden and in open country throughout the summer months and is the only one found from one end of the country to the other. Because this species is so numerous in the fields in England during harvest time, the whole group has become known as harvestmen. To see this

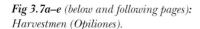

Fig 3.7a–e (below and following pages): Harvestmen (Opiliones).

a The European Harvestman, Phalangium opilio, *is common in gardens and fields.*

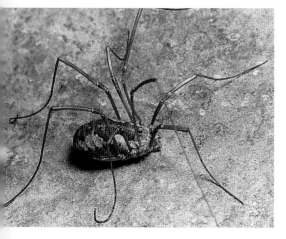

compact oval body racing along on its eight tremendously long and slender legs is an astonishing spectacle in itself, but actually this species is one of the most modest in a group which specialises in seemingly useless ornamentations.

As well as the common name of harvestmen, Opiliones are sometimes called daddy-long-legs – a name more appropriately used for long-legged craneflies. Harvestmen are distinguished from spiders by the absence of silk-producing organs and poison glands, but the most obvious feature is that their compact body is not divided into two parts as in spiders (Fig. 3.7b). Some harvestmen are mite-like but these species always have grooves across the abdomen separating the segments whereas in mites the abdomen is never segmented. While a few harvestmen are blind, most have a single pair of simple eyes spaced on each side of a raised eyemound. In a few groups, however, the eyemound is absent and the eyes are flush with the carapace. The chelicerae are alway *chelate*, which means that there is a fixed and a moveable finger at the end of the second segment. The pedipalps are leglike but in some groups they are large and swollen and armed with numerous stout spines.

Fig 3.7b Rakaia pauli, *a mite-like harvestman. Note the single compact body with transverse grooves marking segmentation. (See also Fig 3.7e.)*

Feeding habits

Scavenger is an appropriate word to describe the harvestmen's general feeding habits, as most will eat any dead invertebrate they come across. Many also prey upon small insects, the long-legged forms grasping them with their cheliceral nippers and the short-legged ones with their formidable spined pedipalps (Fig. 3.7c). Prey is detected, not with the harvestman's eyes, but by the tips of the second pair of legs which are always probing ahead and to the side as the creature moves along. In captivity harvestmen thrive on an extraordinary range of food, from meat and dead insects to bread, fat and milk. Since they do not have poison glands they might be considered as easy prey for their more formidably armed neighbours, but after a short encounter most of these would-be predators back away. This is because they are repelled by a fluid which exudes from a pore on each side of the harvestman's carapace. This fluid ensures that even the most pugnacious of spiders leave them alone. Harvestmen are mainly nocturnal creatures which hide in the dark during daylight and, because they easily dehydrate and die, they choose places where the humidity is high. For this reason they are mainly forest dwellers and to find them during the day you need to search the leaf litter, inside decaying logs and under loose stones and logs on

Fig 3.7c Nuncia *sp. (Triaenonychidae). These brown smooth-bodied harvestmen are the commonest forest group. The strong spined pedipalps are used in prey capture.*

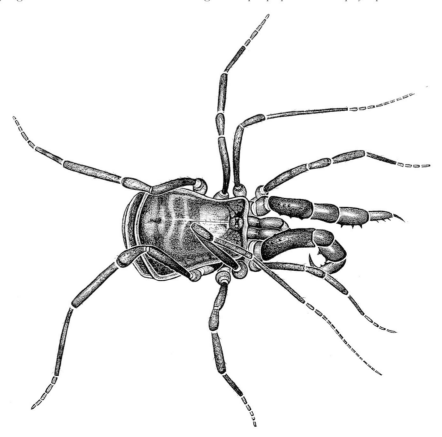

Fig 3.7d The grotesque, long-legged harvestman (suborder Palpatores) are a common group thoughout forests where they frequent low shrubs and ferns. Male Megalopsalis *can be recognised by their huge chelicerae seen here projecting in front of the body and bent beneath revealing the nippers at the ends.*

Fig 3.7e Blind mite-like harvestmen such as Rakaia *live in forest leaf litter. Glands, which excrete a noxious substance used in defence, open out through two mounds, one of which is seen here at the front of the carapace.*

the forest floor. A few live outside forests in tussock country and above the bushline under stones, while some have adapted to a life spent entirely in the darkness of caves.

Reproductive activity

Unlike many other arachnids our native harvestmen are most undemonstrative before mating and no true prenuptial behaviour has been recorded. Mating in our long-legged harvestmen has never been observed, which is a pity because the extraordinary enlargement of the male chelicerae in most species (Fig. 3.7d) must surely have some significance during this procedure. Although the behaviour of the introduced *Phalangium opilio* has been recorded this involves little more than the male assuming a standing position in front of the female and extending its long intromittent organ forward to enter the genital opening on the under surface of the female abdomen. Our short-legged harvestmen follow a similar procedure but as the intromittent organ is much shorter the male usually grasps the palps of the female with his palps and lifts both of their bodies up so that they are quite close together when the intromittent organ is extended.

Mating behaviour in the mite-like cyphophthalmi is completely unknown, but studies of their internal reproductive systems suggest that it must be quite different from that of other harvestmen. Cytological investigations reveal that the male testes produce a series of minute spheres, each sealed with an outer layer of flattened cells. Each sphere is filled with viscous material which appears to come from the protoplasm of the outer cell layer, with sperm being distributed just under this outer layer. The fine structure of these compact bundles suggests strongly that they are spermatophores, or sperm carriers, which, unlike the procedure in the two other harvestmen groups, are transferred into the female's oviduct by some indirect means in much the same way that sperm is transferred in a number of other arachnids, such as the false scorpions or even the spiders.

After fertilisation, eggs are laid via an ovipositor which, when not in use, is sheathed inside the oviduct. Ovipositors are very long in both the long-legged and the short-legged harvestmen but short and compact in the cyphophthalmi. Mostly, eggs are laid here and there around where they live, in leaf litter, rotting logs and even in the soil, and usually take about three weeks to hatch. The young are colourless, differ little in shape from the adults and are able to fend for themselves within a short period. As mentioned below, some short-legged species lay all their eggs in one place and guard them until they hatch – but this behaviour is not common.

The harvestmen are divided into three suborders which are easily recognised:

1. Cyphophthalmi – mite-like harvestmen
2. Laniatores – short-legged harvestmen
3. Palpatores – long-legged harvestmen

Cyphophthalmi

These are all small and the largest species in New Zealand does not exceed 5 mm in length. These drab, dark-brown harvestmen live a hidden life amongst leaf litter and moss on the forest floor where they probably feed on small insects such as springtails. All New Zealand species are blind and the mounds on each side of the carapace, which might be mistaken for eyes, are actually outlets for stink glands (Fig. 3.7e). The sexes are alike except for two striking modifications in the male. The end segments of the abdomen are depressed in the middle so that each side shows up as a hump, and the tarsi of the fourth legs are provided with a stout spur from which a substance is exuded. This is produced by a number of special glands shaped like a bunch of grapes and located within the tarsal segment. The exact function of these features is not known but perhaps they have a role in mating behaviour. The shape of the tarsal spurs and abdominal modifications differ in each species and so are important characters for defining them. The two New Zealand genera are easily separated because the tarsi of the fourth legs in males are subdivided in *Neopurcellia* but remain as a single segment in the more widespread genus *Rakaia*.

Laniatores

The best known harvestmen in New Zealand, about 160 species of Laniatores are described while others await description and naming. They are renowned for the profusion of seemingly useless spines, knobs and pustules and we wonder about the value of these excrescences. One solution springs to mind, however, if you are looking for these animals in the habitat they frequent. They are extraordinarily hard to see against a background of debris and fallen vegetation. The common name for this group is 'short-legged harvestmen' although this is puzzling because none have particularly short legs. In fact, some of them are quite long legged. Perhaps they should be called the 'shorter-legged harvestmen' to distinguish them from Palpatores which have extraordinarily long legs (see below). While the number of Laniatores species is large for a country the size of New Zealand, the great majority belong to a single family, Triaenonychidae, which is also found in Australia, South Africa, Madagascar and Southern South America. Fourteen other species are placed in the Synthetonychidae, a family found only in New Zealand.

Laniatores possess extremely stout pedipalps which are armed with spines and used as grasping arms (e.g., *Soerensenella*, Fig. 3.8a). A hard, hinged plate on the undersurface just behind the legs covers the genital opening within which lie the retracted intromittent organs. By contrast, Palpatores usually have slender leg-like pedipalps which lack spines, characteristically long and slender legs, and a soft flap covering the genital opening.

Two males to one female

When the New Zealand Laniatores were first studied some years ago, it was discovered that, in many species, two different kinds of male were associated with the female. One of these would be much like the female in appearance but the other, as if flaunting his maleness, had much longer, stouter pedipalps and longer chelicerae. The reason for two types of male is still obscure, particularly as experiments show that, as far as the female is concerned, she is not impressed by this apparent show of strength and will mate with whichever male happens to be in the right place at the right time. In fact, the widespread presence of two forms in so many species shows that neither has a particular advantage because if it had, as generation after generation passed, the disadvantaged form would disappear. Interestingly, a similar situation has since been revealed in the Australian Laniatores. Quite apart from these particular cases it is usual for the spines on the body and the palps of the male to be conspicuously stronger than in the female.

Fig 3.8a–f (below and following pages)
Harvestmen continued

***a** (below left):* Soerensenella prehensor *is typical of Laniatores which includes the bulk of our native species. Common in forests, they are also found in more open habitats such as the seashore, mountains and deep caves. They are often hosts for parasitic mites, some seen here on the body. Note the strongly spined pedipalps.*

***b** (below right):* Algidia *is commonly found clinging to the underside of rocks or logs where it is very difficult to see. In this species, three thick spines project forwards from the carapace.*

Fig 3.8c (above left): Pristobunus *is another cryptically ornamented harvestman which, by attaching soil and debris to its body, merges completely with the background. Two eyes are situated on a mound which projects over the front of the carapace.*

Fig 3.8d (above right): Karamea, *one of the very few harvestmen to exhibit maternal behaviour, is seen here with its eggs and young in various stages of development.*

Common harvestmen

The commonest short-legged harvestmen are drab yellow-brown animals with smooth bodies and low rounded eyemounds. Those belonging to the genus *Nuncia* are usually found in bush and tussockland from one end of New Zealand to the other, as well as the Subantarctic, Auckland and Campbell Islands (see Fig. 3.7d). A number of jet black species found in the forest and occasionally in caves belong to the genus *Hendea*. These harvestmen usually have a prominent spine high on the eyemound and another on the hind carapace. Some species have colonised caves and actually lost their black pigment and, as often happens in cave animals, have legs that are relatively longer than non-cave species. For reasons we do not yet understand, the eyemound and carapace spines are also longer in cave forms. For example, in *Hendea spina* from the Nelson caves, the spines are as high again as the body. In the common forest genus *Prasma* the normal eyemound spine is present but the single carapace spine is replaced by two pairs while the carapace itself is closely covered with beadlike pustules. In other genera such as *Algidia* and *Pristobunus* the carapace pustules are arranged in intricate patterns. Moreover, all *Algidia* species have flattened bodies with three to five strong spines projecting forward from the front of the carapace and one or more spines or knobs on the top of the eyemound (Fig. 3.8b). In general this is a hardy group which can survive for a time without water, so some species live in the open under stones. Nevertheless, a species such as the attractive green *Algidia viridata*, which inhabits wet moss deep in the Westland forests, quickly dies of dehydration if removed from this moist environment.

Forest-litter species

The numerous species of *Pristobunus* are all forest-litter dwellers. The spines and tubercles which clothe their dull brown bodies make them look like bits of debris so that they are rarely seen until they move (Fig. 3.8c). Indeed, the eyemound – the most conspicuous part of these harvestmen – consists of a massive cone, its margins armed with a row of prominent spines projecting over the front of the carapace. Females of all short-legged harvestmen deposit their eggs here and there in small clumps as they move about, but a number of species show a greater measure of parental care by guarding their eggs during development.

Egg-guarding behaviour

In the South Island, a number of species with particularly massive pedipalps belong to the genus *Karamea*. Most of these also have a huge spine projecting from the eyemound but the comparable North Island genus, *Soerensenella*, has only a small, erect spine on its eyemound (see Fig. 3.8a). Reproductive behaviour among species in these two genera is similar. As the eggs begin to mature in the ovaries, females cease to feed and settle down in the darkness beneath a stone or log. It is then that males seek them and mating takes place. Before long, the female deposits a small bundle of some twenty to thirty eggs on the surface beneath her and crouches over them like a broody hen. In normal circumstances these eggs would hatch out in about three weeks and her vigil would be over, but a few days later she lays a further batch and at regular intervals continues to lay even more. The eggs are fertilised as they pass down the oviduct to be laid so that when the first batch hatches there are a number of other batches at various stages of development (Fig. 3.8d). Apparently the female does not feed while guarding eggs, but this state of affairs does not carry on until she dies of starvation for the emergence of the first batch of young signals the end of egglaying. Thus her vigil lasts for approximately six weeks, or twice as long as it would have been if she had laid all her eggs at once.

Before completing this brief survey of short-legged harvestmen mention should be made of the pygmies of the group – the Synthetonychidae (Fig. 3.8e). These fascinating little harvestmen, the largest of which has a body of little more than 2 mm, are found only in New Zealand where they live in damp forest moss. Apart from their small size these harvestmen can be recognised by the compact body and absence of an eyemound. Two eyes are placed flat in the middle of the carapace, while the palps are long and slender with virtually no spines. Yet, although these relatively common and widespread harvestmen were described over forty years ago, nothing is known about them other than that they are unique to New Zealand.

Fig 3.8e The minute Synthetonychia, *which rarely exceed 1.5 mm in body length, are known only from New Zealand where they live in forest moss and litter.*

Palpatores

Quite the most comical arachnids are the Palpatores, or long-legged harvestmen. It just does not seem possible that the long thread-like legs, only 1 mm thick and twenty times its body length, can support the animal to which they are attached – yet they not only manage, but manage extremely well (Fig. 3.8f). In spite of their ungainly appearance they are remarkably agile creatures and can put on a surprising burst of speed when disturbed. What is more astonishing is that they are still quite mobile with only three or four legs. Often these harvestmen are seen with missing legs, always broken off at the end of the first joint next to the

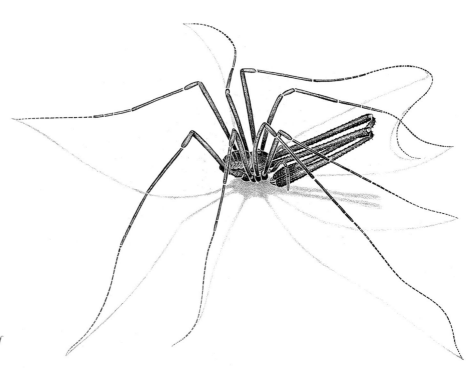

Fig 3.8f The extraordinarily long chelicerae of the male Megalopsalis, *seen here bent in the middle, may have a part to play in mating behaviour.*

Fig 3.8g A female Megalopsalis *with an attractively patterned abdomen which contrasts with the mainly black colouration of the males. Note the very small chelicerae (cf. male chelicerae, Fig 3.8f).*

body. Here, the joint is especially weak so that a slight pull on the leg causes it to break; thus the leg is left behind with a would-be predator while the harvestman escapes. With this useful and apparently often-deployed built-in escape mechanism, it is surprising to find that harvestmen are one of the few arachnid groups not able to regenerate a lost limb.

Male–female differences

Most long-legged harvestmen from other countries do not show a great difference between the sexes, but in many New Zealand species, as well as close relatives in Australia and South Africa, differences between male and female are striking. In fact, males and females of the common native genera *Megalopsalis* and *Pantopsalis* are so unlike each other that, for many years, each sex was thought to be a separate species. In these genera the male is usually black, with only a few white or reddish patches on the body, and the carapace is often covered with short spines. By far the most striking features, however, are the huge, ungainly two-segmented chelicerae towering high above the body like a huge crane. In strong contrast, the female is a most attractive creature with a smooth, compact body, beautifully coloured with various patterns of yellow-brown, green and red while the chelicerae are so small that they hardly reach above the carapace (Fig. 3.8g). Of the many species which live in New Zealand only a few are named and little is known of their habits, but it is thought that the large chelicerae of the male perform some role in mating behaviour.

Zeopsopilio novaezealandiae (Acropsopilionidae), a tiny member of the Palpatores found living in forest leaf litter and moss, makes up for its small size by its grotesque shape. The front half of its body supports an oversized eyemound on each side of which is a large eye. The pedipalps are intermediate between those of the more typical Palpatores and Laniatores, being slender but armed with spines and, as with the latter group, are used to capture their minute springtail prey. Today these harvestmen are represented by very few species, and persist as relicts in New Zealand, Australia, South Africa, southern South America, Japan and the United States. Surprisingly, they have also been found beautifully preserved as fossils in Baltic amber from Europe, having been trapped in gum oozing from trees growing there some 20 million years ago. Therefore, it seems that in the not-very-distant geological past Acropsopilionidae were to be found worldwide but, like so many other forms of plant and animal life, have persisted more commonly in southern continents and islands such as New Zealand.

CHAPTER FOUR

TRAPDOOR SPIDERS AND THEIR KIN
Mygalomorphae

Trapdoor spiders and their kin belong to the Mygalomorphae and are easily recognised by their large size and solid build (Fig. 4.1). The presence of four booklungs on the underside of the abdomen, as well as the up and down movement of the fangs when they capture prey, are further distinguishing features.[1] This special orientation of the fangs is easily seen when the spider is turned over on its back because they lie more or less parallel to each other when at rest. An active spider can be induced to strike at a stick if it is jiggled in front of it, revealing that this spider must first raise itself up to provide clearance for the downward strike of the fangs. This rearing-up action is in strong contrast to the behaviour of non-mygalomorph spiders which have transverse fangs and are able to bite without performing this manoeuvre. The four booklungs can be seen as pale patches on the underside of the abdomen, one pair near the waist and the other pair about the middle (see chapter 1 and Fig. 1.19a).

Trapdoor spiders have come to be known by this name because many of them dig burrows in the ground and close them with a hinged door or lid. However, in some species the burrow mouth is left open while other groups make silken tunnels under stones or logs and even on trees. Burrows may be simple or branched and are lined with silk but there are several variations which we will discuss in this chapter. True trapdoor spiders (Idiopidae) possess a rastellum or serrated comb on the front margins of their chelicerae and these are used for digging. Once the hole is big enough for the spider, it may be completely enclosed with silk after which the spider bites through the silk around

Fig 4.1 *The female tunnelweb spider* (Porrhothele antipodiana) *is large and solidly built, characteristic features of the Mygalomorphae. In body length, this spider can reach 25 mm.*

[1] A single exception is the world's smallest mygalomorph, *Micromygale diblemma*, found in moist leaf litter in Panama forests. The spider, about 0.8 mm, has no booklungs and only two tiny tracheal stubs. Gaseous exchange is probably accomplished by cutaneous respiration.

Fig 4.2 (right): Burrow of a true trapdoor spider (Idiopidae) with its hinged door (or lid), which is almost invisible when closed. These burrows usually reach a depth of some 20 to 25 cm.

Fig 4.3 (far right): Some trapdoor spiders, such as Misgolas huttoni *(Idiopidae), make burrows lined with silk but without a door. Such burrows are said to be open. The open burrows of* Misgolas huttoni *are usually found on the forest floor and are often closely associated with those of* Aparua *(Nemesiidae).*

the entrance leaving a small section as the hinge. This forms the trapdoor (Fig. 4.2). For these spiders, burrows and tunnels make ideal homes. They are roomy, can be adapted to the spider's needs, are insulated against the extremes of heat and cold, and are moisture-proof and relatively secure. They serve as a refuge against enemies, a lair for catching prey, a mating chamber, and a nursery for their young. Indeed, trapdoor spiders can be credited with 'inventing the door' long before humans populated the Earth. Perhaps *Homo sapiens* got the idea from these spiders.

Name changes

When we wrote about these spiders in the 1970s, two familiar mygalomorphs, *Porrhothele* and *Hexathele*, were associated with another New Zealand genus *Aparua* and these three genera were placed in the Dipluridae, a family with a worldwide distribution. Recently, the extensive studies of Robert Raven of Australia have radically changed our views on the relationships of many mygalomorph groups. *Porrhothele* and *Hexathele* are now placed in a new family, Hexathelidae, together with a number of other, mainly Australian and South American spiders, but also some from the Northern Hemisphere. *Aparua*, thought to be closely related to, or even the same as, the Australian *Stanwellia* is transferred to the family Nemesiidae, which is also strongly represented in the Southern Hemisphere. True trapdoor spiders, previously placed in the Ctenizidae, are now grouped under a new family name, Idiopidae, while the former generic name, *Cantuaria*, has given way to *Misgolas*, a name for a similar Australian species which was established earlier than *Cantuaria* and therefore has precedence. Migidae has been retained for the tree-trapdoor group.

Identifying trapdoor spiders

The precise identification of trapdoor spiders to species relies considerably on the form of the reproductive organs – the palpal organ of the male and the internal genitalia of the female – but in fact the initial field identifications are often much simpler. Each species is generally restricted in its distribution so that if it is known that a certain species is found in a particular area it is very likely that this is the one you will find. A quick check can be made on the coloration and particularly the abdominal pattern of the spider both of which tend to be characteristic for each species. A final check can then be made of the genitalia.

There are two distinct groups of trapdoor spiders in New Zealand. Species found on the forest floor are usually open burrow species of which *Misgolas huttoni* (Fig. 4.3) is typical, while the numerous species inhabiting our open grasslands, particularly in the South Island, have lids closing their burrows. Burrows with lids are always difficult to find because particles of the surrounding debris are incorporated into the silk (see Fig. 4.2), but open-mouth burrows of the *huttoni* group are readily located once the eye has become accustomed to seeing the silk-lined opening amongst the litter. Burrows of most species often extend down into the soil for some 20 to 25 cm and are difficult to dig out amongst the roots and soil. To find the occupant it is wise first to push a slender twig or straw gently down the burrow before excavating because the silk lining at the top becomes much sparser lower down and it is easy to lose your way (Fig. 4.4).

Trapdoor spiders of the open country have readily adapted to our pasturelands. As long as these pastures have not been extensively ploughed, populations, particularly in inland Canterbury, can be surprisingly high. Because in general these spiders remain in their burrows, their presence is not necessarily noted except for the occasional errant male found in the house or shearing shed. The concentration of burrows is usually greatest along the fenceline and beside the road in pastureland, but is more scattered in untouched tussock country.

Four mygalomorph families in New Zealand

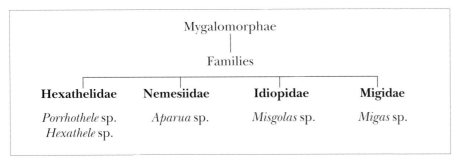

Fig 4.4 *Cut-away view of the deep burrow of* Hexathele petriei *(Hexathelidae) showing the spider at the bottom. Most of the time, the burrow is left open but it may be closed with silk while the spider moults.*

Idiopidae and Nemesiidae are ground tunnel-dwellers, some of which close their burrows with lids, Migidae are tree-trapdoor spiders, and Hexathelidae are known as tunnelweb spiders. Their relationships are shown in the diagram above.

Although the habits of individual species of these four families fall neatly within three particular categories of behaviour as well as the places where they live, there are exceptions. Moreover, the common use of the term trapdoor spiders for Idiopidae can be confusing because not all of them shut their burrows with a lid. In addition, a few migids also construct burrows-with-lids in the ground and some hexathelids also burrow in the ground although none close their tunnel with a lid. Fortunately each of the four types of spiders themselves are very distinctive and it is easy to place them in the correct family.

Many of the spiders placed in *Porrhothele* and *Hexathele* have been known for a long time. In fact *Porrhothele antipodiana* (see Fig. 4.1) has pride of place among our spiders as being one of the first ever to be collected in New Zealand. It was collected in 1827 by the French naturalists Quoy and Gaimard during the voyage of the French corvette *Astrolabe* and described in 1849. *Hexathele* was established in 1871 by the authority on mygalomorph spiders at that time, an Austrian arachnologist, Anton Ausserer. He named *Hexathele hochstetteri* after his fellow countryman, Ferdinand von Hochstetter, who carried out much of the early geological exploration of New Zealand.

Hexathelidae

Hexathelids are the largest of our mygalomorph spiders and, at least by weight, are the biggest of all New Zealand spiders. Their body length may exceed 3 cm. Some of them have an unfortunate habit of wandering into houses where they get trapped in the bath or sink, thus causing consternation in the household next morning. These are invariably males looking for a mate. Two genera are found in New Zealand and all can be recognised by the extremely long posterior pair of spinnerets which project from the abdomen like two small flexible fingers. In spite of their size, and the fact that they are related to the notorious funnelweb spider of Sydney, very few people get bitten – and those few victims

Fig 4.5a A female Porrhothele antipodiana *sits watchfully at the entrance of its silken tunnel. The front legs rest on the silk platform spreading out from the tunnel so that they can detect any movement on the web.*

suffer little more than soreness and inflammation at the site of the bite.

The best known hexathelids belong to the genus *Porrhothele* – large spiders with a bright yellowish or orange carapace and a bluish-black abdomen sometimes marked by a chevron pattern on top. These spiders construct bulky silken tunnels beneath a loose-fitting stone or log, or within a hollow tree, or in the crevice of a broken rock face (Fig. 4.5a). The commonest species is *Porrhothele antipodiana* which can have a tunnel up to 25 cm long and 3–4 cm wide. The tunnel opens out into a wide swathe of silk which probably serves to alert the spider to the approach of an enemy or its next meal. Once an insect is captured, it is quickly dragged back into the tunnel to be devoured. The remains of these meals may be found neatly sealed off with silk at the end of the burrow and so it is not difficult to discover the spider's favourite diet. Although it seems that these spiders attack and eat most arthropods, the bulk of their prey is always beetles, millipedes and slaters. However, a detailed study by Don Laing of Wellington shows that *Porrhothele antipodiana* is also adept at catching another rather unusual prey item, the common garden snail *Helix aspersa*.

Snail-eating behaviour

Snails are generally thought to be unlikely prey for spiders because of their slimy exterior and ready habit of withdrawing into their shells. But Don Laing often found snail remains in *Porrhothele antipodiana*'s tunnels so he decided to investigate. His observations resulted in an astonishing story of specialised predation by this spider. To begin with, Don Laing asked himself: 'How can a spider kill and eat a snail which has retreated into its shell?' He found that the answer lay in the spider's tenacity in the face of a snail's defences. First, the spider touched the active snail with its palps, recognised it as prey, stabbed it and – most importantly – kept its fangs embedded as the snail tried to retreat into its shell. The spider's persistance in remaining attached to the snail's body, despite the snail's defensive production of copious quantities of a foamy mucus that smothered the front of the spider, was the key to its success. Finally, after the snail ceased to resist, the spider pulled its limp body partly out of the shell and embarked on a feeding binge often lasting up to 15 hours.

Reproductive activities

Despite the fact that these spiders are found all over the country, no one has seen their mating behaviour. While we did not see the final stages, we were able to watch a male as he approached a female partly outside her tunnel. As he neared the silken platform he raised the front pair of legs so that the oddly excavated portion of the metatarsus and the swollen and spiny tibia were directed towards the female, simultaneously lifting up the front of his cephalothorax to expose his fangs and palps (Fig. 4.5b). With the front pair of legs vibrating, he moved slowly forward until he reached her but unfortunately the necessary responses from the female did not eventuate and so they remained head to head for some twenty

Fig 4.5b (below left): The male Porrhothele antipodiana *raises his front legs before he approaches a female. This posture is rather similar to his agonistic posture. Note the enlarged tibia and hollowed-out metatarsus of the front legs.*

Fig 4.5c (below right): With the silken retreat pulled aside, a female Porrhothele antipodiana *crouches near her eggsac from which spiderlings are beginning to emerge. Behind the eggsac are the bundled up indigestible remnants of the prey she has eaten.*

minutes and nothing else took place. If the female is receptive, probably the male then grasps her so that her fangs are held firmly between the stout tibial bulge and hollow metatarsus while he inserts the tips of his simple bulblike palpal organs into her genital opening to fertilise her. For this to be successful, she would need to raise the front of her body so that he could grasp her fangs and support her while he reached forward with his palps. This head-to-head mating position (see Fig. 2.14a) is common to most mygalomorphs. It is most likely that mating takes place inside the female's burrow.

In early or middle summer the female constructs a single bulky eggsac containing from two to three hundred loosely wrapped eggs which she attaches to the wall at the end of the tunnel (Fig. 4.5c). The spiderlings hatch in about thirty days but remain with the mother inside her retreat for some time before leaving to construct their own similar but miniature homes close by. Little is known of the lifespan of these spiders but it seems that two or three years pass before they are mature. If the spider is a male, he has only a few months to live. For the female, however, maturity does not mean her life is drawing to a close because she can moult again, even though adult, and continue to grow. Some females have been kept in captivity for four years after rearing their first brood, so they are known to live for at least six years.

Six spinnerets in Hexathele

Superficially, many species of *Hexathele* look like *Porrhothele* and the only sure way to identify them is by counting the spinnerets. *Porrhothele* has only four while *Hexathele*, as its name suggests, has an additional pair. In the North Island, the commonest species of *Hexathele* is large, with a yellowish-brown carapace and a prominent chevron pattern on the upper abdominal surface. This spider, *Hexathele hochstetteri* (Fig. 4.6) and other similar species, all construct loose silken tubes under logs and stones in much the same way as *Porrhothele*. A number of dark coloured, and sometimes black, species are also found in the North Island. Some, such as *Hexathele maitaia* from Nelson, live in typical silken tubes under fallen logs but others construct burrows in the ground. Typical of these subterranean species is *Hexathele huka* (Fig. 4.7a, b) from Wairarapa, a spider which excavates a deep burrow, lines it lightly with silk and attaches a few threads radiating from the entrance to act as warning triplines.

The South Island species of *Hexathele* range from those which still show traces of the abdominal chevron pattern to others where the chevrons have joined until they can barely be seen. Moreover, some live in silken tubes while others construct tunnels in the ground. There are also two strikingly different

Fig 4.6 *The pale but distinctively patterned abdomen is characteristic of* Hexathele hochstetteri *(Hexathelidae) which has a body length of about 20 mm. This spider makes a silken retreat very much like that of* Porrhothele.

Fig 4.7 Hexathele huka, *a burrowing hexathelid, is shown here to illustrate how abdominal patterns change during growth.*
a*: A sub-adult female with a very conspicuous abdominal pattern.*

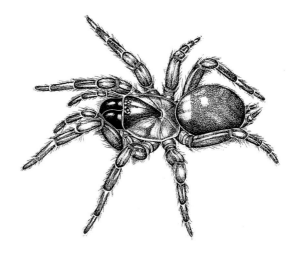

b: The adult Hexathele huka *with more subdued markings.*

***Fig 4.8a** (above top):* Hexathele petriei *is a burrowing hexathelid like* Hexathele huka, *with only a few threads spreading out from its open-mouthed silk-lined burrow (see also Fig 4.4).* **b** *(above): Here,* Hexathele petriei *guards her eggsac at the bottom of the burrow.*

***Fig 4.9a** (below) The hinged 'wafer' lid of* Misgolas toddae *(Idiopidae) can hardly be seen when closed.* **b** *When open, the burrow below is revealed.*

yellowish-brown species where the chevrons have been reduced to faint remnants on the hind portion of the abdomen. One of these, *Hexathele petriei* (Fig. 4.8a, b) from inland Otago, is found in the dry open country typical of this area where it lives in tunnels in the ground. In 1886, P. Goyen, one of the early school inspectors in Otago, described this species and its habits. He suggested that all these spiders leave their burrows to hunt for prey, but this is doubtful because pitfall traps we set near a concentration of burrows only caught mature males. Our results support findings for other mygalomorphs – that is, only males forsake their burrows and then solely for the purpose of locating a mate. So juveniles and mature females always remain within their own burrows and simply reach out from the entrance to seize insects passing by. Surprisingly the closely related *Hexathele rupicola*, which lives in the riverbeds of inland Canterbury, shows no inclination to burrow, but instead makes a typical silken tube amongst the stones that accumulate on river banks.

Idiopidae

Typical or true trapdoor spiders of New Zealand are currently placed in the genus *Misgolas*, all of which make burrows in the ground. About forty species are known and these can be separated into two groups based on the presence or absence of a hinged lid. So although this means that some trapdoor spider species actually live in open burrows, most of them in fact do make thin wafer-like lids (Fig. 4.9a, b). These lids are very different to the thick plug lids constructed by some overseas trapdoor spiders. All *Misgolas* spiders are characterised by a group of strong spines (the rastellum) on the front of the chelicerae and these are used in making burrows. The four spinnerets are small and, unlike other New Zealand mygalomorphs, can barely be seen from above (Fig. 4.10a).

Digging behaviour

When burrows are being excavated, the palps, the first pair of legs and the chelicerae are all used to dig and scrape the earth away. Grains and small lumps of earth are gripped between the partly closed fang and basal segment of the chelicerae, brought to the surface and dropped a short distance from the opening. Small crescent-shaped mounds of discarded earth, which appear at intervals as the spider enlarges its burrow, usually provide the clue to their presence. But even when you know a burrow is nearby the spiders are still very difficult to find. As the excavation proceeds, the walls are firmed by saliva and pressure from the fangs, and finally the internal silken lining is spun. At this stage, the lining covers the entrance, enclosing it completely. Using its fangs, the spider bites through

Fig 4.10a Misgolas napua *(Idiopidae) common near Palmerston in the South Island, constructs its burrow-with-a-lid in open grassland. The tips of two of its four spinnerets can just be seen from above.*

Fig 4.10b *(below): After the final moult,* Misgolas assimilis *has acquired his 'mating' palps. Here, these prominent palps can be seen in front as he sets out in search of a mate. Spiders of this group are generally found in open country and pastureland.*

about two-thirds of the entrance margin, leaving the rest as a hinge. This design not only ensures that the door does not flip open but also that it closes as soon as it is released. This might be a disadvantage if the spider got shut out but providentially it never leaves its burrow. Lurking near the entrance, just beneath the partly opened lid, the spider waits patiently until some insect ventures near enough to be caught. Males behave in the same way until they are mature, when they abandon their homes and set out in search of a mate, and some at least do not eat again.

Surprisingly, most males undergo their final moult sometime during the early winter months of April to June, after which they set out in search of females and often wander into houses (Fig. 4.10b). The complete mating sequence has not been described for any of our species but from periodic observations their behaviour seems to be fairly uniform. As the male approaches the female's burrow he holds his first legs straight out in front and vibrates them rapidly at regular

Fig 4.11 A pale abdomen and dark mottled pattern characterise this species, Aparua kaituna *(Nemesiidae), named after Kaituna Valley, Canterbury, where it is found. All known* Aparua *species live in open burrows in the ground, generally on the forest floor.*

intervals on the door. Presumably this drumming acts as a signal to the female. It is not clear whether a responsive female then opens the door to let him in or whether, in the absence of any adverse reaction, he opens it himself, but the actual mating probably takes place within the tunnel.

Distribution

Long-term studies of population and distribution patterns in *Misgolas toddae* were begun in the 1960s by a husband and wife team, Brian and Molly Marples, who marked individual spiders and their burrows in selected areas. They found that these large spiders can form high-density populations, with as many as 23 spiders being found in a square metre. Yet the distribution is patchy for reasons not yet understood. Here and there, clay banks may be inhabited, and burrows occur even in stony or moist or tussock-covered situations, while on the West Coast of the South Island burrows with trapdoors are found in dense, wet bush. While observing a clay bank population, the Marples noticed that these spiders usually sat at the entrance to their burrows holding their doors ajar. They were intrigued to discover that the sound of a car door being slammed sent a 'Mexican wave' across the bank as one spider after another hastily shut its door.

These studies were interrupted when the Marples went back to live in England but on returning to the area about twenty years after initiating this work, they were surprised to find many of the previously marked, mature females living in their original burrows. It seems, therefore, that these long-lived females seldom, if ever, leave their burrows; only mature males do so, in search of mates. Observations of similar species in France showed that when spiders seize their prey, they hold on to the rim of the burrow with their hind legs as they reach out for it. The lid is thus held ajar, ensuring an easy backward return. It is likely that *Misgolas* behaves in much the same way.

Nemesiidae

By the 1940s a considerable number of hexathelids ('diplurids') belonging to *Porrhothele* or *Hexathele* had been discovered, and it was believed that no others of that size could have been overlooked. It was therefore a great surprise when, in 1944, Valerie Todd (now Valerie Todd-Davies), who was studying hexathelids for her master's thesis, found an entirely different species living in open-mouthed burrows in the forest floor near Wanganui. This species she placed in a new genus called *Aparua* from a Maori word meaning hole (Fig. 4.11).

Distribution and habits

Today, *Aparua* species can be found from the Three Kings Islands in the far north to Stewart Island in the south and are reputed to be the most widespread group of mygalomorphs. These spiders resemble true trapdoor spiders, and all construct open-mouthed burrows similar to true trapdoor spiders and hexathelids. Most species have a distinctively mottled brown abdomen but some

Fig 4.12a (below left): This lichen and bark encrusted door belonging to Migas goyeni *(Migidae) conceals a retreat built into the loose exterior of a New Zealand tree fern. In the camera's flashlight, the door is much more noticeable than it is in nature. **b** (below right): With the lid propped open,* Migas cantuarius *peers out from its 15-mm-long tunnel. Built on the outer surface of bark, this well camouflaged retreat has a back door and a front door.*

Fig 4.12c Migas saxatilis *is a migid which makes its retreat on cliff faces. Its abdomen is a purplish-brown with dark patterning. The male (top) is about 8 mm long, somewhat smaller than the female.* ***d*** *The female* Migas saxatilis *is from 10 to 11 mm long and has a much larger abdomen than the male.*

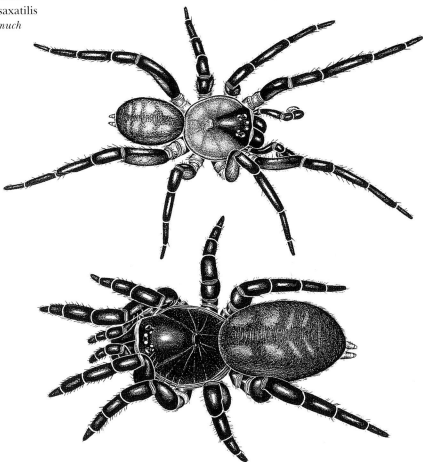

are so pale that the abdomen seems almost white. However, long posterior spinnerets separate this group from true trapdoor spiders, and the mottled abdominal pattern separates them from all hexathelids. As in most burrowing mygalomorphs, the eggsac is attached to the wall near the bottom of the burrow. When the spiderlings emerge they do not seem anxious to leave and continue to share the burrow with the mother spider for some time, no doubt also sharing her catches. This reluctance to leave may last a year or more as occasionally quite large juveniles are found still occupying the nest with a newly laid next-season's eggsac.

Migidae

Habitats

The name 'tree-trapdoor spider' is usually reserved for spiders belonging to the Migidae, although not all of them live on tree trunks. Generally rather squat and less than half the size of true trapdoor spiders, most species (such as *Migas cantuarius* for example) are dark brown or blackish. They build their vertical tunnels directly onto the outer surface of bark on a tree but sometimes one is built within a crevice so that only the door is visible (Fig. 4.12a). Made of a strong and resistant silk into which is incorporated bits of the adjacent bark, the finished retreat merges completely with the surroundings. While most tunnels have a single door, every now and then one is found with two, one at each end (Fig. 4.12b). However, when *Migas* opens a door, it remains open until the spider shuts it. This may be because the hinge is much smaller than in other groups so that the door can be opened wider, added to which is the effect of gravity. This allows *Migas,* unlike *Misgolas* for instance, to leave its retreat, albeit briefly, to catch prey. Tunnels are especially hard to see on tree trunks, and we find that the best way to locate one is to run a hand lightly down the trunk until the soft hiding place is felt.

Some tree-trapdoor species have forsaken what was probably their original habitat – the forest – for cliff faces, particularly near the sea, and some even make burrows on the ground. An example of a cliff-face dweller is *Migas saxatilis* (Fig. 4.12c, d). Being so much smaller, they seem at first glance rather like immature true trapdoor spiders. Very little is known about the life cycle of migids. However, we believe that they mate during the summer because once males become mature, females are soon found guarding eggsacs within their retreats. The eggsac is attached to the wall of the burrow and may contain as few as twenty eggs or, in most species, as many as one to two hundred eggs. When spiderlings leave the mother's home they show little desire to explore their new world but settle nearby on the same tree trunk or cliff face.

A trapdoor spider enemy

One of the greatest enemies of trapdoor spiders is *Priocnemis (Trichocurgus) monachus,* New Zealand's largest pompilid wasp. These parasitic wasps need to provide each of their growing larvae with a large spider for food and it is for this reason that trapdoor spiders are targeted. Those mygalomorphs that live in tunnels without doors are most at risk. Once these spiders are about two-thirds grown, they fulfil the wasp's requirements.

Anthony Harris, a well-known New Zealand entomologist, has carried out extensive studies of these spider-hunting wasps. His observations reveal that once the wasp locates a trapdoor spider, it stalks it with frenzied zeal. The fleeing spider eventually turns to face its pursuer, rearing up as it does so. A struggle ensues but the spider, after being stung about the body several times, collapses although it is still alive. As the spider is usually much larger than the wasp, it has to be dragged backwards to the already prepared wasp's nest. This is a hole in the ground with as many as fourteen cells. Each cell is provisioned with one trapdoor spider upon which the wasp lays an egg. When the egg hatches, the larva feeds upon the inert but still living body of its host until it is ready to pupate. Only at the last minute does the larva eat the spider's vital organs, at which stage the spider dies.

Survival strategies

While all this suggests that trapdoor spiders have little chance against these wasps, Don Laing of Wellington has found that *Porrhothele antipodiana* has several behaviours that help it survive when the wasp comes hunting, and which help explain why spider wasps are unable to exterminate them. For example, the spider may (i) ignore the wasp moving over its sheet web outside the tunnel, (ii) be in a completely inactive state, (iii) hide in a side tunnel if the wasp enters, (iv) fight the wasp in its tunnel where it has an advantage, (v) happen to have built a silk cover over the tunnel entrance, or (vi) emerge and flee. This last reaction provides the wasp with an almost certain chance of making a capture. Laing's studies showed that the spider was more likely to survive if it followed any of the other options and, more particularly, if it remained inactive at the end of a tunnel. Of course, the wasp itself sometimes falls victim to the spider. This means that the tunnel is an important haven for these large trapdoor spiders

CHAPTER FIVE
LIVING FOSSILS
Araneomorphae

The 'living fossils' discussed in this chapter are grouped with the Araneomorphae (see diagram on p.5) because of marked changes in several important characters and behaviours compared with the Mygalomorphae. For example, araneomorphs (sometimes called True Spiders) move their fangs from side to side (diaxially) rather than up and down (paraxially) as in the mygalomorphae. Another distinguishing feature is that araneomorphs have one pair of booklungs and tracheae whereas mygalomorphs have two pairs of booklungs and no tracheae. Chapters 5 to 16 are about those families which belong to the larger group, Araneomorphae.

Of course every rule has its exceptions. So today, among one hundred-plus araneomorph families worldwide, there are three families – Hypochilidae, Austrochilidae and Gradungulidae – which have diaxial or semidiaxial fangs but still possess the four booklungs which characterise the Mygalomorphae. It is generally agreed that these families represent the last survivors of the kind of spiders which flourished hundreds of millions of years ago. To arachnologists, their existence today has the same scientific significance and arouses the same degree of interest that the tuatara engenders among reptile specialists.

The first exception to the rule
Up until 1955 only one of these three families, the Hypochilidae, was known. Species belonging to this family had been found in Asia and the United States, where there has always been a strong interest in spiders. Their four booklungs linked these hypochilids with primitive mygalomorphs. But mygalomorphs are substrate-based and restricted to burrows or silken tunnels. Their mobility is limited because the downward strike of their fangs means that prey has to be dragged backwards to their burrows. Hence prey has to be seized either from the burrow or in close proximity to it. However, the four-lunged hypochilids are lighter-bodied and longer-legged. Moreover, they are cribellate spiders (see Fig. 1.8e) and build distinctive snares to catch their prey. Their fangs are semidiaxial so that they can lift and carry prey. But the retention of booklungs, considered to be more subject to water loss than tracheae, means that hypochilids must seek humid habitats. Hence they build their domed and lacy webs under overhanging rocks close to the ground and often near running water.

At that time, the evolutionary progression seemed clear. The change from downward-striking fangs to transverse fangs allowed spiders to scoop up their prey, hold it securely and, if necessary, transport it to a safe place for consumption. Once that happened, they were no longer obliged to live within the confines of a solid substrate. Silk use was already well developed, being known from Devonian fossil spiders, but lighter bodies and longer legs were required to maximise the advantages of aerial webs. So hypochilids bridged the gap between mygalomorphs and araneomorphs. But this was only part of the story, as we shall see.

An amazing discovery
There was little thought in the mid 1950s, when one of us was just beginning a serious study of New Zealand spiders as a change from harvestmen, that we might be on the brink of a discovery which would overnight, as it were, bring

Fig 5.1 *Gradungula sorenseni* is the spider that led to the establishment of a new family, Gradungulidae. First found in Longwood Forest, Orepuki, Southland, New Zealand, this is therefore the type locality of both the species and the family. Notice the attractive body pattern and the long proclaw visible on one of the front legs. Body length 14–16 mm.

the New Zealand spider fauna to the attention of spider workers all over the world. A check of the spiders gathered from beech forest near Orepuki in Southland by Mr Jack Sorensen, a prominent local naturalist, showed that one particular spider appeared to have some characteristics typical of hypochilids. This spider, about 14 to 16 mm in length, looked very much like a wolf spider except that its eyes were in two more or less straight rows and its chelicerae were semidiaxial (Fig. 5.1). But the really significant aspect was that four pale patches on the undersurface of the abdomen indicated it had four booklungs, and this was later confirmed from its internal anatomy.

By this time hypochilid spiders were known from four species, each living as small populations in widely separated parts of the world: China, Tasmania, South America and the United States. All were large spiders with long slender legs, and they captured their prey in webs coated with hackled silk produced by the cribellum. The presence of a cribellum and the hackled-silk snare, as well as four booklungs and semidiaxial chelicerae, were accepted as the principal characters of these primitive spiders. To our surprise, however, the spider from Orepuki did not have a cribellum and, according to notes by Jack Sorensen, did not construct a snare either; in other words, it was a hunting spider. Soon we were on our way to Orepuki to look for more of these living fossils and to study them in their natural habitat. Once there, we climbed the hill to reach the forest a few hundred metres from the road, and there they were – hiding under logs on the forest floor. Why no one had found them before was a puzzle because, as we later learned, the spider was not only relatively common at Orepuki but in the west and south coast forests of the South Island and Stewart Island as well.

Clearly, like hypochilids, these spiders did not live in burrows or silken tunnels, yet they still had the four booklungs so diagnostic of mygalomorphs but unknown then in araneomorphs. Despite these common features, this strange New Zealand spider did not have a cribellum, built no snares but hunted its prey – a major change from the sit-and-wait habits of other four-lunged spiders. Not only that, spiders with these particular features were not known anywhere else in the world. Without a snare or silken lair, how did these hunting spiders catch their prey? As we shall see, they had acquired a potent new 'tool' which aided their free-living lifestyle and facilitated the capture of prey.

History of a new family of spiders

Special features

Eventually this new species from Southland was named Sorensen's Gradungula (*Gradungula sorenseni*) in honour of its discoverer. Furthermore, because so many characters differed from earlier known representatives of Hypochilidae, it was necessary to establish a new family, Gradungulidae. Apart from the four booklungs, one of its most distinguishing features is that the inner superior claw on the first two pairs of legs is almost twice as long as its counterpart in hypochilids and indeed, in almost all other spiders. These two characters alone make it easy to identify in the field. The name Gradungulidae means graded claws and so we know that all members of this family have a large proclaw and two progressively smaller claws like this (Fig. 5.2).

Further discoveries

Before the description of Sorensen's spider was published, however, a strange coincidence occurred. Amongst spiders sent to us for identification by T.E. Woodward of Queensland University was an immature specimen unquestionably related to *Gradungula sorenseni* particularly since it also possessed those curious claws. Of course, this was a different species, *Gradungula woodwardi*, and so when the new family was established it was recorded from both Australia and New Zealand.[1] The fact that these two species, which had been separated by a

---

[1] Later, a new genus was established so this spider is now *Tarlina woodwardi*.

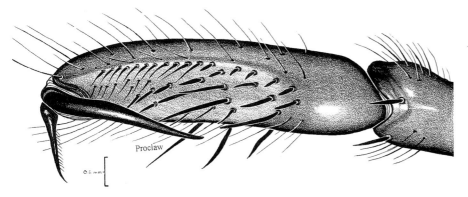

Fig 5.2 The tarsus of a front leg of a gradungulid spider showing the large proclaw folded in towards the concave spiked groove. Long sensory hairs at the distal end of the tarsus may be involved in the detection of prey. The retroclaw is about half the size of the proclaw while the third claw is reduced to a small stout curved spine, not visible here.

wide expanse of ocean for many millions of years, could be so similar caused little surprise because it is often found that survivors of earlier geological periods evolve very slowly. Perhaps the most astonishing part is that they came to light at much the same time.

This marked the beginning of an important step forward in our understanding of the evolution of these early spiders and sparked off an exciting period of discovery. Not long after, an enthusiastic band of speleologists, who were systematically exploring New Zealand caves, found a remarkable long-legged spider living in caves in the Nelson area and this was soon identified as a second representative of gradungulids in this country. Much bigger than the forest species, this spider also had the long legs typical of overseas hypochilids. Indeed, taking account of the leg span, it is the largest spider in New Zealand. Commonly known as the cave spider and officially called *Spelungula cavernicola*, its name reflects the interests of the group who discovered it and the habitat in which it lives. More recently, Anthony Harris of Dunedin discovered an entirely different species (*Pianoa isolata*) in a small isolated patch of forest bordering Piano Flat in Southland. At the time he was observing the capture of spider prey by pompilid wasps. This species lives on the forest floor in much the same way as *Gradungula sorenseni*.

Australian representatives of this family

Meanwhile, interest in the group was developing in Australia and, to everyone's further surprise, twelve different species of gradungulids were discovered, mostly in small populations along the east coast and separated from each other by distinct geographical boundaries. Ten of these were ground-hunting spiders like *Gradungula sorenseni* with six spinnerets (Fig. 5.3a) but the most exciting finds were of two separate species, one from the Carrai Bat Cave in New South Wales and the other in the tropical forests of Northern Queensland. Both these latter species possessed a cribellum (Fig. 5.3b) and constructed simple snares which were, however, very different from the webs of hypochilids.

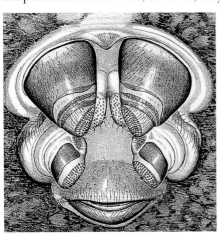
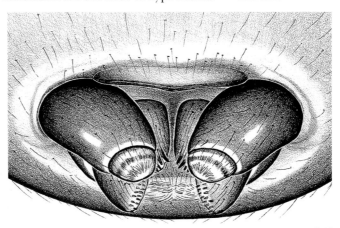

Fig 5.3a (below left): The six spinnerets of a hunting gradungulid such as Gradungula sorenseni. *b (below right):* Compare with the six spinnerets and cribellum (see p. 9) of the snare-making Progradungula carraiensis from the Carrai Bat Cave, New South Wales, Australia.

It so happened that as long ago as 1883, a four-lunged spider had been found in caves in the Mole Creek district in Tasmania and given the name of *Theridion troglodytes* (Theridiidae). This long-legged spider was known to make an extensive horizontal web but because of its placement in Theridiidae, it did not attract much attention. However, when Willis Gertsch revised the Hypochilidae in 1967, it was renamed *Hickmania troglodytes* and placed in this family. Similar spiders were known in Chile and parts of Argentina and because of genitalic differences, structural variations and their large horizontal webs (so markedly different from those of hypochilids) another family, Austrochilidae, was established for all of these spiders. This means that the distribution of the three families now accepted is: Hypochilidae in USA and China; Austrochilidae in Tasmania and South America; Gradungulidae in southern New Zealand and the east coast of Australia. So, after much confusion and name changing, arachnologists felt that they had finally reached consensus on the grouping of these unusual spiders into families which reflected their relationships and evolutionary history as well as ancient geological land movements, and so it remains to this day.

Hypochilidae

Lampshade webs

Our first sight of a *Hypochilus* web was in the Great Smoky Mountains National Park, USA, in 1976. With our American colleagues, Herb Levi and Fred Coyle, we clambered up to an expansive rocky area near a trickling stream, a typical habitat of this spider, *Hypochilus pococki*. Under an overhanging rock in a space of some 50 to 60 cm in height, we were soon looking at a most unusual web structure. Commonly described as a 'lampshade' web, we saw that the upper cone consists of a woven cribellate network over a dry silk framework while the lower circular edge flares out. This whole structure is held under tension by a series of auxiliary lines fastened to an encircling frame line which, in turn, is anchored to the lower substrate.

Hypochilus gertschi is another large cribellate spider with a mottled dark and faintly yellowish dorsal colouring, inconspicuous against the rock above. It hangs

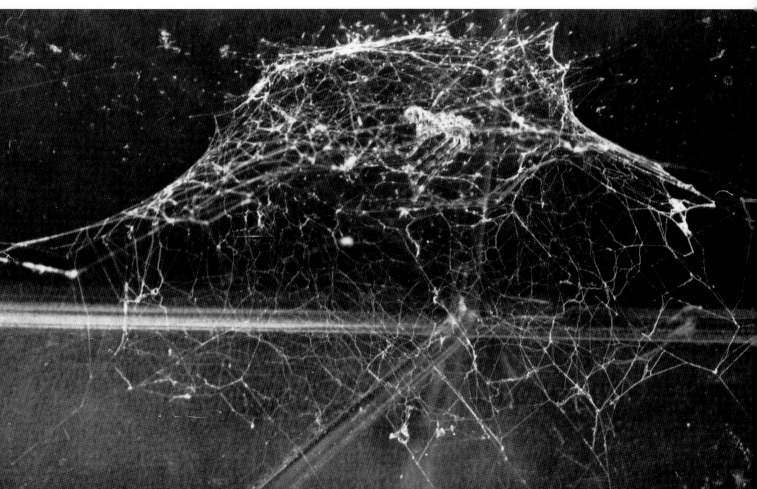

Fig 5.4 A typical lampshade web built by Hypochilus pococki *(Hypochilidae) in laboratory conditions. The spider in the centre of the web has just caught prey and its vigorous shaking action has blurred the picture in this area. Great Smoky Mountains, USA.*

from the centre of its lampshade web, holding on to it with the third and fourth pairs of its long slender legs. Bill Shear's studies show that prey consists mainly of flying insects such as tipulids or other flies, and these are treated as prey only if they come into contact with the sticky part of the lampshade. Once alerted to prey, the spider quickly grasps the threads nearest to the struggling insect with its first and second pairs of legs, pulls the entangled bundle toward itself, bites it and injects venom several times. In less than five minutes, the struggling subsides and the spider begins to feed.

Courtship and mating behaviour in this group was observed for the first time in *Hypochilus pococki*, in the course of a laboratory study by Kefyn Catley of Cornell University, USA. In its lampshade web the female waits for prey and mate in her usual posture in the centre of the web (Fig. 5.4). Upon encountering a female web, recognised perhaps by olfactory cues, the wandering male tugs at the cribellate silk and bobs up and down with his body. These actions eventually induce the female to approach him and mutual leg-stroking begins. This stroking increases in intensity and, some three minutes or so later, the female suddenly drops her abdomen at an angle to the web, and becomes quiescent. The male moves towards the female with his palps extended forward and inserts them, one at a time, into her genital opening. Eggsacs are suspended in the web, the female herself fetching debris from below to attach to them, as camouflage.

Austrochilidae

Horizontal webs

Austrochilids are separated from hypochilids not only because of very different web structures but also because the calamistrum, which combs out the cribellate silk, has only a single row of metatarsal bristles. Two genera, *Austrochilus* and *Thaida*, are found in Chile and adjacent Argentina while one, *Hickmania*, is found only in Tasmania. These spiders always live in very moist conditions but little is known of their behaviour. All of them build highly visible horizontal webs spreading out from funnel-shaped retreats hidden in rocks, tree roots or other inaccessible places. Here, spiders hide during the daytime, only venturing out at night to hang beneath their webs.

Although the Tasmanian cave spider, *Hickmania troglodytes*, is found in caves, it will also build its extensive horizontal web – sometimes more than a metre long and half a metre wide – in any cool, dark, damp cavity. The female, about 19 mm in length, has a reddish carapace and a dark, relatively unmarked abdomen. Its very long and slender legs are about four times the length of its body. Pear-shaped eggsacs are hung from the web by a narrow stalk. In caves, the eggsac is pure white but in hollow logs, for example, it may be covered with debris.

Gradungulidae

Extra long superior claws

Gradungulid spiders can be readily distinguished from both hypochilids and austrochilids by the large superior claws of the first two pairs of legs. Remember, though, that each leg has three claws, two superior (upper) claws and one small inferior (lower) claw. Of the two superior claws on the four front legs, it is the proclaw or 'inner' claw which is enlarged and lengthened. This proclaw can be bent towards a concave groove on the ventral surface of the distended tarsus and this groove is armed with bristles and stout spines. The superior claws can be moved independently and are often used as nippers. It is these four elongated proclaws which provide the spider with 'new and improved' hunting and climbing tools.

Three genera are found in New Zealand, one (*Gradungula*) from the damp moss-covered forest floors of western and southern South Island and Stewart Island, the second (*Pianoa*) from Waikaia Forest, Southland, and the third

(*Spelungula*) between the twilight and deep zones of Nelson and West Coast caves. Only one species is known for each genus. Australia has four genera with twelve species, ten of which are ground hunters like *Gradungula* and *Pianoa*, one (*Progradungula*) which is found in caves and the other (*Macrogradungula*) from Northern Queensland forests, the latter two being cribellate species.

Ground-hunting gradungulid species are all ecribellate – that is, they do not have a cribellum and calamistrum so do not produce hackled silk, nor do they spin snares to catch their prey. It should perhaps be mentioned here, that many spider groups (discussed in other chapters) make silk snares from glands other than the cribellate gland. When the family Gradungulidae was established, it was believed to be an ecribellate family but the finding of two cribellate species in Australia with the long proclaws which proclaimed them as Gradungulidae, was another 'amazing discovery'. The first such revelation happened while we were at the Queensland University in Brisbane, looking through a collection gathered from the Carrai Bat Cave when, amidst great excitement, it was realised that a new gradungulid species found there was cribellate.

The Australian cave spider

This cribellate spider, *Progradungula carraiensis*, is so named because, in an evolutionary sense, it is believed to have preceded *Gradungula*. This is because a spider is a spider because it spins silk, so the very earliest creature qualified to be a spider was an eight-legged silk-spinning animal. And so, since *Progradungula* makes a snare to catch its prey, one of this spider's ancestors is considered to be the harbinger of free-living members of the group. Although the Carrai Bat Cave in New South Wales is the only recorded locality for *Progradungula carraiensis*, there have been tales of similar spiders in caves elsewhere in the region. It is also possible that this spider is not entirely restricted to caves since the steep forested slopes and limestone outcrops surrounding them offer suitable sites for web construction.

A prey-catching web

A medium to large spider with very long legs, *Progradungula carraiensis* has a fawn coloured abdomen with a darker central stripe and chevron markings. The proclaw of the first and second legs is strongly developed with a length almost half that of the tarsus (Fig. 5.5). Webs are built beneath the overhang of

Fig 5.5 *Micrograph of the powerful proclaw of* Progradungula carraiensis, *the Australian cave spider. Like the proclaw, the smaller superior claw has a row of teeth which greatly improves the spider's grip on prey.*

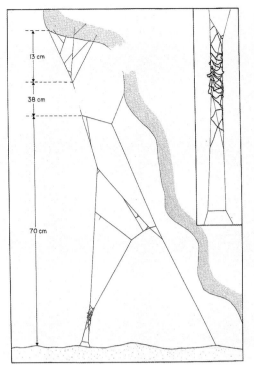
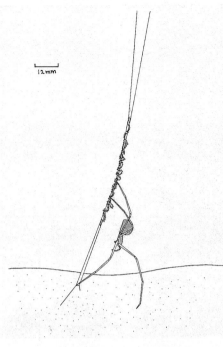

a cave wall and stretch to the ground below, sometimes reaching a length of more than a metre. At the upper level beneath the overhang, a network of support threads serves as a retreat for the spider and this is attached to the main scaffolding threads. These threads are anchored to the base above which a small catching ladder, overlain with a layer of cribellate silk, is constructed. The spider adopts a head-downwards hunting posture with its second, third and fourth legs grasping the catching platform. Its long first legs hang free but near enough to the ground for the spider to be able to detect and seize potential prey (Fig. 5.6).

Prey-catching is described by Mike Gray of the Australian Museum as follows: If a likely food item strays across the cave floor near the spider's front legs, it is instantly seized and cast into the sticky silk ladder where it becomes entangled. Still held by the legs, the spider bites its prey and when struggling ceases, the spider moves to a head-up position and begins to wrap the prey with silk. The wrapped bundle may be eaten on the spot but is sometimes carried to the upper retreat. Gray notes that the only prey he saw being caught were tineid moths. These virtually flightless insects commonly wander on the cave floor and apparently alert *Progradungula* to their presence by vibration. This 'bite-wrap' behaviour brings to mind the distinction made by orbweb spiders when moths are intercepted by their webs. In a 1976 study of these spiders in Panama, Mike and Barbara Robinson found that, instead of the wrap-first, bite-later method used for most prey, orbweb spiders always bite moths at once and wrap them afterwards (see chapter 11). Is this a behaviour pattern that began with *Progradungula*?

Fig 5.6 The web of Progradungula carraiensis *(Gradungulidae) as it might be built under an overhang in the Carrai Bat Cave, New South Wales, Australia. The spider holds onto the web above the sticky catching ladder (see inset for details) and hangs downwards so that its front legs reach the floor. (From Forster & Gray, 1979).*

The New Zealand cave spider

The first official discovery of *Spelungula* was in 1957 at an elevation of about 550 metres in the upper Oparara river area, Karamea, in the northwest region of the South Island, and other reports followed soon after. It quickly became apparent that the Honeycomb cave system in the Oparara Valley was home to these large spiders, many of those seen still being juveniles. Ida Cave, Star Draft Cave, Wonder Sump Cave, Y Cave and Neil's New Passage were among those where its presence was noted. A mature female collected in 1973 from Ida Cave, and a male in 1983 from Wonder Sump Cave (both sites in the Oparara Valley, Karamea) were used to formally describe and name these spiders as *Spelungula*

Fig 5.7 (above) The New Zealand cave spider, Spelungula cavernicola *uses its long proclaws to cling to the rough underside of the cave roof. Body length 20 mm. (above right) The proclaw and the smaller superior claw. Note the strong bristles on the undersurface of the tarsus.*

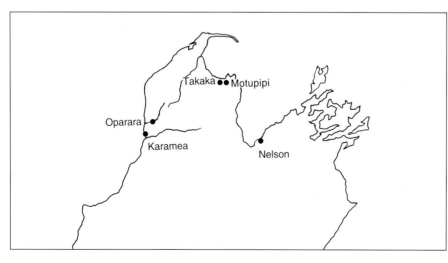

Fig 5.8a (far left): Spelungula cavernicola *holds a captured cave weta (*Gymnoplectron *sp.) with three of the front legs. Notice, however, that the other front leg has stretched forwards to seize a harvestman (*Hendea *sp.). Its proclaw has closed over the harvestman's hind leg, clamping it tightly in the 'crook' of this claw. The spider's four hind legs are sufficient to retain its grip on the cave surface even with the added weight of the weta.* **b** *(left): A close-up of* Spelungula *feeding on the weta. The left front proclaw is gripping a joint of the weta's hind leg while strands of silk entangle its antennae. The pedipalps assist in restraining and manipulating the weta while fangs masticate and mouthparts exude digestive juices, a barrage of weaponry against which the weta is helpless. (Photographed by Ian Millar in the Honeycomb Cave, Oparara Valley, see map.)*

cavernicola by Forster and colleagues in 1987. A preliminary survey of populations in this cave system by Ian Millar and Trevor Worthy in 1983 noted the preference of these spiders for living in areas just beyond the twilight zone although they are sometimes found within this zone or even in deeper parts of the caves. Such a transition area, although dark, is still subjected to external conditions which cause fluctuating temperatures, and changing levels of humidity. The deepest zones are noted for their constant temperatures and humidity as well as higher levels of carbon dioxide.

This striking ecribellate spider, *Spelungula cavernicola* (Fig. 5.7a), is differentiated from *Gradungula* and *Pianoa* by its larger body size (about 20 mm), much longer legs and very different habitat. The brown abdomen, densely covered with hairs and black bristles, has a faintly patterned appearance, quite distinct from the chevron markings of *Pianoa*. In some cave populations, however, paler spiders have been found, thus raising the possibility of a second species, although this has yet to be investigated. It has been confirmed that *Spelungula* is more likely to be found just beyond the twilight zone (a dimly lit region which may extend into a cave for up to 20 metres) into the dark zone (where no light penetrates) but is said to be cave-adapted rather than troglobitic. True troglobites, like the *Lycosa* species found in perpetually dark lava caves in Hawaii, are blind and usually pale in colour. *Spelungula* is a dark-coloured spider and there is no indication of a reduction or loss of eyesight. Moreover, this spider has often been reported to be feeding on cave wetas (Figs 5.8a, b) (Ian Millar, pers. comm.), an insect group said to be twilight-zone dwellers but which are also found in areas frequented by *Spelungula*. In addition to the spiders that Ian Millar saw and photographed at that time, he counted six large pear-shaped eggsacs suspended by thin stalks (Fig. 5.9a) from the cave roof, evidence that this was a breeding colony.

Spelungula *behaviour*

Further detailed observations on *Spelungula* populations in the Honeycomb Cave system by Andrew McLachlan of Christchurch have added to our knowledge of this species. He noticed that smaller spiders (large spiders less frequently) often hang from the roof on draglines and that some feed (Fig. 5.9 b, c) or moult in this

Fig 5.9a (below left): A typical pear-shaped eggsac of Spelungula cavernicola, *suspended from an 8 cm long stalk, is firmly attached to the roof of the cave. Recently laid, its surface is sprinkled with cave detritus. Box Canyon Cave, Oparara Valley, Karamea, New Zealand.* **b** *(below centre): This half-grown* Spelungula *has lightly wrapped a large weta (*Gymnoplectron *sp.) with strands of silk and is biting its catch in the head region. Stardraft Cave, Oparara Valley, Karamea, New Zealand.* **c** *(below right): With digestive juices flowing,* Spelungula *holds the weta firmly as it feeds. The silk thread from which the spider hangs is held by a fourth leg. (Photographs Andrew McLachlan)*

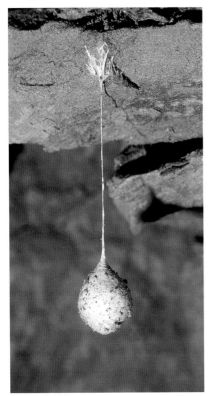

position. In 500 metres of cave passages that he monitored regularly, numbers of spiders varied from less than fifty to more than a hundred, apparently partly due to periodic factors such as spiderling emergence. Nevertheless, this finding is encouraging for it indicates that these spiders are present in quite viable and active populations. It was also confirmed that cave wetas form the main bulk of their prey but that spiders sometimes feed on blowflies which enter the caves.

An important new observation was made by Barry Chalmers, a caver from Karamea, whose attention was first attracted by the sudden movement of a *Spelungula* snatching at a weta and then grasping it with the four front legs. Letting itself down on a dragline of some 30 to 40 mm, the now almost quiescent weta was wrapped lightly in silk before being devoured by the spider. Perhaps the absence of a struggle by the weta resulted from the venom injection but another possibility is that by dropping on a dragline *Spelungula* has developed a ploy which immediately removes its victim from the substrate, the basis of its usual mobility.

Motupipi Cave population

More of these spiders were found in the Motupipi Cave in the Takaka Valley, Nelson, a cave system described by Ian Millar (pers. comm.) as being in a narrow 'wall' of scrub-covered limestone standing above alluvial flats near the Dry River, further noted as a partially isolated limestone outcrop surrounded by farmland. It is in this cave system that the paler spiders have been found, and as this population is some 80 km east of the Honeycomb Caves of Karamea, this isolation makes a second species more likely. Unfortunately, although eggsacs have been found, no adults have been available for study, so key features such as genitalia cannot be compared.

Despite one sighting of a large *Spelungula* on a tree trunk in the vicinity of Honeycomb Hill by Phil Wood of the Buller Caving Group, it is not known yet whether these spiders venture out of caves regularly, whether the males wander from one locality to another, or even whether their young 'balloon', but unless they do so, it is obvious that each of these cave populations could be quite isolated. According to evolutionary theory, the less often potential mates come into contact with each other, the less likely they are to interbreed. Given a sufficiently large but isolated population, *speciation* (the gradual development of a new species) is likely to occur. Should a population become too small, however, inbreeding may lead to less vigorous individuals and the ultimate extinction of this species. For this reason smaller populations need to be protected and monitored. We note here that *Spelungula cavernicola* is one of New Zealand's few protected spider species, a status conferred on it by the Department of Conservation because of its rarity, its ancient origins, the fact that it is the only species in this genus and the possibility of its extinction.

Ground-dwelling forest gradungulids

The two ground-dwelling species, *Gradungula sorenseni* and *Pianoa isolata*, are found only in the South Island and then only in native forests. *Gradungula sorenseni* is the more widespread species being found all along the West Coast, Fiordland and parts of the southern South Island as well as Stewart Island whereas *Pianoa isolata* is restricted to a small isolated patch of rugged bush country known as Waikaia Forest. The rich mixture of moss, leaf litter, and decaying logs is a significant feature of all these forests, with deep layers of up to 50 cm in many places. Indeed, the depth of this layer is an important indicator of the well-being of the forest beneath which it lies. Here, in this nursery for the new growth of our native trees, are to be found millions of minute invertebrates including a great variety of spiders and insects, their eggs and larvae, as well as mites, symphyla, snails and slugs, millipedes and centipedes.[2] Within

Fig 5.10 *A female* Pianoa isolata *with two eggsacs, revealed when a rotting log was pulled apart. The white eggsac with the long stalk was probably laid more recently because it is not yet coated with debris.*

[2] Surveys suggest that up to 100 million invertebrates exist in a hectare of rain forest.

such tiny thriving nutrient-recycling worlds, the young of both *Gradungula sorenseni* and *Pianoa isolata* find their food and grow. Of course, these spiders too fall prey to some of the many animals living in and above the litter layers.

The adult *Pianoa isolata* is a medium-sized spider, about 12–13 mm long, whose pale fawn colour is softened by a slightly pinkish 'bloom'. This nocturnal hunting spider, easily distinguished by the chevron pattern on its abdomen (Fig. 5.10), four booklungs and four elongated proclaws, wanders the moss-covered leaf-litter floor of the forest in search of food. Young spiders mostly stay within the moss and leaf litter hunting for the tiny midges, grubs and other minute creatures that form their prey. Once over half-grown, juveniles are more likely to hide in and under decaying logs during the day and hunt only at night. Adult spiders, whether male or female, are usually found under decaying logs but often hide deep within them.

Changes in the use of silk

Earlier we mentioned that the most primitive spiders living today are the liphistiids and that *Liphistius* species live in ground burrows lined with silk. Silk use progressed from this stage to the more extensive webs now associated with ground-dwelling spiders such as *Porrhothele*. We have seen that the first spiders to live above the ground are typified by the hypochilids and austrochilids and that they make elaborate silk snares to catch their prey. In the Gradungulidae, however, there has been a dramatic change of lifestyle. Although *Progradungula* and *Macrogradungula* make simple prey-capture webs, other species in this family are hunting spiders which have abandoned the use of silk for this purpose. *Pianoa isolata*, for example, is a 'living fossil', an ancient representative of those first spiders which adopted a free-living, hunting existence. This may have been brought about, in particular regions, by the growth of dense rain forests and thick mossy floors as a result of heavy rainfall over prolonged periods. Extensive web structures might have been quite impractical so those spiders which evolved to 'make do' without them, or found refuge in caves, are among the ones that have survived to this day.

Prey capture

Two features are particularly important to *Pianoa* in the capture of prey. The first is the ability to sense the whereabouts of potential prey immediately in front of it, an awareness brought about by the possession of sensory hairs and trichobothria on the tarsi of the front legs (see chapter 1). The second is the development of the large proclaw on the front legs. Captures usually occur when both the prey and spider are on the move. With a lightning flash, the nearest front leg reaches out and seizes the prey with its long proclaw. This flexible claw snaps down into the hollow tarsal groove where, in conjunction with stout opposing spikes, the prey is held in a vice-like grip. Immediately it is conveyed to the spider's mouth where it is bitten repeatedly and, as the victim is manipulated by the fangs and palps, digestive juices flow copiously over its body. All this takes place in a matter of seconds, and only the close observation of many captures has ascertained this sequence.

If it is disturbed, the spider may drop its prey and run away, or it may run to a sheltered spot carrying the prey in its fangs. Examination of the dropped prey shows that the fangs and claws have inflicted massive wounds and that the victim has been killed quickly. Feeding commences at once while the prey continues to be masticated by the fangs. Usually a meal is finished within fifteen to twenty minutes, and a small indistinguishable bundle of chitin is all that remains. A variety of prey is caught by *Pianoa*, the main provisos being that it is less than the size of *Pianoa* itself and that it is moving. Other hunting spiders are amongst the more common victims, which also include ground wetas, different kinds of flies, and moths. Beetles are avoided and *Pianoa* either runs quickly away from larger creatures or feigns death. Sometimes, when spider prey is involved, a struggle

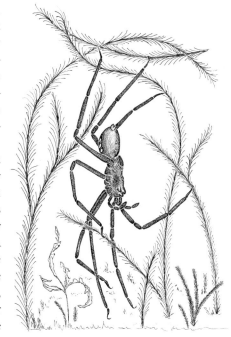

Fig 5.11 Suspended from moss strands by its hind legs, Pianoa isolata *waits for prey to come within reach of its long proclaws. In this illustration, the moss vegetation has been greatly reduced, for clarity.*

ensues, but *Pianoa*'s longer legs and raptorial claws usually prevail.

Alternative tactics

This ground-based style of hunting 'on the hoof' is obviously efficient but, surprisingly, a second 'sit-and-wait' strategy has been observed. When the spider is amongst a bed of loosely aggregated strands of moss it may hang head downwards with its third and fourth legs and sometimes the second, holding on to two or three strands of moss while its front legs hang down (Fig. 5.11). Prey capture has been witnessed several times with the spider in this posture, which strongly resembles that of *Progradungula* but with strands of moss being substituted for the web. The technique used to catch prey is almost identical to the ground-based captures – the spider scoops up the prey with one of its raptorial claws and transfers it to the fangs. Prey may be eaten on the spot or carried elsewhere. Then the proclaws are used, particularly by the juveniles, to hook onto twigs and tendrils as they climb through moss and undergrowth.

Early development

Mating behaviour has not been observed in this species. However, we know that *Pianoa* females attach their spherical eggsacs by narrow stalks to the upper roof of moss-covered cavernous logs or rotting stumps. Within about three weeks each developing spiderling casts off the egg membrane inside the eggsac, emerging a few days later to shed the first moult skin. This takes place within 24 hours, just outside the eggsac. Several times in the wild we have found empty eggsacs with some 18 to 25 pale skins hanging in a cluster nearby and, in captivity, we have been able to verify these observations. This means that the very pale first instar spiderlings emerge from the eggsac before undergoing their first true moult, whereas in most other spiders this moult happens within the eggsac.

This potted taxonomic and biological history of three primitive families is understandably brief but as new discoveries were made and interest in the group grew, spider enthusiasts in New Zealand, the United States and Australia (Forster, Platnick and Gray) collaborated in a review that was published in 1987. On a worldwide basis, this doubled the number of genera known at that time to a grand total of twelve and boosted the species count from eleven to thirty-two.

CHAPTER SIX

FREE-LIVING SPIDERS

Many diverse groups of spiders have no fixed abode such as a hole or a web, and simply wander about in their quest for prey. Such free-living spiders, often called vagrants or hunters, include groups which utilise a wide range of predatory behaviours. For example, most jumping spiders actively search for prey but other free-living spiders, such as gradungulids, rely on chance encounters with suitable food organisms, while many groups, like the crab spiders, employ sit-and-wait tactics. Sometimes these different behaviours may be found within a single family, but since most of our readers first want to know 'what sort of a spider is this?' the chapters in this book deal primarily with groups of families in a way that provides a basis for identification.

In the next five chapters we describe those 'wandering' families that are found in New Zealand. Some of them, including nurseryweb and wolf spiders, are also abundantly represented in overseas countries, but the Huttoniidae, for example, is found only in New Zealand. It is widely accepted today that most vagrant spiders evolved from web-building ancestors, which means that a long time ago they gave up making snares and began to move about in search of food. That this is a reasonable hypothesis is shown by the fact that, in many of these families, a few species still construct a snare for the capture of prey. Examples are found amongst tropical nurseryweb (Pisauridae) and wolf (Lycosidae) spiders, but not in the New Zealand representatives discussed in this chapter.

Wolf spiders: Lycosidae

The chances are that the spider you see dashing across the garden path or between the plants on a sunny summer's day is one of the common wolf spiders (e.g., *Lycosa hilaris*), for it is one of the few fast-moving spiders which hunts in the open during daylight (Fig. 6.1a). On the other hand, if you are out in the garden at night with a torch, you may find out one reason why they are called wolf spiders. Tiny pinpoints of light often appear here and there as you shine the torch directly in front of you. These may arise from the eyes of wolf spiders for the torchlight is reflected by a special layer of crystalline cells called a tapetum at the back of their eyes. The eyes of wolves also have a reflecting layer which makes them glow in the dark, so perhaps that explains how these spiders got their name.

Many different species

A close look at these spiders from different parts of the country shows that, although they are very similar in appearance and habit, on the basis of their genitalia they are, in fact, different species. The initial impression of a rather drab brown or greyish animal is dispelled on examination with a magnifying glass. The strong legs of the female are usually banded and mottled with yellow, green, black and brown. The carapace is attractively patterned with yellow or white hairs, relieved by two broad black bands which do not quite meet in the normally yellow centreline. The short, furlike pile on the abdomen is set off on the front by a vivid yellow or orange stripe down the centre, and this is accentuated by a cluster of black hairs along the margins. The male is similar

Fig. 6.1a The female striped wolf spider (Lycosa hilaris) *is about 6 mm in body length and its colour and pattern make it difficult to see amongst the plants where it roams.*

Fig. 6.1b A male Lycosa hilaris *pauses for a few seconds before he makes his next dash.*

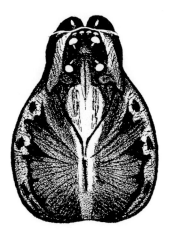

Fig. 6.2 Carapace of a wolf spider showing the four small eyes in a row at the front, and four large eyes spaced apart on the upper carapace.

in appearance to the female but has a smaller body and is longer legged, while the stripe on his abdomen is often much wider and longer (Fig. 6.1b).

Wolf spiders are always fast and agile runners, but if you watch one closely you will notice that it always runs in short bursts and then pauses, almost as if to get its breath back before dashing off again. In fact, it is quite likely that the spider does need to release carbon dioxide and allow air to enter through the spiracles from time to time (see chapter 1). Although its muscles function anaerobically, Rudiger Paul and his associates found that organs such as the central nervous system, heart, Malpighian tubules and midgut gland work continuously and as a result are aerobic. Since the structure and habits of most wolf spiders are fairly uniform, an account of the common striped wolf spider from our garden in Dunedin is used as a general guide to the behaviour of many other species found in New Zealand.

Striped wolf spiders

Wolf spiders are true hunting spiders because, with very few exceptions none of which occur in New Zealand, they capture their prey 'on the hoof'. Studies carried out by Jerome Rovner testify to the keen sight of these spiders so it is likely that our own spiders are also able to detect movement visually and that this is one means by which they react rapidly to objects in their surroundings. Experiments by F. Magni also showed that wolf spiders navigate by utilising polarised light from the sky but they do not have the jumping spiders' ability to distinguish shapes. Moreover, wolf spiders also send and receive vibratory, tactile and olfactory signals.

The relative sizes and positions of the eight eyes on the carapace are useful characters by which to recognise wolf spiders (Fig. 6.2). They are grouped in three rows, the four front (anterior) ones on the forward sloping face of the carapace being small and in a more or less straight row. The large median pair (PM eyes) have a field of vision to the side and the front, while the large hind pair (PL eyes) are well back and directed to the side and rear. Rovner found that the two large PL eyes detect a moving object and initiate turns of up to 160° so that the spider faces the source of the movement. If this stimulus is prey, the spider makes a dash towards it, an action primarily directed by the PM eyes.

Prey capture

If you watch a wolf spider near its potential prey you will see that even when prey is behind it, the spider quickly turns towards it. Once an insect a few centimetres away is detected, the spider exhibits the sudden speed and certainty of action which has doubly earned it the name of wolf spider. Like lightning, the gap is closed and the spider is next seen clasping its prey with the four front legs while the fangs plunge in to inject the paralysing poison. After a few frantic jerks the insect gives an occasional twitch then the spider withdraws its legs.

Feeding begins while prey is transfixed by the fangs. Depending on size it takes the spider from 5 to 30 minutes to reduce an insect to the soggy little ball of indigestible crushed cuticle and wings that is dropped and left behind as its quest for further food continues. So what does a wolf spider eat? To find out what web spiders eat, you can collect the vestiges from their webs or examine the ground below for prey remnants. But because the wolf spider is a wanderer we could not do this. Instead we kept some in captivity and fed them with a variety of likely food items from their usual habitats. The answer is simple – a wolf spider eats practically any invertebrate that is smaller than itself, except for slaters (woodlice) which, unfortunately, are the most common animals in our garden.

Although *Lycosa hilaris*, like most wolf spiders, may find refuge in silk-lined cocoons under loose stones or logs, they do not occupy permanent burrows as many overseas species do, but instead hide in cracks and crevices or the base of clumps of grass. The silken cocoons, however, are constructed for special purposes such as moulting or during mating, egg-laying and emergence of the spiderlings. Spiders at each of these phases can be found in these snug little retreats.

The way a female wolf spider cares for her eggsac and spiderlings is a special feature of this family. After fashioning her eggsac she carries it behind her, attached to her spinnerets (Fig. 6.3). Few other spiders behave in this way although, unexpectedly, two quite different native New Zealand spiders sometimes carry their eggsacs like this. One of these is a minute, not-yet-described, cobweb spider (Theridiidae), often found in crevices in the ground, while the other is a small, long-legged spider (Synotaxidae), recently described from the South Island. If the female in either of these two species, is disturbed in her web, she quickly attaches the eggsacs to her spinnerets and moves them out of danger.

Eggsac construction

Our observations show that the wolf spider's eggsac is usually constructed at night. In early evening the gravid female finds a sheltered spot and lays down a thick sheet of silk. Some 4 cm in diameter and roughly circular in outline, this sheet is firmly attached to the ground by numerous threads around the margin. The spider lays her eggs in the centre of this sheet by standing on tiptoes and lowering her abdomen. Eggs ooze out in a glutinous mass from the gonopore under the abdomen and settle as a small pinkish ball of eggs. Immediately, the spider begins to lay down an upper sheet which will cover and seal the eggs within a cocoon. At first the cocoon looks like a lightly poached miniature egg, but as more silk is laid it becomes a white biconvex disk (Fig. 6.4a).

Fig. 6.3 Wolf spider females always carry their eggsacs behind them, attached to their spinnerets. Here, Allotrochosina schauinslandi *keeps her abdomen high so that the eggsac is held above the ground. This species is restricted to damp habitats within native forests and is found throughout New Zealand.*

The positioning of this upper layer of silk is by no means haphazard, and with careful observation it becomes clear how she achieves such a perfect result. After depositing and sealing the eggs, she stands over the mound on the tips of her legs and then, by bending her abdomen, she touches the eggs with her spinnerets. She then backs down the dome of the egg mass, keeping her spinnerets on the sloping surface. As soon as the spinnerets cease to move downwards and reach a level surface, a new behaviour pattern is triggered, because it is then that she begins to exude silk. The silk is laid with an upward, stroking movement of the abdomen. After the first strand is laid, she lifts her abdomen, bends it first to the left and then to the right to stroke up two more strands so that at each position she lays three overlapping strands. After three strokes have been completed, she moves further up towards the centre of the sheet and goes through the same procedure. This is repeated until the spinnerets reach the top of the eggmound; then she stops, turns slightly in another direction and edges backwards to the margin of the sheet where the sequence is repeated. It takes much longer to describe her activities than it does for her to perform them. Each set of actions, from the margin of the sac to the apex, only takes some 40–50 seconds, but as each layer is very thin the whole process is repeated again and again so that two or more hours may pass before the eggsac cover is completed. The end of the task is apparently recognised by the spider when the slope down the egg mass is fully bridged by the silk and this is marked by the abrupt cessation of these movements.

The final processes

It was the next stage which had us most puzzled for here was the female with an egg-filled disc firmly attached to the ground. How was this to be transformed into the familiar, free, globular eggsac that bounces from her spinnerets? Actually it proved to be quite straightforward, and this is how she does it. Standing astride the sac so that her fangs are over the margin, she uses them to tear at the edges of the attached sac. As some threads are freed, she slips her palps under the margin and by lifting herself up to the very tips of her legs, tugs hard to free it further. Sometimes the four front legs are also slipped under the sac while she levers herself with the two hind pairs. As one portion becomes free she moves to another section and quickly pulls this free. When most parts have been detached she grasps the whole sac with her two front pairs of legs (Fig. 6.4b). With a sudden tug she pulls it free and holds it firmly with her legs under her body. Now it becomes clear how the eggsac attains its ball-like form because she revolves it round and round with her legs while depositing further silk (Fig. 6.4c). It is also evident why lycosid eggsacs have a distinct ridge around them, as this is the seam where the upper and lower sheets are joined. Moreover, it is here the eggsac eventually breaks open to release the spiderlings for the next stage of their lives.

Once the eggsac is complete, the female bends her abdomen to touch it with her spinnerets. With a few dabs of silk she dashes away with the eggsac carried behind and concentrates on the task of catching food. The sight of one of these active spiders makes you wonder how the eggs, and eventually the spiderlings, can survive this treatment for the month or so ahead. However, the silken cover is resilient and because the mother has now acquired a singleminded need to have something attached to her spinnerets the sac is rarely lost. This instinct is particularly strong for the first week, so much so that if the sac is removed and given back after a day or so she will grasp it with such a sense of urgency that it is almost as if she is saying 'thank goodness my long-lost eggsac has been found'. A few simple experiments soon show that little more is involved than an urge to carry a small roundish object. She will just as readily accept another spider's eggsac including that of quite different spiders and even a small ball of cottonwool or cork as long as it round and not too heavy. With time, however, this instinct becomes weaker and if the eggsac is then removed, it is quite likely that she will abandon the cares of motherhood completely.

Early development

It is not possible to watch the development of the eggs through the opaque silken cover, but the sac can be removed and a small hole cut through the silk to see what is happening. To follow the development of the eggs for any length of time the eggsac must be kept in fairly humid conditions. Upon returning the eggsac to the spider, she will examine it carefully and seal the opening before attaching it to her spinnerets again. About five weeks elapse between egg-laying and emergence of the spiderlings. But spiderlings actually break out of the egg membrane in less than four weeks and so, for six to seven days, the helpless spiderlings remain quietly within the eggsac. After undergoing their first true moult they are then ready to emerge.

By now, the once gleaming white eggsac is a dirty grey and the silken cover is stretched tight as the spiderlings jostle within the sac. In all our observations the

Fig. 6.4a (below left): This female wolf spider (Lycosa hilaris) has completed the task of laying eggs and is now covering them with silk. **b** (below right): Tugging hard, the female pulls her eggsac away from its attachment base.

Fig. 6.4c *(above left): The eggsac is held beneath the spider as she rotates it and adds more silk.* **d** *(above right): After the eggsac has been carried around by the female for about three weeks, it bursts open and the small pale spiderlings make their way out.*

sac opens at the seam, although in some overseas species the female actually bites it open. This we never saw and although the female, at times, pecked at it with her fangs, the eggsac always broke open by itself, perhaps due to the pressure exerted by the spiderlings moving about inside. At this stage the spiderlings are colourless except for the conspicuous black pigment around their eyes. Emergence is slow, so it takes two to three hours before the first spiderlings move out (Fig. 6.4d) and it may be next morning before the sac is completely empty. Just before emergence, however, the female usually spins a silken retreat where she remains until all spiderlings have moved out of the sac and amassed on her abdomen. The number of spiderlings ranges from 80 to 135, so the mother has to cope with a great many of them clinging tightly to her abdomen or to each other for the next stage of their lives.

The ride of their lives

The first task faced by spiderlings after clambering out of the eggsac is to make their way safely to the mother's back. Fortunately, the female is normally quiescent at this time and those that climb out on the wrong side often touch a leg as they wander about and use this as a guide to reach her back. Because there is rarely enough room for all spiderlings to grasp a hair or bristle, two or three layers accumulate, so that a female running about with her brood aboard looks rather top heavy (Fig. 6.5). But she shows little or no maternal solicitude and her passengers are ignored unless they stray forward over her eyes and then, with a flick of her front legs, she brushes them aside. So it is not long before the load is reduced to a more manageable size, for she is unconcerned when some of her offspring lose their footing and fall by the wayside. Unless they are lucky enough to grasp a leg and climb up again, they are left behind. Spiderlings do not feed during this time nor do they attack one another, but in a week or so, as their next moult approaches, they rapidly disperse and from then on have to fend for themselves. So what are the advantages of this 'piggyback' behaviour? Perhaps their chances of survival in that first week are assisted by the quick reactions of the mother to danger. Or maybe the female's instinctive maintenance of critical humidity requirements during this vulnerable period prevents her young from becoming dehydrated.

At home in the open country, the herb fields and the tussock country, the striped wolf spider is one of the few spiders to have adapted to life in our pastures and gardens. Another wolf spider occasionally found hiding in damp places under boards or clumps of grass, is the brown wolf spider, *Allotrochosina schauinslandi* (see Fig. 6.3). This species, our only regular forest dweller, depends on a constantly high humidity to survive, soon dying if kept in a dry container, for instance. In its forest habitat, it can be found in daylight hours hiding

Fig. 6.5 *This undescribed wolf spider has a full load of spiderlings on her back as she sets out to hunt for food.*

Fig. 6.6 A male mountain wolf spider clasps the female firmly as he mates with her. Note the difference in colour and pattern of the male and female abdomens.

under or within rotting logs or amongst the leaf litter on the forest floor. It only comes out to hunt at night. The bluish-tinged eggsac, decidedly smaller than that of the striped wolf spider, has a more distinct ridge marking the junction of the two silk layers. The female tends to carry the sac beneath her raised abdomen when moving about.

Courtship and mating in an alpine Lycosa

During courtship, the male in this mountain species raises his front legs into the air and then, while approaching the female, he bends them in a U-shape and vibrates them so fast that it is hard to see the legs moving at all. If the female is not receptive she waves her front legs in unison above her head, at which point the male usually retreats. Leg movements by either the male or the female act as a threat or defensive signal when they meet at inopportune times and immediately cause the challenged spider to depart. Adult spiders rarely attack each other although they may seize and eat a younger spider. If the courting male's advances are accepted, he creeps over the female's head and reaches below her abdomen to insert his palpal embolus into one of the epigynal openings, first on one side and then with the second embolus on the other side (Fig. 6.6). This entire activity is extremely rapid and is completed in a few minutes.

Riverbed species

Many quite distinct lycosid species are found amongst the stones in riverbeds, particularly in the South Island, although relatively few are named. One of these is the slatey grey *Lycosa arenivaga*, frequently seen in the heat of the day running across rocks and capturing small insects sheltering between them. *Lycosa* often falls victim to a spider wasp, *Priocnemis nitidiventris*, which darts about among the stones until it locates a spider (Fig. 6.7a), which the wasp then stings and drags back to a safe hiding place while it prepares its burrow in the sand. After pushing the spider into the burrow, the wasp lays a single egg on its body and when the larva emerges, it feeds on the live but comatose spider until it is ready to pupate (Fig. 6.7b).

Many species of drab wolf spiders live in riverbeds while others are found along beaches where rivers flow out to sea. Other habitats, and probably the source of many riverbed populations, are the screes above the bushline where many mostly dark-coloured species live. However, despite the similarity of lycosids from one riverbed to another it is likely there are a number of species involved. For example, there is a larger fawn-coloured spider with a distinct yellow stripe on the hind part of its abdomen which hides beneath the stones during the day and hunts at night. One of these, *Pardosa goyeni*, is found in Central Otago. Neither this spider nor *Lycosa arenivaga* make a permanent burrow, but in early summer females are found in compact horizontal burrows scooped out beneath stones and situated well above the normal waterline. It is in these retreats that the eggsac is constructed and the young are hatched.

Fig. 6.7a (below left): Priocnemis nitidiventris, *a parasitic wasp, seizes and stings this riverbed wolf spider as it tries to hide.* **b** *(below right): The paralysed spider has been removed from the wasp's burrow, and it can be seen that the wasp larva has grown fat as it fed upon the spider's internal tissues.*

Most wolf spiders are of sombre colouring and are hard to see against their background, but a number of undescribed seashore wolf spiders have truly mastered the art of camouflage. In these species, the pale cream background colour dotted all over with dark patches ensures that the outlines of the body and appendages are effectively obscured. Walking along the beach above the high tide level, our attention may be drawn first to a scurry on the sand, but out of the corner-of-an-eye we catch a further movement like a puff of sand. A closer look enables us to distinguish the outline of the spider, motionless against the sand with its legs outspread, thus blending perfectly with the mottled sand beneath it (Fig. 6.8a). We find these spiders along most of the sandy coasts of the South Island and Stewart Island, and note the subtle change in the spider's own colours as the overall shade of the sand varies from locality to locality. Compare this to the striking colour and pattern of another wolf spider, as yet unnamed (Fig. 6.8b).

Fig. 6.8a (above left): Motionless upon the sand, this seashore wolf spider is hard to see. The different colours of the sand grains are mimicked by the hues and patterns of the spider's body and legs. b (above right): By contrast, this handsome wolf spider does not blend with the background on which it rests. This spider was found in Dunedin but little is known of its habits.

Lycosid burrowers

Overseas, many lycosids make and live in burrows in the earth, but we have seen only two instances of this in native species. One, a fairly large undescribed species living in open-mouthed burrows in Northland, may be closely related to some Australian burrowing lycosids. The other is a much smaller spider rather like the common striped lycosid from Central Otago. This species, named as *Lycosa bellicosa* by Goyen, inhabits grassland where it digs a more or less vertical burrow some 10 to 15 cm deep and extends the opening with a rim of silk into which it incorporates a few pieces of the surrounding vegetation. From its position at the opening of the burrow the spider dashes out as an insect approaches, captures it and returns to the retreat to feed. The female attaches her eggsac in the usual manner but remains in the tunnel, preying on passing insects. On a fine day she comes regularly to the mouth of the burrow and sits with her head down so that the eggsac, held by her spinnerets, protrudes from the burrow into the sun. In captivity, this behaviour is easily elicited. Within a short time of bringing a lamp close to the mouth of the burrow the eggsac is seen being pushed through the few threads which normally draw the opening together.

Nurseryweb spiders: Pisauridae

Large, gleaming white nurserywebs on the tips of low shrubs such as gorse, broom or manuka are familiar to everyone (Fig. 6.9) but, although most people know these webs are made by a spider, how many have actually seen the spider itself? Because it is nocturnal by habit, you will more readily see the architect of this web if you go out at night with a torch. Only then will you find the female *Dolomedes minor* spider clinging to her nurseryweb having crept out of her daytime hideaway to guard it during the night. She is a magnificent creature whose body seems clothed with the finest velvet. Her legs are brown and the carapace is patterned

Fig. 6.9 At night, the female Dolomedes minor *spider can be seen guarding her nurseryweb, but during the day she hides somewhere below it.*

on each side with three longitudinal bands, first brown, then yellow and finally black where it reaches the narrow yellow stripe down the centre. This yellow stripe continues down the middle of the abdomen flanked by a brown band extending back to the spinnerets, and marked only by pairs of small white spots. The sides and lower surface of the abdomen are paler but speckled with brown. Although the body is only about 2 cm long the females always seem huge because the legs span 5 cm or more. The much smaller male, with its noticeably slender legs, does not make webs and its main use of silk is the dragline.

Distribution

Spiders belonging to the Pisauridae are found all over the world. Because they are normally associated with water they are often called fisher or swamp spiders, but the widespread New Zealand nurseryweb spider (*Dolomedes minor*) is one of the exceptions. Although it is as frequently found in swamps as in dry scrubland this native spider does not require free water to thrive, so its webs are a familiar sight along the sides of roads and on scrub-covered hillsides. Wherever there is low cover, particularly scrub, grasses or sedges, *Dolomedes minor* is found but as it avoids shaded areas it is never found in deep forest (Fig. 6.10). Although favouring the lowlands it is also found high up in the mountains where its nurserywebs are seen binding the tips of the tussock together and swaying in every breeze.

However, two other New Zealand species are more dependent on an aquatic habitat. The best known one, *Dolomedes aquaticus*, is found on the shores of some lakes as well as on the stony banks of rivers and streams which run through open country. These spiders were well known in Central Otago in the goldmining days of last century. Large numbers lived beside the water races built by miners for sluicing and so were called 'water spiders' – a name which still persists. This is largely because Goyen perpetuated this name in an account of this spider's habits in 1887. A second species of aquatic nurseryweb spiders, discovered only recently, is restricted to streams passing through forest and is referred to as the forest water spider. We will come back to these two groups of spiders later.

Prey capture by Dolomedes

Despite the abundant use of silk for nurserywebs, these spiders do not construct snares for they are hunters that catch their prey in a manner similar to that of wolf spiders. Very sensitive to vibration, they pause with the two front legs stretched forward as an insect approaches. Little happens until the insect is very close or even actually touching the spider, but the ensuing action is extremely rapid. With a quick flurry, the spider pounces on top of the insect, clutching it

Fig. 6.10 *Daylight reveals the striking colours of this* Dolomedes minor *found near Fox Glacier on the West Coast of the South Island.*

tightly under the cephalothorax with all four pairs of legs. Fangs are thrust in at once but as the prey's struggles decrease the legs are withdrawn and the victim is then held firmly by the fangs and palps (Fig. 6.11). As digestive fluids ooze from the mouth, the fangs open and close and the chelicerae move out and in, crushing the body of the prey. The insect is rapidly reduced to a small moist ball which is turned round and round as fluid is alternately exuded from and sucked into the mouth. Eventually all that is left is a bundle of indigestible fragments, which is then discarded. Depending on the size of the catch, 10–30 minutes may elapse from capture time to completion of the meal, after which the spider spends another 30 minutes grooming itself. It begins by rubbing the palps together, then draws each one between a moistened fang and its basal segment. This is repeated several times after which the spider goes through this procedure first with the front legs and then the other legs in turn. To finish, the spider sweeps each palp across the eyes with a movement just like us when we brush the hair from our eyes. With its toilet completed, the spider sallies forth once more to search for food.

Fig. 6.11 Dolomedes *feeds upon her prey, in this case a blowfly, which is being manipulated by the palps and chewed with her fangs.*

Reproduction

It is late spring or early summer when female spiders are seen, each carrying beneath the cephalothorax a huge eggsac held there by her fangs. Because of its size the spider has to move about on tiptoes. These eggsacs are carried for at least five weeks before the spiderlings emerge – but because the female uses her fangs for this new task, how does she manage to eat? What happens is that at the same time as prey is seized with the legs, the fangs release the eggsac which is quickly pushed back under the body by the tips of the palps thus leaving the fangs free for action. Sometimes the spider releases the eggsac entirely but keeps it close by so that at least one leg is touching it lightly. At the slightest disturbance the eggsac is scooped up with the fangs, clasped firmly beneath her body and she dashes nimbly away carrying both the sac and prey without any noticeable reduction in speed or agility.

The nurseryweb

Towards the end of five weeks, the eggsac loses its pristine whiteness and begins to turn grey. The mother may be seen pecking at the surface with her fangs although not actually breaking through the silk cover. Perhaps a week or so before this, spiderlings cast off their egg membrane and are now due for their first moult. Once these *ecdyses* (moults) are completed they have reached the next stage of their lives. But how the mother knows when spiderlings are ready to leave is a puzzle. Does she detect movement in the sac or has she a biological clock which signals to her when the requisite time from egglaying to emergence has elapsed? Whatever the mechanism is, her whole behaviour now changes. She ceases to hunt, and climbs to the top of some nearby shrub or clump of grass, firmly attaches the eggsac and spins the bulky nurseryweb around it (see Fig. 6.9). This always happens at night and, although the eggsac is now disposed of and protected within the web, she still does not desert her young but clings faithfully to the side of the nest. Here she remains for another week, creeping down to hide at the base of the shrub during the day but resuming her vigil as darkness comes. The spiderlings emerge from the eggsac shortly after the web is completed and climb freely within the 'nursery', laying dragline threads as they move about inside. Surrounded by this sturdy web, they are protected from the weather and from hungry predators who may perceive the web but cannot penetrate the silk.

Spiderlings remain here for about a week, during which time the nurseryweb becomes a little tattered and growing plant shoots may even burst through the silk. It is said that the mother spider actually opens up the web to allow the spiderlings to escape but since empty nurserywebs are usually pitted with small holes, it seems more likely that most of them make their own way out. Even after

Fig. 6.12 *Spiderlings remain near the nurseryweb for a week or so and their combined draglines fashion an extension to their original home. Shortly afterwards, they balloon away to distant fields.*

the spiderlings emerge they do not necessarily move away and many are found near the nurseryweb for some time (Fig. 6.12). However, by the second week most spiderlings have gone, leaving behind the dilapidated remnant of the once shining white web, the inside of which is peppered with the tiny crumpled cast-off skins left behind by the second instar spiderlings.

Water spiders *(Dolomedes aquaticus)*

Water spiders are slightly bigger than the closely related nurseryweb spiders but their colouring is more subdued so that they blend in with the riverbed stones amongst which they are found. In the common Otago species *Dolomedes aquaticus*, the legs are fawn, the carapace is strikingly patterned with a narrow yellowish band down each margin while the inner area is blackish brown. In this species, however, a barely visible band runs down the mid-carapace in contrast to the conspicuous yellow band of the nurseryweb spider. A broad greyish band extends down the upper surface of the abdomen which also lacks the yellow streak of the nurseryweb spider. The male is much paler and, being more slender in the body and legs, is more agile than the female.

During the day these spiders shelter under stones near the water's edge and so are rarely seen, although under the right conditions they are present in large numbers. As they are nocturnal feeders, a cautious visit to a riverbed with a torch may reveal one clinging motionless to a stone with the front legs stretched out on the surface of the water. These limbs are able to rest on the water because numerous fine hairs on the undersurface of the metatarsi and tarsi (scopulae) prevent them from breaking the surface film (Fig. 6.13a). If there is an incautious step or a stumble by the observer, the spider vanishes. Now, no longer on the surface of the water, the spider has run down the side of the rock below the surface and there it can be seen, upside down, gleaming in the light of the torch. Its silver sheen results from a film of air caught between the hairs on its body (Fig. 6.13b). Spiders are able to stay under water for at least 30

Fig. 6.13a *Dolomedes aquaticus clings to a stone while its other legs rest on the water. If anything edible touches one of these legs as it floats or swims past, it will be seized, impaled by the fangs and eaten.*

Fig. 6.13b *If disturbed, the spider runs down the stone under the water where it can be seen covered with a silvery sheen. This is because air has been trapped by the spider's hairs as it submerged.*

minutes, sustained by the air trapped around the tracheal and booklung spiracles, a respiratory mechanism known for various spiders as a physical gill. If the spider is given a slight prod while it is clinging beneath the water it will release its grip on the rock and come flashing to the surface, often on its back. Although upside down, it rights itself with a quick flip. The water immediately runs off and the spider rests, high and dry, on the surface.

Prey capture on the water
Water spiders are primarily sit-and-wait predators, this strategy being carried out at the water's edge. The front legs act as sensors, with sensory hairs such as trichobothria, slit sensilla and metatarsal lyriform organs the means of vibration detection. However, instead of detecting movements of the air or ground-borne vibration, the spider waits for ripples on the water. If the prey is not moving, spiders show little or no response unless it actually floats so close that it touches one of the legs. At once the victim is scooped up, the spider clambers out of the water and seeks shelter while devouring its meal. However, during the scooping-up manoeuvre the spider may lose its footing and its prey, and slip into the water. When this happens, the spider extends its legs and floats downstream and, as soon as it touches the shore, clambers out and takes up its predatory stance once more. At other times, the spider may float on the surface until it encounters a waterborne insect which is immediately grasped with its four front legs. Then it either floats to shore or actively sculls to the river's edge using the second and third pairs of legs.

If these spiders are disturbed while floating on the water they lift their bodies up and run across the surface almost as rapidly as if they were on dry land, an activity also observed in a North American species, *Dolomedes triton*. In fact, *Dolomedes aquaticus* have often been observed using this method of locomotion, even jumping to cross rougher water encountered now and then. We were curious to see whether they actually eat prey while floating because we reasoned that digestive fluids would be largely diluted by water before breaking down the body tissues and the meal would be a very watery one. However, in all instances, the spider came back to shore before beginning its meal.

Mating rituals
As our efforts to observe these spiders courting and mating in the wild were unsuccessful, we decided to put them together in a special cage. The first attempts went sadly astray because as soon as the male was introduced the female pounced on top of him and he became her next meal. We tried satiating the female with flies before introducing the male, but the same thing happened again, although a further trial succeeded. Aggressive female behaviour, despite theories proposed in many popular natural history books, is by no means usual and as this could not happen so consistently in nature we knew that there must

be some factor we had overlooked. We even considered that, like the European nurseryweb spider *Pisaura mirabilis*, it might be necessary for the male to offer a nuptial gift to his mate. In this latter species, the male suitor first captures an insect and after carefully wrapping it in silk, carries it in his chelicerae as he approaches the female. By raising himself on the tips of his hind legs and balancing on the end of his abdomen he offers this carefully wrapped bundle to the female who, if she is receptive, proceeds to eat it as mating takes place. Unfortunately, our male *Dolomedes aquaticus* did not appear to be in any way interested in this technique.

As the only successful mating was with a newly moulted female, perhaps females always mate immediately after their final moult and subsequently regard any approaching male as just a meal. In this instance, the successful mating we did observe was preceded by a very short courtship. The male slowly approached the female as he alternately waggled the front pair of legs up and down. She remained still as he neared and touched her gingerly, after which he crept directly over her carapace and bending over her abdomen inserted the embolus of one palp. Within a few seconds he withdrew this palp, moved over to the other side of the abdomen and inserted the other palpal organ. The whole process was over quickly but it still remains to be seen whether this single observation is typical of our water spiders.

Eggsac construction in pisaurids

The method of eggsac construction by New Zealand pisaurids is unknown but is probably similar to that described for related European species by Pièrre Bonnet. In these species the female first spins a few threads from which she hangs. Next, a thick silk platform is woven and then her body weaves circles around it until a small silk dome is completed. At this point she hangs upside down from the dome, her body held at an angle with silken threads. The eggs are deposited into the cuplike interior of the dome as she presses her abdomen underneath it after which she quickly secures them in place with an extra swathe of silk. Further silk is added until the eggs are completely enclosed and the sac begins to take its final form. The threads which hold the sac are cut so that the female now hangs down from her hind legs while holding the sac firmly with her palps and legs. The sac is rotated as the final covering of silk is added; then with the eggsac held firmly with her fangs, she moves about freely for five weeks or so until the spiderlings emerge (Fig. 6.14a). However, the female catches her prey and feeds in the same way as the nurseryweb spider and even moves over water as easily as she did before.

Curious to know whether the water spider constructed a nurseryweb in the same way as *Dolomedes minor*, we spent a year in futile searching. To begin with we knew that there were typical nurserywebs on shrubs along the banks or beds of rivers and streams but we soon discovered that they always belonged to *Dolomedes minor* and not the water spider. One day, a large sheetweb strung between stones on a riverbed caught our eye (Fig. 6.14b) and under this we found a female *Dolomedes aquaticus* crouched next to her eggsac, which was attached to the side of a boulder. In contrast to the 'upward' instinct of the nurseryweb spider, apparently the female water spider moves 'downwards' into a crevice or between rocks and boulders beside the riverbed when her young are ready to hatch. Here, she attaches her eggsac and spins her 'nurseryweb' above and between the rocks, to protect the eggsac. Once you know where to look, these 'nurserywebs' are easy to find.

Fig. 6.14a *A* Dolomedes aquaticus *female with her huge eggsac held below her body by the fangs. Despite its size, she runs nimbly across the water and readily catches prey.*

A new Dolomedes species

When we first wrote about these spiders only two species were known in New Zealand – *Dolomedes minor* and *Dolomedes aquaticus* – which we have described above. Since then a third species of *Dolomedes* has been found and this spider is recognised by its greyer, more subdued coloration marked with a distinctive dark-speckled pattern on the upper abdomen (Fig. 6.15). It is further differentiated by its much longer legs. As it has not yet been formally described, it is known here as *Dolomedes* sp. This 'species' is associated with streams which run through native bush or dense vegetation and is found under stones or roots by the water's edge in much the same way as *Dolomedes aquaticus*. But, despite their aquatic habits, these spiders build typical nurserywebs in low shrubs, herbs and grass alongside stony streams and also occasionally beside creeks flowing through pastures from which forest has been cleared. The nurseryweb itself is similar to that of *Dolomedes minor* although the webbing is not as dense.

In 1979 David Williams made a study of this new spider's ecology and habits and found that, like *Dolomedes aquaticus*, it also sits on a stone with its anterior legs resting on the water. However, it not only catches prey which makes contact with these legs, but also races some distance across the water in pursuit of its quarry. This sounds very much like the behaviour of the North American *Dolomedes triton*, which is known to respond to water surface waves generated by insects trapped by individual 'menisci' at the air/water interface, a finding linked to a series of studies on these spiders by Bleckmann. While adult aquatic insects are the main prey of *Dolomedes* sp., flying insects and ground invertebrates are also caught, detected in the main by tactile and airborne vibratory signals. Moreover, this *Dolomedes* sp. is known to catch small fish, mainly cockabullies, which swim beneath its legs, almost close enough to touch them with its fins. As with other prey, this spider retreats to solid ground to predigest and imbibe its large catch.

Fig. 6.14b (above): Before the eggs hatch, the female builds a nurseryweb amongst the stones beside the river or lake.

Fig. 6.15 (below): This third Dolomedes *species is more likely to be found in creek beds shadowed by native forests or dense vegetation. It appears to be more stream-adapted than* Dolomedes aquaticus, *since it is adept at running across water to seize its prey and does at times catch small fish.*

Fig. 6.16a *(above left): The South Island fernbird brings a* Dolomedes minor *female to feed her hungry chicks.* ***b*** *(above right): A* Dolomedes *eggsac, packed with protein, makes a good meal for growing fernbird nestlings. (Photographs Wayne Harris)*

A Dolomedes *predator*

Almost all free-living spiders are food for birds and lizards and even fish but a major predator of *Dolomedes aquaticus* and *Dolomedes* sp. in southern New Zealand is the South Island fernbird. In a long-term study of the breeding ecology of this bird, Wayne Harris showed that these spiders and their eggsacs make up a large proportion of the diet of nestling fernbirds (Figs 6.16a), although other spiders are also taken. These birds, small inconspicuous passerines with poor flying capabilities, occupy areas with dense assemblages of low vegetation and significant amounts of surface water, habitats favoured by these two *Dolomedes* species. Fernbirds exhibit considerable tactical skill in their capture of *Dolomedes* for they first shake the shrub bearing a nurseryweb, so that the spider drops to the ground. Immediately, the fernbird follows and, as the spider rushes for cover, the bird tracks its path and seizes it. Eggsacs are also provided for the young nestlings (Fig. 6.16b), ones taken from the nurseryweb itself being full of ready-to-hatch spiderlings.

Impact of spider predation on insect pests

Many lycosids and pisaurids are common in pasture and open country, preying on insect pests which damage crops and grasses. These include beetles, moths, flies and their larvae as well as grasshoppers, crickets, plant suckers and the like. In New Zealand, no long-term studies on the impact that spiders have on insect pests have been undertaken. However, an overseas expert, A.L. Turnbull, estimated that spider populations in meadows and pastures in England could be as high as $130/m^2$ and that each spider might consume 0.1 g of prey every day. Only local studies can determine the validity of such potential predation in New Zealand. There is clearly a need for ecological and behavioural studies in this country because biological control methods can supplement and may eventually supercede the chemical management of pests.

CHAPTER SEVEN
CRAB SPIDERS

Crab spiders are always rather flattened in appearance, an effect enhanced by the sideways position of the legs, which are kept close to the ground very much like the familiar seashore crab. All are free-living vagrants and although they are generally classified as hunting spiders, they have poor eyesight and their vibratory systems are apparently short-range. These spiders wait motionless for prey, which is grabbed when near enough for the strong pair of front legs to be brought into action. Then the victim is crushed before being bitten with the fangs. The Sparassidae and Thomisidae are found in New Zealand, but only the Thomisidae is native here.

Giant crab spiders: Sparassidae

Members of this family are large and crablike, hence their name giant crab spiders. They are not native to New Zealand although the banana spider, *Heteropoda venatoria* (Fig. 7.1), periodically arrives here with imported fruit – usually bananas – and often manages to survive fumigation. Some years ago this spider was a very common visitor, but today it is less often seen because control measures are now more rigorously applied. Despite their more frequent importation in the past they never became established. Those that did arrive were often holding grey, lens-shaped egg-sacs beneath the cephalothorax, each containing hundreds of eggs, but fortunately the young spiderlings did not thrive in our climate and rarely lived beyond their second moult. In most tropical countries these giant crab spiders are a familiar, and often welcome, sight as they scuttle over the walls, capturing and eating many obnoxious insects including cockroaches, which for many are their favourite food.

Fig. 7.1 Better known as the banana spider, Heteropoda venatoria *(Sparassidae) is relatively harmless. In the past it was often found in New Zealand amongst imported bananas – hence its common name. The position of the legs and its rapid sideways scuttling movements are similar to those of a crab, which is why the group is known as crab spiders.*

Fig. 7.2a *(left): The Avondale spider,* Delena cancerides *(Sparassidae), introduced from Australia many years ago, is now well established in parts of Auckland, especially Avondale after which it is named. Reports suggest that it is now spreading further afield.* **b** *(right): Although a close-up view makes the Avondale spider look rather fearsome, it is not known to bite.*

Avondale spider

One member of the Sparassidae now firmly established in the Auckland district is the huge and hairy Australian Triantelope *Delena cancerides*. It made its first appearance in the Avondale suburb of Auckland some forty to fifty years ago and is now widely known as the Avondale spider (Fig. 7.2a, b). These relatively harmless spiders with their 5 cm leg-spread are a familiar sight in many Australian houses where they are not only tolerated and but also appreciated for their diligent hunting of household pests, and perhaps one day they may enjoy a similar role in this country. It was this spider, collected in Avondale by Grace Hall of Landcare, Auckland, which came to fame in the film *Arachnophobia*. It so happens that this native spider of Australia is a prohibited export from that country, but the American producers of the film were able to obtain a supply from the Avondale population which, of course, has no such protection.

A natural habitat for *Delena cancerides* in Australia is under the peeling bark of eucalyptus trees or sometimes in foliage. They leave their hiding places at night and can usually be found on tree trunks where they catch their prey. Often known as huntsman spiders, they are opportunistic predators, eating a variety of flies, moths and even frogs and lizards. When disturbed *Delena* moves with astonishing rapidity, frequently with a sideways or backwards motion. In the wild it disappears under bark or logs in a flash, and even in captivity it can climb on glass and cling underneath, able to resist efforts to dislodge it. This is achieved by a brush of short black hairs (scopulae) on the underside of the tarsus and metatarsus of the first three pairs of legs and these hairs have 'end-feet' (see chapter1) which, with the aid of fine surface moisture, anchor the spider to the substrate. These legs are pressed against the ground when the spider sidles or climbs but when it runs it raises itself on the tips of the legs.

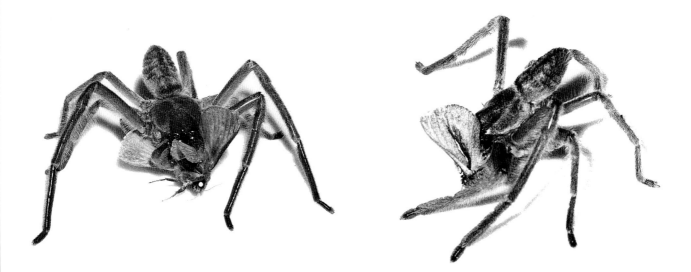

Fig. 7.3a (above left): A fully grown Delena cancerides *is shown here with large prey, in this case a moth. Using its palps to hold the moth, the spider bites it repeatedly. Notice that it has lost its right front leg and the ruptured area has healed. The spider is not handicapped by this loss. **b** (above right): Here,* Delena *circles with the moth. The head region is low on the ground, the abdomen raised high and the legs angled appropriately as the spider lifts and lowers itself while it circles, first in one direction, then in the other.*

Predatory behaviour

Delena is a sit-and-wait spider and only when a housefly, for example, comes into contact with the front legs is there a flurry of activity during which all the legs surround the prey enclosing it in a basket-like trap. The fly is rapidly bitten, the spider impaling it with its fangs and immediately covering it with digestive juices. Another fly is seized in like manner and the process repeated, the second fly being quickly blended with the first by the subsequent mastication. The spider completes its meal in about fifteen minutes and a few scattered shreds of legs, wings and cuticle are all that remain of the flies. After a meal, the spider grooms and cleans itself thoroughly by wiping all its appendages through moistened chelicerae.

A strange ritual

When a large moth is the offering, it too is seized and masticated (Fig. 7.3a). Moths, however, are much larger prey and the spider reacts to large items in a rather curious way. First, it holds its prey off the ground by stilting, with the palps and forelegs lightly restraining the prey. The tarsi quiver several times very rapidly while the prey continues to be bitten and manipulated. About two minutes after capture, the whole body of the spider seems to quiver, it raises its body even higher and then makes a clockwise revolution with the abdomen first being raised in a sweeping arc, next the head sinks to the ground and the tarsi are flattened, (Fig. 7.3b), and then the abdomen is lowered. There is a pause, then the spider rotates again in an anticlockwise direction, the head and abdomen being raised and lowered as before. Another pause, and these head and abdominal movements are repeated three or four more times, all while the feebly struggling moth is being masticated. As suddenly as it began, the 'circling' stops, the spider then taking about an hour to complete its meal.

This odd performance with larger invertebrate meals was observed on several occasions and we wondered if fine silk was being laid down on the base. It was clear, however, that if silk was involved, it was not being applied to the prey. After the spider finished its meal, we would dust the area with cornflour and look carefully for silk under magnification. None was seen. This ritual was never observed when small prey, such as one or even two house flies or small spiders were involved, but if three or more houseflies were caught at once, the circling occurred. The first time this behaviour was observed was in the wild in Australia, so the above experiments were undertaken to see if an explanation could be found.

By chance, however, when the spider was on a piece of bark wedged upright in its cage and it caught a blowfly, circling occurred and this time silk was observed, the end result being an uneven ring of silk around the spider. Does the solution to this curious behaviour lie in the fact that when spiders catch prey on the vertical trunk of a tree, they have to stand on tiptoe and so cannot use their tarsal scopulae to grip? Perhaps the silk provides a hold should the spider slip! But surely by then it would be too late to be useful. Is it a boundary mark to prevent others from sampling their meal? If so, why produce silk on a vertical substrate and not on a horizontal one? And if there is to be no silk, why go through this performance at all? This mystery awaits a solution.

Antagonistic behaviour

At first, interactions between male spiders were thought to be caused by mistaken identity – that a male could not distinguish between another male and a potential mate, and so began to 'court' as if he was in the presence of a female. However, it is now known that interactions between males are common procedures and, as with mammals for example, are designed to ascertain a male's fitness within the species. Because different species have their own rituals, it also helps a male spider to recognise its conspecifics – that is, members of the same species. Moreover, as males generally wander in search of a mate, interactions between males are probably quite common and may also help to keep individuals widely spaced, thus ensuring that females have plenty to eat when they begin egg production.

When two male *Delena* spiders meet each other, their interaction is not a placid affair. In the observations described here, Male 1 was the older of the two, being mature when collected, whereas Male 2 moulted in captivity. The two males quickly made contact and much leg intertwining occurred while they circled around each other. Each male was seen to be pushing against the other, with Male 1 pushing harder and Male 2 losing ground. Suddenly, Male 1 thumped its body on the ground twice, a sound clearly audible to the observer. This double thump was repeated at intervals during the interaction while circling, leg intertwining and pushing continued, a performance lasting some six to seven minutes. But Male 2 did not perform any body thumps and eventually disentangled himself and ran away, although Male 1 continued to body-thump for another two minutes. The assumption is, that Male 1 established his dominance and so was able to seek a mate, but more observations are needed to determine whether such spider behaviour is a regular occurrence and hence similar in function to that of higher animals.

Small crab spiders: Thomisidae

Crab spiders are all able to move sideways or backwards with considerable ease and speed, and those smaller versions known as the Thomisidae are no exception. In New Zealand there are at present three genera, *Diaea*, *Sidymella*, and *Cymbachina*, although a revision of this family may lead to further genera being established. All three groups are cryptically coloured, *Sidymella* mimicking the colour and texture of the logs, tree trunks and detritus (Fig. 7.4) where they are

Fig. 7.4 This strange looking spider belongs to the genus Sidymella *(Thomisidae). It is well camouflaged against the bark of a tree.*

likely to be found; *Diaea*, whose colours merge with the leaves and flowers (Fig. 7.5a, b, c) where they lurk; and *Cymbachina*, which blend with the lichens in their habitat. Although described as wanderers, these spiders are more likely to ambush their prey, remaining largely hidden from sight against the background with which they blend.

Despite the fact that these spiders spend much of their time above the ground and are plentiful on the foliage and trunks of trees and shrubs, they are seldom seen as often as other spiders. *Diaea* are always very small, sometimes being described as pygmy-sized spiders, but they are likely to abound in your garden. If you hold a tray beneath a shrub which you shake vigorously, a myriad of mostly tiny creatures will drop into it. After a minute or so, a number of small *Diaea* will begin to move about (Figs 7.6a, b). A magnifying glass is needed to reveal their various colours and patterns. One of the commonest species in New Zealand is *Diaea ambara* (Fig. 7.7), a thomisid which may have a variety of hues and markings, in addition to changes that occur at maturity. This particular species, originally found by Urquhart in Whangarei Harbour and named by him as *Philodromus ambarus*, was given its present name by Elizabeth Bryant.

Fig. 7.5a *(above left): Sheltering within a moisture-laden leaf,* Diaea *(Thomisidae) is concealed from prying eyes.* **b** *(above centre): Inconspicuous against the flower that it matches, this yellow* Diaea *remains alert for prey or enemies.* **c** *(above right): Sometimes known as flower spiders, this* Diaea *lingers near a bloom likely to attract insects.*

Fig. 7.6a *(above left): While many* Diaea *are green, markings and colour vary as seen in this immature male.* **b** *(above right): This female* Diaea, *almost entirely red, shows up against the green leaf on which it was photographed. In other surroundings it would be invisible.*

Fig. 7.7 *(left): This male* Diaea *was shaken from moss in the forests near Franz Josef Glacier on the West Coast of the South Island. His maturity is marked with red palps and forehead which may be the means by which the female recognises him.*

Fig. 7.8 *After tying the female to a leaf with silk, the male mates with her. However, the female remains docile and these precautions would hardly restrain her if she wanted to move, so perhaps the real function of the silk bonds has yet to be discovered.*

Male tactics while mating

Little is known about the behaviour of *Diaea ambara* but a holiday in Central Otago gave us the opportunity to make some observations of the bright-green variety commonly found in shrubs. When we had found a suitable shrub where *Diaea ambara* males and females were present we settled down to watch. It was not until dusk that we noticed a male whose activities suggested he was seeking a mate. Soon he came to a leaf on which a female rested with her front legs outstretched in a typical 'sit-and-wait' hunting position. She made no movement as the male approached and halted at her side. After a short pause, he extended his front legs, touching first her legs and then her body while she retracted her legs and crouched. To all intents and purposes, the female appeared passive but to our surprise the male turned tail and ran. Reaching the edge of the leaf, he immediately dropped on his dragline, hanging with legs outspread 2–3 cm below. After several seconds, he climbed rapidly up the thread, gathering it up as he did so and, on reaching the top, he dashed quickly to where the female still waited and then immediately climbed on her back. Here, he kept moving his abdomen from side to side across the carapace and legs of the female, behaviour that left us wondering!

With the aid of a magnifying glass, we were able to confirm our suspicions; he was lashing her down securely to the surface of the leaf. Within a minute or so she was firmly trussed. He then moved to the end of her abdomen and crawled under one side (Fig. 7.8) and mating began. This involved the male moving from side to side, alternately applying his palps for half an hour or more, before he departed in haste. There was no question of him removing the restraining threads and his hapless mate was left to struggle free as best she could. But she shrugged off the threads with a few rapid movements and ran off, dragging the strands behind her and so, despite appearances, the silken bonds were little more than a gesture.

If, however, a female is unwilling to mate, or is threatened with an enemy, her legs are likely to be raised high in a defensive position (Fig. 7.9a). Despite such a deterrent, these spiders are not safe from all possible threats. Here (Fig. 7.9b) a parasitic mite has attacked a *Diaea* from behind and is sucking up its body fluids. The spider is not likely to live long under such circumstances.

Fig. 7.9a *(below left): If a female* Diaea *is unwilling to mate, she is likely to raise her legs in the air like this. She might make a similar gesture if disturbed or a possible enemy approaches.* **b** *(below right): A red parasitic mite has attached itself to* Diaea *and is sucking up the body fluids of this hapless spider.*

Fig. 7.10 (far left): Diaea albomaculata, *a species with a distinctively patterned white and green abdomen, is shown here with her large eggsac inside a flax leaf.*

Fig. 7.11 (left): The colour and markings of this unusual thomisid, Cymbachina albobrunnea, *mimics that of the lichen where it lives. Although it was once believed to be rare, the discovery of its habitat led to it being found throughout New Zealand.*

Other species

The more common spiders that we call *Diaea* have variable colour patterns rather than being uniform green or yellow. Typical of these is *Diaea albomaculata* (Fig. 7.10), which we found guarding her eggsac within a silken retreat inside a curled flax leaf. We do not know whether *Diaea albomaculata* and other species related to it share the mating behaviour of *ambara* but it is very likely that some of them do.

A lichenicolous spider

An unusual thomisid whose body and legs are mottled with black and white and a patch of green on the carapace (Fig. 7.11) was only rarely seen until it was realised that its colour pattern resembled the common black and white lichens. These lichens are often found on the trunks of trees and also on old fence posts, so once this habitat was revealed a number of spiders were found although nowhere do they seem to be abundant. First noticed by Urquhart a hundred years ago near Auckland, this species has now been found from one end of the country to the other. It is known today as *Cymbachina albobrunnea* but nothing is known of its lifestyle and habits.

The blunt-ended spider

The grey to reddish-brown crab spiders with the blunt-ended abdomen are all placed in the genus *Sidymella* (Fig. 7.12). Among the twenty or so species known from Australia and New Zealand, the carapace and abdomen are sometimes smooth but more often granular and leatherlike. While these spiders are sometimes seen on tree trunks, they are most often found on the forest floor where they live amongst the leaf litter. There they wait patiently for prey to wander close enough to be scooped up with their spiny pairs of legs. Apart from the dull and drab coloration, which is readily accounted for when one finds them so admirably concealed, the peculiar square shape at the end of the abdomen has

Fig. 7.12 Much larger than Diaea, *this oddly shaped spider with the triangular abdomen belongs to the genus* Sidymella. *Although these spiders are relatively common, their rough texture and cryptic coloration make them hard to find.*

led to much speculation because unusual modifications generally only persist in animals if they have some function. While we may not have the complete answer to this question, we have observed that these spiders make use of their shape in an extraordinary way.

Acrobatics

By and large, when spiders become stranded on their backs, as frequently happens, they push with their legs on one side until they gain sufficient purchase to turn over sideways and so right themselves. But *Sidymella* achieves this same result in its own peculiar way. When thomisids are disturbed, their main defence is to drop on the end of a thread of silk where they dangle with legs outspread before reaching the ground. *Sidymella*, however, always seems to land on its back. If you watch one dangling from a leaf you will see that the first two pairs of legs are spread-eagled outwards, while the rear legs are held close to the abdomen with one of them grasping the dragline. But as the spider falls, its three-dimensional triangular abdomen weights its dorsal surface towards the ground so that this part touches the ground first.

As soon as the spider reaches a firm substrate it severs the dragline and remains lying upside down with its legs curled over its body, looking for all the world like a piece of debris. Here, the spider may remain completely motionless (Fig. 7.13a) for a minute or so but eventually the front two pairs of legs are stretched out and then drawn back under the cephalothorax (Fig. 7.13b). After a further short pause, the front two pairs and the hind pair of legs are stretched to lever the front of the body off the ground. Because the front two pairs are very long and the hind pair short, the spider now rests at an angle with the abdomen supported on its angular edge. A sharp push with the front pairs of legs raises the body to a vertical position steadied by the fourth pair of legs but with the remaining three pairs free and outstretched to balance it (Fig. 7.13c).

This acrobatic stance, with the spider resting squarely on the hind portion of the abdomen, seems perfectly stable; here it may remain for 30 seconds or so, until, by a flick of the legs to the front of the body, the centre of gravity is changed and it topples over to land the right way up, whereupon it is ready to move away (Fig. 7.13d). The first time we saw this complicated manoeuvre we thought it must surely have been an accident – but no, after repeated experiments, we now know that this happens every time. With this unusual preview of the habits of these spiders we looked forward to recording some spectacular aspects of their mating behaviour but this we found to be most prosaic.

Fig. 7.13a (opposite): After falling to the ground, Sidymella *(Thomisidae) lies on its back for 1–2 minutes before starting to move.* **b** *Using its front and side legs, the spider cautiously levers itself into an upright position.* **c** *Once upright, and supported by the blunt end of its abdomen, the spider – now perfectly balanced – pauses briefly.* **d** *By shifting the first and second pairs of legs towards the front, the spider falls forwards and is now ready for its next activity.*

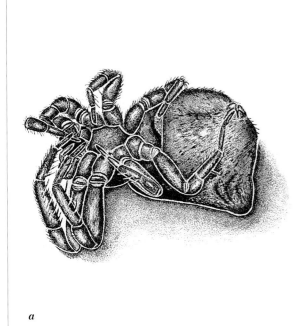

a

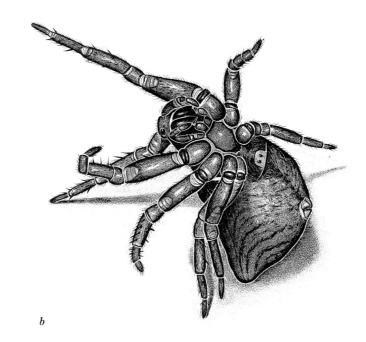

b

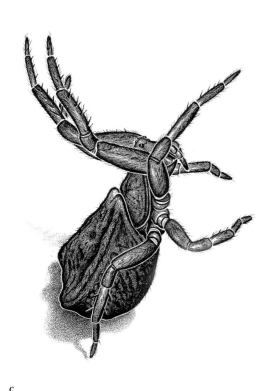

c

d

Mating behaviour

Although the male is much smaller than the female, his approach seems devoid of any preliminary behaviour to ensure that his advances will be accepted. He moves straight up to the female as she rests motionless and, lightly tapping her with his front pair of legs, creeps immediately onto her back. There, while drumming with his palps, he moves around until he faces the end of her abdomen and after a short pause creeps under her abdomen so that he is facing the same way as his mate, but upside down. After applying his palp to one side, he moves back on to the top of her abdomen and creeps beneath the other side to apply the other palp. He may cross from one side to the other four or five times and up to half an hour may elapse from start to finish. The end comes as abruptly as it starts, as he just climbs off and departs. Shortly after mating the female usually locates a dried up leaf and moving beneath it lays down a sheet of silk on which she deposits the eggmass. A thick layer of finely spun silk covers the eggs so that an irregular and slightly raised eggsac remains attached to the surface of the leaf (Fig. 7.14). The female guards the eggsac for some time but apparently does not remain with it until the young emerge.

Fig. 7.14 Sidymella *lays its large eggsac on the underside of a dead leaf lying on the ground. This leaf, turned over so it could be photographed, closely matches the colour of the spider herself.*

CHAPTER EIGHT
HUNTING SPIDERS

Several families of free-living spiders or vagrants roam about searching actively for prey in the habitat where they live. These include the Oxyopidae (lynx spiders), Clubionidae (hopping spiders), Corinnidae (fleet-footed spiders), Miturgidae (prowling spiders), Gnaphosidae (stealthy spiders), Lamponidae (white-tailed spiders), Mimetidae (pirate spiders), Malkaridae (shield spiders) and Cycloctenidae (scuttle spiders). With the exception of Oxyopidae and Corinnidae which are diurnal, the other groups are nocturnal hunters although this does not deter them from seizing prey during the day should the opportunity arise.

Lynx spiders: Oxyopidae

Like the Zodariidae, lynx spiders are usually associated with warmer climes, where they may attain considerable size and also be gaily coloured in greens and reds. New Zealand is only a marginal home for oxyopids and our single species is relatively small and drab. Specialists for life in the foliage of plants, lynx spiders hunt actively during the day and are often seen running and leaping about in pursuit of small insects on the leaves. All are distinguished by the shape of the carapace, which is elevated and appears large in relation to the pointed abdomen, while the slender legs are conspicuously armed with stout spines and bristles protruding at right angles to the surface of the legs (Fig. 8.1).

Named after the keen-sighted lynx, a member of the wild cat family, these spiders too, are reputed to have good eyesight, although this has not yet been investigated. However, their daylight hunting habits support this reputation. The eight eyes are always well separated from the chelicerae and are arranged in three rows of 2:4:2 (Fig. 8.2).

The only named New Zealand species is *Oxyopes gregarius*, described by Urquhart from the specimens collected on his farm in Karaka, south of Auckland. This straw-coloured spider is common in the North Island and parts of the South Island (see Fig. 8.1) where it is found in shrubland and along forest margins. It has also adopted the urban habitat and is often seen hunting on the foliage of garden hedges. In some parts of the South Island and particularly in Central Otago there lives a similar but silvery-grey oxyopid which may represent

Fig. 8.1 (below left): A female Oxyopes gregarius *(Oxyopidae), plump with eggs, hunts on leaves and flowers in her search for food. The long bristles on the legs, a distinguishing feature, can be clearly seen.*

Fig. 8.2 (below centre): A frontal view of the head of a lynx spider (Oxyopidae) depicts the eyes, well placed to detect movements or shadows around it.

Fig. 8.3 (below right): After the eggs are laid in a curled-up leaf, this Oxyopes *females's abdomen returns to its normal elongated shape (cf. Fig 8.1). She will remain on guard until egg development is well advanced.*

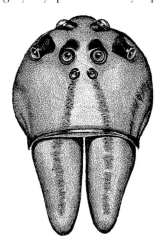
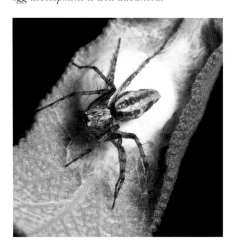

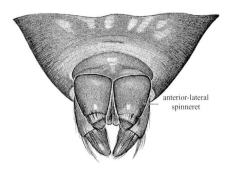

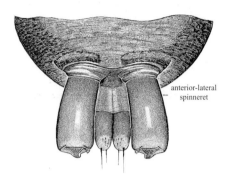

Fig. 8.4a *(above top): In the female clubionid spinnerets shown here, the anterior-lateral ones (in front) are short and close together.*
b *(above): The anterior-lateral spinnerets (seen here on the outside) of gnaphosids are broad and long and widely separated.*

a colour variety or another species. The flattened eggsac is laid in spring or summer, usually on the surface of a leaf and guarded for a period during which the female straddles the eggsac (Fig. 8.3).

Hopping spiders: Clubionidae

Distinguishing features

As the result of a recent revision, there are now twelve species in the cosmopolitan genus *Clubiona* (Clubionidae). Although some clubionids found in New Zealand may be confused with Gnaphosidae, the spinnerets help to separate the families. In female clubionids, the anterior spinnerets are squat and close to each other (Fig. 8.4a) while in gnaphosids they tend to be elongate and separate from each other (Fig. 8.4b). Most of these species live in open areas on low shrubs, tussock or flax or under stones, but a few species are restricted to the forest where they inhabit shrubs and ferns or forest floor litter. Their general colour is pale brown but species can be divided into two groups, one whose abdomen is covered with dark spots (Fig. 8.5a) and a second with a dark band down the midline of the abdomen and chevron remnants to the rear (Fig. 8.5b).

Perhaps the most noticeable habit of many species of *Clubiona* is the way they move about on foliage. Every now and then the quick forward run is interrupted by a sudden hop. This hopping is quite different from the 'jump' of jumping spiders because the body and legs just seem to rise up from the substrate while the spider itself does not move forward more than one or two body lengths, and then not in any particular direction. Perhaps the purpose of hopping is to deter or deceive would-be predators or other enemies.

Mating inside a cocoon

Courtship consists merely of the male waving his legs just before leaping or climbing onto the docile female. More often, however, the male makes a silken cocoon beside that of a female awaiting her final moult. Once the female is mature, the male makes a hole through which he enters her cocoon. We have often seen males mating with females in their retreats. Here, too, the eggsac is constructed and females are commonly found in curled up leaves or, as in the case of the swamp-dwelling *Clubiona cambridgea* (Fig. 8.5c), in curled-up flax fronds, where they seal themselves within the cocoon until the spiderlings hatch.

Fleet-footed spiders: Corinnidae

In our 1973 spider book, we did not mention the active black and white ground spider *Supunna picta*, but, if we had, it would then have been placed in the Clubionidae. Since that time an entire section of this world-wide family, mainly those which mimic ants, has been separated from it and established as a new family, the Corinnidae. Moreover, this family now also includes *Supunna picta*, which, in its native Australia, is a wasp mimic.

Fig. 8.5a *(below left): The spotted* Clubiona *(Clubionidae) shown here is a fast runner whose movements are interrupted now and again by a little hop up in the air.*
b *(below centre): Two striped* Clubiona *spiders have found a sheltered place in which to mate.*
c *(below right): The sheltered space for mating is an ideal place for laying eggs and here, the female* Clubiona cambridgea *guards her emerging spiderlings.*

A Cuvier Island discovery

It so happened that in 1943, one of us (RRF) on radar watch on the remote Cuvier Island in the Hauraki Gulf, just off the coast of Auckland, found a live male *Supunna picta*. At that time, though, it was not realised that this was another addition to New Zealand's immigrant Australian fauna. Not for some fifteen years was this spider, which in the 1940s must have been well established in Auckland, identified as a common Australian spider. Subsequently it began to be reported from all over the North Island. The first South Island record came from Marlborough in 1969 and it took a further twenty years or so for this spider to slowly expand its range to Nelson and the West Coast, to Canterbury and Central Otago, and finally to the East Coast and Southland.

A number of species of *Supunna* are known in Australia but only *Supunna picta* (Fig. 8.6) appears to have been introduced to New Zealand. These spiders, some 7–10 mm in body length, are readily identified by their black colour contrasting sharply with numerous white patches and the pale orange hue of the front pair of legs. In this country these spiders are often seen hunting in the sunshine, mainly in the drier open country and riverbeds, and more rarely in shaded areas, while occasionally some are found in fields and pastures and even gardens. Very fast moving spiders, they run in short bursts and then freeze with all their legs spread out on the ground, making them almost impossible to detect. The lenticular eggsac is usually attached to the undersurface of a stone or on a rock face and is abandoned by the female shortly after it is laid.

In Central Otago we have found *Supunna* under stones or at the base of scrub where they often shelter within a loose network of threads. Here they have been observed with a variety of prey, mostly insects and at times even spiders. From time to time people contact us after catching a glimpse of this unusual spider but the description of its appearance and swift erratic movements make it easy to identify.

Fig. 8.6 *This distinctive Australian spider,* Supunna picta *(Corinnidae), is easily recognised by its black and white markings and orange forelegs.*

Prowling spiders: Miturgidae

Widespread group

The New Zealand representatives of this family are bulky brown or grey spiders which are found throughout the country (Fig. 8.7). Most species live in the forest but some are found in screes above the bushline and even persist in areas where the forest cover has been destroyed. In many places they may be found creeping about the house. All are nocturnal ground hunters which do not construct snares. At night they prowl the forest floor or tree trunks but at daybreak they retreat to a shelter under a log or debris on the forest floor. Despite this daytime concealment, in some areas these spiders frequently fall prey to the large black or red spider wasps, often seen struggling backwards as

Fig. 8.7 *The female* Miturga *(Miturgidae) is a medium to large spider with brown markings, and is a night-hunter.*

they drag their large victims to already prepared burrows. The male *Miturga*, usually more slender and with longer legs than the female (Fig. 8.8a), actively hunts for a mate when he reaches maturity and so is more often seen than the female. After mating, the female prepares a chamber, often lightly lined with silk, beneath a log or stone on the forest floor or even inside a rotting log (Fig. 8.8b). A single round eggsac is laid and guarded by the female until the young hatch and disperse.

A smaller and more slender representative of this group is *Cheiracanthium stratioticum* recently assigned to this family from the Clubionidae. This spider is shared with Australia, from where it is believed to have come. *Cheiracanthium* is common in the North Island but is seldom seen much further south than Canterbury in the South Island. While most of them are usually pale brown in colour, some have green or red pigmentation (Fig. 8.9a, b). They live mainly on grass clumps and tussock where they construct conspicuous retreats. At first glance these look like small pisaurid nurserywebs but, unlike nurseryweb spiders, *Cheiracanthium* are usually found within the web either moulting or guarding eggsacs. Although, in their Australian homeland, some species of *Cheiracanthium* have a reputation for giving a nasty bite, the one living here has only once been suspected of any serious envenomation.

***Fig. 8.8a** (top left):* Miturga *males, like the one shown here, are slimmer than females and have longer legs. His orange-hued palpal organs can be clearly seen on the underside of the cymbium (see chapter 1 for details).* **b** *(top right): When this rotting log was moved, a female* Miturga *with her eggsac was exposed. The very wet and crumbly log was carefully replaced to protect the female and prevent the eggsac drying out.*

***Fig. 8.9a** (below left):* Cheiracanthium stratioticum *(Miturgidae) is the only species in this genus known in New Zealand. This colourful, red-striped female clings to a twig while grooming. Here she cleans her third leg while pulling it through her partly extended fang.* **b** *(below right): This male* Cheiracanthium stratioticum, *found in Foxton in the North Island, is marked with green pigmentation.*

Stealthy spiders: Gnaphosidae

New placements

Recently a number of groups of spiders have been removed from the worldwide family Gnaphosidae and placed in new families. Apart from the Gnaphosidae itself, only one of these new families, the Lamponidae (see below), is found in New Zealand. Four of the ten gnaphosid genera we know to be here are each represented by a single species and are considered to be recent immigrants from Australia. The other six genera, indigenous to New Zealand, each have a number of species. One of these genera, for example, is *Hypodrassodes*, which is also found in New Caledonia where more than twenty species are known.

Many gnaphosids are dull grey or blackish spiders which lead rather retiring lives within small silken retreats under rocks and bark during the day, emerging only at night to hunt their prey nearby. They may live in the same retreat for some time, returning to it after each night's foray. Although most of them run rapidly when disturbed, their normal movements consist of a stealthy progression similar to the shuffling gait of some haplogyne spiders. These spiders generally just seize their prey as they come upon something suitable. If this turns out to be too big to handle, or an enemy, they deal with it in a very different way. Rapidly turning away and raising the abdomen high in the air, they spread their long spinnerets widely and trail a wide band of silk behind them. This either entangles the offender or prevents it following its would-be victim.

A common gnaphosid

One of the most common and widespread of these gnaphosids in New Zealand is the silvery-backed *Anzacia gemmea* (Fig. 8.10) which is usually found hidden within small silken retreats on the underside of stones. During the summer months the female *Anzacia* remains for some time within her retreat keeping watch over her flattened eggsac until the spiderlings emerge. Many gnaphosids share with *Anzacia* the silvery gleam of the abdomen which, under strong magnification, is seen to stem from the presence of numerous flattened hairs. These form a close covering of light-reflecting scales very similar to the scales on a butterfly's wing. *Anzacia gemmea* is usually found where there is little plant cover, such as on shingle banks and screes, riverbeds and beaches, and numerous retreats are grouped together. They seem to prefer dry habitats and are never found in wet forest conditions.

Fig. 8.10 In this silken retreat, an Anzacia gemmea *(Gnaphosidae) female is seen with her newly hatched spiderlings. The silvery sheen on its back may help to disguise this spider as it hunts among the stones and screes where it lives.*

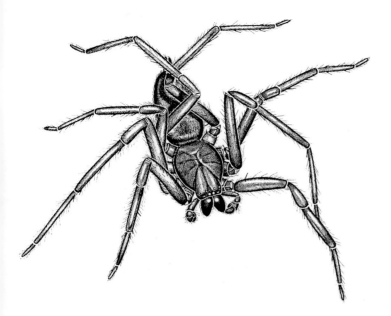

***Fig. 8.11a** (above left)*: Hemicloea rogenhoferi *(Gnaphosidae) has a flattened body and very long legs which are kept angled over its body. On the run, however, these legs are unfolded and the spider reveals its speed.*
***b** (above right): The eggsacs of* Hemicloea rogenhoferi *can be found under bark and stones, with the female in close attendance.*

Eucalyptus spider

The curiously flattened *Hemicloea rogenhoferi* (Fig. 8.11a) is an Australian immigrant originally recorded by A.T. Urquhart as *Hemicloea plautus* and, as long ago as 1895, was found from Te Karaka in the North Island to Dunedin in the south. In 1917, Compte de Dalmas proposed two other names, *Hemicloea alacris* and *Hemicloea celerrima*, for specimens he collected in New Zealand on the assumption that they were native. However, it is now accepted that there is only one species in New Zealand and that this came here from Australia during the last century.

Interestingly this spider, with its flattened body and laterigrade legs, takes advantage of its native habitat, for it is regularly found beneath loose bark on Australian eucalyptus trees growing in this country. Clearly, its flattened shape and crossed legs are adaptations to this slim hideaway under bark. However, it has also found a satisfactory hiding place under the loose bark of other trees and beneath stones and debris on the ground. Like most introduced spiders it does not live in New Zealand's forests but prefers drier habitats. Common in the North Island and parts of the South Island, it is only rarely seen south of Dunedin.

Nothing is known of their mating habits. Eggs are laid in early summer on a thin silk sheet under bark or beneath a stone and then sealed with a smooth cover through which the eggs can be seen (Fig. 8.11b). The eggs hatch in about four weeks but the spiderlings do not emerge from beneath the silken cover until they have moulted, normally in another four weeks. They generally reach maturity before winter and are ready to begin the cycle again as summer approaches later in the year.

Native gnaphosids

Two striking native gnaphosids are often found under the loose bark of some native forest trees such as *Fuchsia*, usually well within the forest. These are *Scotophaeus pretiosus* (Fig. 8.12) and *Notiodrassus distinctus* (Fig. 8.13), two spiders which clearly show the effect of being clothed with modified hair scales. The shiny *Scotophaeus* is densely coated with flattened scales over both the carapace and abdomen, whereas *Notiodrassus,* although lacking such scales, has slender hairs which allow the dull brown pigmentation of the skin as well as the black bands down the carapace and abdomen to show through.

White-tailed spiders: Lamponidae

Many Australian species

Until recently *Lampona* and its close relatives belonged with the gnaphosids, but now they are placed in a separate family, Lamponidae. Within this new family some 170 species are known, with 70 species being recognised as *Lampona*, a genus whose native home is in Australia as well as some of the islands to the north. This cylindrical-bodied grey-to-black spider with the conspicuous white patch on its 'tail' – hence its name, the white-tailed spider – is now widely known in New Zealand (Fig. 8.14a). It has become notorious in recent years, both here and in Australia, for the supposed consequences of its bite (see chapter 17 'Harmful Spiders'). Until the recent revision of this group it was thought that this *Lampona*, known here as *Drassus formicarius* since Urquhart named it thus in 1886, was the same spider named by Koch earlier in 1866 as *Lampona cylindrata*, and to which name it was subsequently returned. Now it is believed that there are two species of *Lampona* in New Zealand, both Australian in origin, the original *Lampona cylindrata* and the newly recognised *Lampona murina* which is very similar in appearance.

At home in New Zealand

Lampona is often found in and around houses where it may be seen stalking and eating other spiders, frequently after luring them out of their webs. These spiders are known as *araneophages* for they subsist almost entirely by eating other spiders and, by a twist of fate, their primary targets are the Australian *Badumna*,

Fig. 8.12 (above left): The silvery covering of light-reflecting scales on Scotophaeus pretiosus *(Gnaphosidae) are apparent in this flashlight photograph.*

Fig. 8.13 (above right): In this male Notiodrassus distinctus *(Gnaphosidae), which lacks the light-reflecting scales of* Scotophaeus *(see Fig 8.12), the markings of the carapace and abdomen can be clearly seen.*

*Fig. 8.14a (right): The white-tailed spider (*Lampona cylindrata, *Lamponidae) is easy to recognise with its dark body and white patch at the end of the abdomen.*

Fig. 8.14b *(above left):* Lampona cylindrata *is seen here devouring a* Badumna longinqua *spider (Desidae), the prey it often lures from its retreat.*
c *(above right): The female* Lampona *lays its eggsac beneath bark or stones and guards it fiercely until the spiderlings hatch.*

the dark-grey house spiders (Fig. 8.14b) which make those untidy webs around our houses. When adult, *Lampona* are mostly grey or blackish, but still retain the white 'tail' mark. The juveniles are distinguished by conspicuous pairs of white patches on the abdomen as well as the patch on the tail, but sometimes paler versions of the abdominal patches persist into adulthood. Eggs are laid beneath bark or stones (Fig. 8.14c) and the young hatch in about three weeks provided the temperature is above 20°C. The fact that the eggs generally do not hatch at temperatures below this indicates the preference of these spiders for the warmth provided by human habitation.

Pirate spiders: Mimetidae

Worldwide group

The Mimetidae are a worldwide group of spiders, very similar in form and behaviour. In New Zealand, native mimetids are also delicately coloured little spiders with distinctive 'cut-off' abdomens patterned in red, brown and black. Their attractive, variegated colour pattern sets them apart from two other kinds of spiders which also have truncated abdomens – the brown thomisid *Sidymella*, and the drab theridiid *Episinus*. If in doubt, a look at the two front pairs of legs will soon establish which is which. Mimetids have a unique and conspicuous arrangement of bristles on these legs, consisting of rows of long erect bristles interspersed with shorter ones (Fig. 8.15), whereas *Sidymella* and *Episinus* do not. These bright little spiders are found in many countries and currently the New Zealand representatives are placed, awaiting revision, in the widespread genus *Mimetus*. Commonly found resting on leaves, these spiders also dangle motionless on a thread, particularly at night.

The common name of pirate spiders has been bestowed on them because they prey on other spiders. Frequently observed feeding on *Diaea*, which they

Fig. 8.15 *Mimetids are known as pirate spiders because of their habit of preying on other spiders. The* Mimetus *shown here reveals the blunt end and long bristles characteristic of this group.*

must simply have grabbed as they roamed about, they are also known to invade webs and seize the occupant, especially the common theridiid, *Achaearanea*. Observations of the British mimetid *Ero furcata* by W.S. Bristowe confirm this reputation. Bristowe kept some of these spiders in captivity and soon discovered that they showed no interest in the various insects that he offered to them. In its cage, *Ero* had laid a few threads across the upper surfaces and hung from them. But a young *Theridion*, for example, introduced into the cage, soon wandered onto these threads. Immediately, *Ero furcata* stirred, then crept towards it, grasped it with its front legs and quickly pulled it close. After biting the femur of the leg of the *Theridion*, the victim collapsed at once, suggesting that *Ero*'s venom was particularly effective. The captive was sucked dry, leaving only the empty husk behind. This feeding behaviour is also common to the theridiids. Perhaps the New Zealand *Mimetus* has similar behaviour.

Shield spiders: Malkaridae

When a group of hard-bodied spiders was first discovered living on the forest floor in New Zealand, they were initially thought to belong to the Mimetidae. This was because they possess the familiar rows of long-and-short bristles on the legs that we mentioned as distinguishing features for that family. In addition, there are similarities in the genitalia, as well as in the eggsac, so they probably are related to the Mimetidae. But unlike the mimetids, the abdomen is not 'cut-off' or blunt at the end, being the more usual, oval spider-like shape (Fig. 8.16a). All known species are dark brown but the really unusual feature is that the abdomen is partly enclosed in a hard, shiny sheath or shield (Fig. 8.16b) which may be a protective device. These spiders move slowly over the forest floor at night and secrete themselves under logs and debris during daylight. It is not known whether, like the mimetids, they are araneophages, but it seems likely that they do feed on other spiders.

Southern distribution

Since that discovery, similar spiders have been found in Australia and South America, so we now know that this family has a typical southern distribution and may yet be found in South Africa. Val Davies of the Queensland Museum made the first formal advance with this group by naming one of the Australian species, *Malkara loricata*, which she placed in a new subfamily, Malkarinae, in the

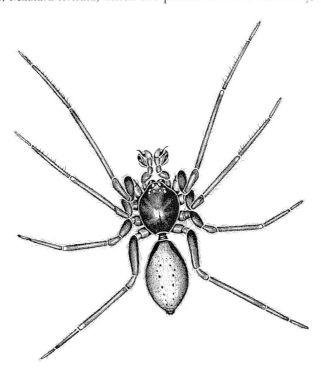

Fig. 8.16a (below left): *Unlike mimetids (see Fig 8.15), the Malkaridae or shield spiders have normal oval, spider-like abdomens. Body length 4 mm.* *b* (below right): *This dark brown little shield spider (Malkaridae) has a hard shiny sheath protecting its abdomen, hence its name.*

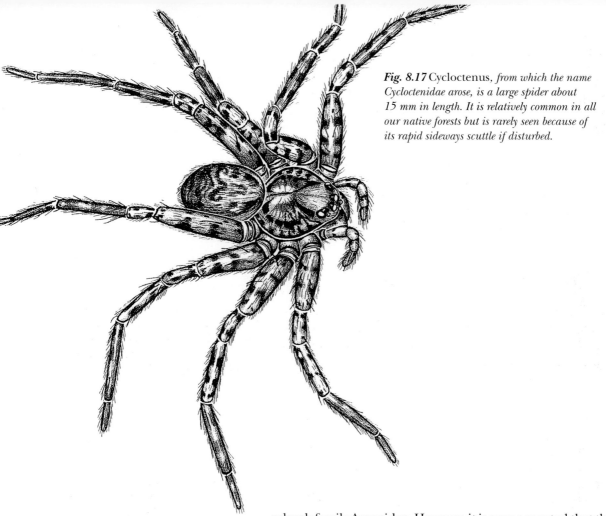

Fig. 8.17 Cycloctenus, *from which the name Cyclocentidae arose, is a large spider about 15 mm in length. It is relatively common in all our native forests but is rarely seen because of its rapid sideways scuttle if disturbed.*

orbweb family Araneidae. However, it is now accepted that they are more closely related to the mimetids than the orbweb weavers. Although little is known about their behaviour, their habit of dwelling in moist confined spaces makes it unlikely that they build orbwebs.

Scuttling spiders: Cycloctenidae

Distinctive eye pattern

Cycloctenidae was established as a new family to accommodate two genera, *Toxopsiella* and *Cycloctenus,* which could no longer be placed in the Toxopidae. They were originally put in this family because of the distinctive arrangement of the eyes (see Fig. 1.14f)) but other features, such as the reproductive organs and tracheal system, distanced them from it. The similarity of the eye pattern in Toxopidae and these two genera may be an example of parallel development, meaning that this arrangement had evolved independently, perhaps because of a particular advantage resulting from such a disposition of the eyes.

The cycloctenids are hunting spiders which do not construct snares and, so far, have only been recorded from New Zealand and Australia. This family is characterised by its simple tracheal system, particular features of the genitalia, and also by the eye pattern, as explained above. Nevertheless, with one exception, none of these spiders appear to rely very much on sight. *Cycloctenus,* however, is one genus that may make use of vision. This is because it always detects movement very rapidly and disappears in a flash around the trunk of a tree or log if you try to catch it. All *Cycloctenus* species are fairly large (Fig. 8.17) and are inconspicuous on fallen logs or tree trunks where their mottled orange and brown colours merge with the background. Moreover, all are laterigrade spiders with their legs directed sideways as in the crab spiders and, being rather flattened too, they are able to sidle quickly in all directions, under bark or into crevices, while remaining tightly pressed to the surface.

Camouflaged eggsacs

The eggsac is a circular shape, but flattened top to bottom, and is usually attached to the underside of a log or stone or even concealed under loose bark (Fig. 8.18). Initially gleaming white, the eggsac soon becomes virtually indistinguishable from its background as the mother spider incorporates nearby debris within a silken outer coating. She often remains near the eggsac until, some weeks later, the spiderlings hatch and emerge through a small hole. Although most of the ten known species of *Cycloctenus* are forest dwellers, a few have been found living under stones high above the bushline, while some species have adapted to life in caves. Although less well known, similar spiders in Tasmania as well as on the east coast of Australia are found, as they are in New Zealand, in forests and caves.

The second genus recorded from New Zealand is the dark-coloured *Toxopsiella* which, unlike *Cycloctenus*, is not laterigrade or flattened so as to provide for a life in narrow spaces, but is found at night freely hunting on the forest floor or sometimes amongst the rocks in screes above the bushline. While most of the known species range in length from 7 to 10 mm and are black and brown there is a smaller species, *Toxopsiella minuta* (Fig. 8.19) with relatively larger eyes and a broad pale band down the abdomen. This spider is common in leaf litter along the West Coast of the South Island. Despite differences in their appearance these spiders share with *Cycloctenus* the unusual disposition of the eyes.

To which family does Plectophanes belong?

Another genus of spiders associated with the Cycloctenidae is the peculiar eight-eyed *Plectophanes*, first recorded in 1935 by Elizabeth Bryant from an adult female she found in a collection from the Canterbury Museum. The family relationship puzzled her as it could not be associated precisely with any other spiders so, in the meantime, she put it into a group of eight-eyed spiders, Plectreuridae, similar in appearance and supposedly having similar habits. This particular group of American spiders spins tubular retreats in cracks and crevices, whereas we now know that *Plectophanes* takes over burrows vacated by insects. However, the plectreurid spiders are clearly a very primitive group and after a male of *Plectophanes* was discovered some time later it was realised that our spiders were more advanced and not in any way related to these American spiders.

This elongate spider, *Plectophanes* (Fig. 8.20), is one of our strangest-looking

Fig. 8.18 A female Cycloctenus fugax *and its eggsac were found on the underside of a piece of fallen bark. Its brown and rust-coloured pattern provides excellent camouflage against this background.*

Fig. 8.19 (below left): This little spider, Toxopsiella minuta *(Cycloctenidae), is only about 4 mm in length. Clearly visible are the four large eyes of the curved posterior row. The four tiny eyes of the anterior row lie below (and cannot be seen).*

Fig. 8.20 (below right): Plectophanes *is a spider that puzzles taxonomists because no one is really sure where it belongs. Its posterior row of eyes is situated on the top of a ridge which projects beyond the chelicerae.*

spiders because the front portion of the carapace is extended in front to form a wide and shallow ridge which bears the eyes. This extension juts out well beyond the chelicerae. The function of this 'visual oddity' remained a puzzle for many years. Because it was present in both male and female, it was not believed to play any part in courtship or mating, as do similar modifications in species where particular features are found only in the male. Up until 1940 only two further specimens of *Plectophanes* had been found, including an adult male from the Port Hills in Christchurch. Particular features of these new discoveries clearly showed that they were not related to *Plectreurys*, but then no one had found out just where the spiders lived.

A fascinating discovery

It remained for C.L. Wilton, who farmed sheep in the Wairarapa but who spent his spare time studying spiders, to discover the spiders at home. They were found living in small burrows vacated by wood-boring insects in the twigs and branches of trees. Observations revealed that a spider positions itself at the mouth of the burrow during the daytime with its eyes projecting out like a periscope (Fig. 8.21). From this hideaway, the spider detects approaching prey, rushes out to capture it and quickly drags the victim backwards into the burrow where it is devoured. Unlike segestriid spiders, such as *Ariadna* and *Gippsicola*, which often occupy adjacent burrows, *Plectophanes* does not lay down triplines and seems to rely on sight rather than vibration to detect its prey. When an insect comes within range there is a flurry of activity as the spider dashes out and just as quickly returns to the protection of its retreat.

Before these observations, *Plectophanes* was one of our rarest spiders but now, in most patches of bush containing the creeper *Muehlenbeckia*, the chances are that insect burrows, lined with silk but lacking triplines, mark the presence of these spiders. With this information, we have been able to record five different species from all over the country and, undoubtedly, there are more to find. Indeed, in our own patch of bush on Saddle Hill near Dunedin, we have located these spiders alongside the more common *Ariadna*, inhabiting empty burrows in galls left in *Muehlenbeckia* stems by the *Morovia* moth.

Even though some sixty years have passed since Bryant originally named these spiders, we still do not know exactly where they belong. As with a number of other unusual temperate Southern Hemisphere spiders, we need much more information before we can confidently determine such relationships. For the present, we list *Plectophanes* with the rare Fiordland *Anaua unica* (Fig. 8.22), both of which are currently associated with cycloctenids. *Anaua unica* is a relatively small, pale-cream spider which looks like a miniature *Cycloctenus* and is found on low shrubs and ferns in the forest.

Fig. 8.21 *(far left): With its 'visual periscope' projecting from a hole in a branch or tree trunk,* Plectophanes *is able to watch for approaching prey.*

Fig. 8.22 *(left):* Anaua unica *has been found only in the vicinity of Lake Te Anau in Fiordland, the origin of its generic name. Its pale colour is heightened by the black eyes and large amber-coloured palps.*

CHAPTER NINE

JUMPING SPIDERS
Salticidae

By far the most fascinating arachnids are the Salticidae or jumping spiders, so astonishingly anthropoid in posture and movements that one is tempted at times to ascribe human thought processes to them. From the moment the small, perfectly fashioned spiderlings emerge from their eggsacs, the basic behaviour patterns are present, each succeeding moult adding only to their efficiency. The only major exception is associated with the development of the reproductive organs, which reach maturity with the final moult. This perfects the mechanism which is triggered, in the presence of a male and female, into a series of courtship displays unequalled in the spider world.

Salticids are found worldwide including, of course, New Zealand – but only one alien species, the Australian *Helpis minitabunda* (Fig. 9.1), has managed to make its home here. We first came across this spider in Auckland in 1972 but since then it has slowly made its ways southwards, being collected in a Palmerston North garden a few years ago. Another attractive jumper, *Menemerus* sp. (Fig. 9.2) arrived here in a crate of bananas, but its tropical homeland conditions could not be matched in New Zealand. Considering that some thirty species from other countries have managed to become established here, and to spread, it is surprising that only one of these is a salticid.

How to recognise a jumping spider

Ranging in size from 2 to 10 mm, these spiders are generally more conspicuous in the North Island. Their frequently bright and attractive coloration is broken up by bands, stripes and speckled patterns, while their bodies and legs are adorned with special hairs and spines. Such eye-catching features are often

Fig. 9.1 *(below left): This long-legged jumping spider,* Helpis minitabunda, *arrived in New Zealand from Australia many years ago and has become established here. It is generally found in tall grasses and leafy shrubs.*

Fig. 9.2 *(below right):* Menemerus, *a tropical salticid visitor from Fiji, has been seen only occasionally in New Zealand. Its bright colours, conspicuous hairs and orange eye-bands are more suited to its home environment.*

Fig. 9.3 *Peering from beneath a leaf, this New Zealand jumping spider (which belongs to the* Trite *group) takes stock of its surroundings. The two big front eyes scan for objects in front of it while the side eyes are alert for movement.*

significant in courtship and, combined with other characteristics, make recognition of this group easy. The distinctive, almost rectangular cephalothorax can swivel at right angles to the abdomen, thus providing the action necessary for the visual efficiency of the eight eyes. Four of these eyes, two large ones flanked by a smaller pair, face forward, prominently situated on the squared front of the carapace, while a tiny pair of laterally directed eyes forms a row behind them. A third row consists of another pair of moderate-sized eyes whose visual range extends backwards, sideways and upwards. It is the size and arrangement of the salticid eyes which provide the most identifiable features (Fig. 9.3).

As their common name suggests, these spiders are able to jump. However, jumping is used mostly for catching prey and for crossing gaps in their path, such as from twig to twig or leaf to leaf. On unbroken surfaces many of these spiders walk or run, their usual way of getting around their environment.

The eyes of jumping spiders

More studies have been undertaken on the eyes of jumping spiders than in any other spider family. Largely because of work carried out by Heinrich Homann of Germany and Mike Land of England, we know that the two anterior-median (AM) eyes of the front row have long eye capsules and retinae that are divided into four layers at the rear and that these two eyes can move in tandem back and forth, up and down, and diagonally – movements which greatly increase their fields of view (Fig. 9.4). The anterior-lateral (AL) eyes and posterior-lateral (PL) eyes are used for movement detection and when stimulated by an object in their surroundings, the spider swivels in that direction. More recently it was shown that the AL eyes also have a role in mediating chasing behaviour by the spider,

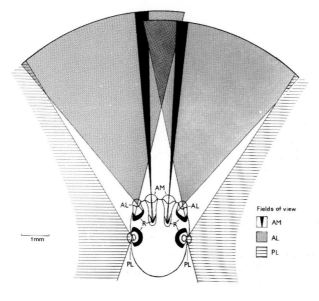

Fig. 9.4 *Diagram of a horizontal section through the cephalothorax of a jumping spider. The long eye capsules of the anterior-median (AM) eyes can move from side to side thus extending their fields of view (black) as shown by the light grey area, but they also move up and down as well as diagonally. As the visual distance of these eyes extends to some 20 cm in front, spiders have a three-dimensional viewing perspective. The anterior-lateral (AL) and posterior-lateral (PL) eyes do not move but their hemispherical retinae allow similar target areas. Note that the fields of view for the PL eyes have not been shown in full. R = Retinae. (See text for details.)*

whereas the AM eyes identify the target and guide the stealthy forward approaches to it. During courtship, the AM eyes identify the target as a possible mate by its size, shape and leg movements.

While the eyes clearly play a large part in the behaviour of jumping spiders, it should be realised that the sense organs of spiders have not all been studied in detail and that the lives of invertebrates are governed by sets of instinctive reactions to various stimuli beyond the normal range of human detection. Nevertheless, it is evident that the eyes of jumping spiders do possess powers beyond those of any other spiders, and that these powers are concerned with the recognition of specific targets, distance estimation and even colour. These abilities have not only enabled the salticids to abandon their snares and simultaneously evolve the specialised actions associated with hunting and mating but have also allowed the spider to hunt in daylight and thus utilise food resources not available to many other spiders. A few groups, such as the semitropical *Portia* genus (Fig. 9.5), have secondarily adapted to a web lifestyle while retaining salticid behavioural skills, most only slightly modified to suit their transient silk environments.

Fig. 9.5 *This jumping spider,* Portia, *is not a New Zealand spider but is found in northern Australia and parts of Africa. It is unusual because it makes a web of its own in which to rest and here it may also catch prey. More significantly, it regularly invades other spiders' webs, capturing insects, other spiders and often the resident spider as well. However, outside webs,* Portia *stalks its food in the usual salticid fashion. In this photograph its three front legs, half hairy and half slender, are spread out in front holding the web. Its hind legs are stretched out to the rear while the bushy orange palps are directed sideways. Notice the prominent anterior-median (AM) eyes and the much smaller anterior-lateral (AL) eyes on either side.*

Behavioural versatility

It is no exaggeration to say that these spiders jump, hop, skip, crouch, stretch, run, leap and dance – no choreographer could wish for a more varied repertoire! Such antics vary between species and have led to a broad separation between hoppers, jumpers and runners, all of which are able to jump but some do so only under specific circumstances. The front pair of legs, often heavily built and in some species noticeably larger are not – as may be thought – used for jumping or seizing prey, although they are often used for holding and manipulating the victim. Instead, evidence shows that they are primarily a signalling device and also possess chemotactile sensitivity. Spiders with one or even two of their front legs missing are certainly not inconvenienced in their normal locomotory activities, and prey-catching proceeds with as much vigour as usual. It comes as a surprise to find that the relatively unspecialised fourth legs are those used for jumping; moreover we have seldom found any of these spiders with a hind leg missing. No doubt this is because they can always detect movement some 10 to 15 cm behind them and turn to face a potential enemy. Whatever the mode of progression, it is marked by a dragline of silk laid down by the spinnerets. Not infrequently, this jumbled network may be gathered up and actually eaten by the spider. Apart from this, silk is only used for the construction of retreats where the spider rests at night and also moults, and to cover the eggs after they are laid. This release from the dependence on silk snares for prey-catching enables the spider to go further afield in its search for food.

Common New Zealand salticids

Of the larger jumping spiders found in New Zealand the most widely distributed species are *Trite planiceps* and *Trite auricoma*, although *Euophrys parvula* (often called the house hopper) is probably the one most often seen, particularly in the North Island. The type species for *Trite* is a New Caledonian spider, and Eugène Simon accepted this for *planiceps* but *auricoma* is likely to be placed in a separate genus. The seashore spider, *Marpissa marina*, can be found hunting on rocks on sunny days along Otago coastlines. There are, however, a large number of undescribed salticids, primarily because this group has not been recently studied taxonomically.

Trite auricoma is a squat golden-brown spider that is broadly grouped as a runner because its jumping is reserved largely for prey-catching although it sometimes leaps across a gap. Often found within the dry rolled-up leaves of flax (*Phormium* sp.), or beneath the dry underfronds of the cabbage tree (*Cordyline* sp.), it is just as likely to be found under stones, in shrubs and low vegetation as

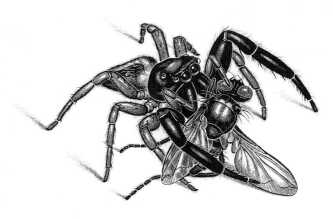

***Fig. 9.6** (above left): Native to New Zealand, the male* Trite auricoma *shown here is mature, revealing this by the band of yellow hairs below its row of front eyes. Females have no such yellow clypeal band. Length 7–8 mm.*

***Fig. 9.7** (above right):* Trite auricoma *stalks its prey and jumps at it from a distance of some 3 to 5 cm. As the spider lands on the victim, its outstretched fangs pierce and crush it while the front legs and palps are used to restrain and manipulate it.*

well as on the ground. Blending perfectly with a background of soil, leaf-mould, dead leaves or pebbles, it can remain virtually unseen until it moves. The adult male is distinguished from the female by a yellow clypeal band between the frontal eyes and the chelicerae (Fig. 9.6), longer front legs and a slimmer body.

Prey catching

Employing a specialised combination of movements, this spider makes full use of its sharp eyesight to stalk and catch prey. If a fly alights within about 15 cm in front of it, the spider is immediately alert. First, there is a slight shifting of the legs and straightening of the abdomen. Then the hind legs are drawn in and the spider slowly and stealthily creeps towards its prey, very much like a cat stalking a mouse. Even if no further movement of the fly takes place, it does not matter – its death is assured as the last 3 to 4 cm is covered with a lightning leap, and outstretched fangs pierce the victim with a tenacious grip from which there is no escape (Fig. 9.7).

If a fly should alight behind one of these jumping spiders, the movement is detected by the posterior eyes. Instantly, the spider swivels around (Fig. 9.8) to face the prey and then resumes his stealthy stalking. Sometimes the fly delays the moment of destruction by moving. This avails him nothing, for now the spider postpones its stalking and follows the victim around with a watchful gaze, turning its cephalothorax with a swivelling action until it is at right angles to the abdomen. Quietly, imperceptibly, the abdomen slides into line with the cephalothorax, the stalking begins again and ends in that final decisive jump.

If, by chance, the prey begins to run away, the spider sets off in pursuit, exceeding the prey's speed so that it can catch up on it, only slowing down as it gets ready to jump. Chasing by the spider is controlled by the anterior-lateral (AL) eyes which, like the posterior-lateral (PL) eyes, are also movement detectors. At the same time, the main (AM) eyes are able to keep track of the image they see. In the end, the prey may pause, with the inevitable result, although sometimes it flies away or moves beyond the spider's reach.

Practice makes perfect

Young jumping spiders emerge from their eggsac after the second moult and have to fend for themselves, since the mother takes no further interest in them. It is important that they try out their instinctive hunting skills as soon as possible, for unless they are successful within a few days, they are unlikely to survive. Observations show that these spiderlings first turn several times to look at movement within their surroundings before they attempt an approach. After they have run towards something moving that might be edible, they jump at it – but generally miss the first two or three times. Then suddenly, success! This is the key to future prey capture. Once a spiderling perfects the movements required to make a catch, then the only improvements are speed and accuracy. However,

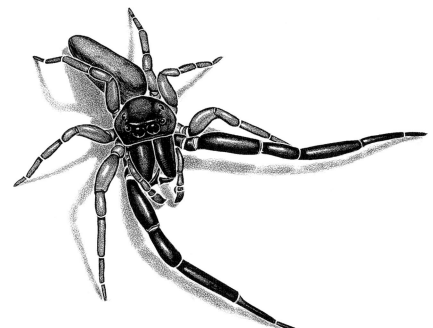

Fig. 9.8 Responding to movements behind it, Trite planiceps *has swivelled to face the source of this movement. If this movement comes from prey, the spider will bring its front legs forward very slowly, straighten its abomen and begin to stalk it stealthily. Length 8–10 mm.*

they must also learn what is good to eat. For instance, spiderlings will stalk and jump at slaters (woodlice), if on offer, but one taste is enough. They seldom make that mistake again.

Identifying targets

There is no doubt that *Trite auricoma* can determine the difference between enemy, prey and a possible mate with uncanny accuracy. Vision is unquestionably involved. Yet a male will use the same initial signal in the presence of either male or female and only when he is facing the female at a distance of not more than about 10 cm will the pattern of his leg waving change. In addition there are variations between threat and inter-male display.

The female, on the other hand, catches her prey in an identical fashion to the male and remains passive if she is sexually receptive to a mate. If she is not receptive, or is confronted by an enemy, her threat posture is instantaneous and forbidding. The cephalothorax is raised, the two front pairs of legs become stiffened and straight, the abdomen is bent at right angles to the thorax while the spinnerets are pressed firmly to the ground. The palps are rounded and the chelicerae extended although the fangs are kept recessed. Perhaps it is only our imagination but even the eyes have a baleful glare. Comical though it appears to us, there is no doubt of its effectiveness for, thus confronted and outstared, the enemy turns tail and flees. A male treated in this fashion forgets his intentions and leaves the scene in great haste. It is a common belief also that the female devours the male upon the least pretext and that if he is not quick to depart after mating he is promptly eaten. We have never seen this, even when we have had the two spiders in a container from which there is no escape.

Agonistic behaviour

Two females brought face to face will at once adopt the threat stance which is always accompanied by sidestepping, first right and then left, although this seldom brings the two antagonists any closer. These jerky movements serve to draw the attention of one to the other most effectively, as if saying 'Look, I am threatening you.' Where two females are concerned, the stance and movements are continued until one gives way, although what determines who departs is difficult to see.

When two male *Trite auricoma* spiders come across each other, they swing round until they are face to face, and first one and then the other will raise its front legs sideways at an angle of about 120° (Fig. 9.9a). These legs are then waved up and down, each rhythmical movement being accompanied by sidestepping, the two spiders always moving in the same direction. At the same time,

Fig. 9.9 *When two* Trite auricoma *males meet, they immediately raise their front legs.* **a** *After looking at each other they approach slowly, legs waving up and down, palps rounded and stepping alternately from side to side.* **b** *Eyes firmly fixed upon each other, they stand on tiptoes, extend their front legs and palps, and attempt to touch.* **c** *Standing high on their back legs, with abdomens bent to the ground, they stretch their front legs and palps and will eventually touch along their length.*

the spiders approach each other cautiously with their eyes firmly fixed on each other. In addition, each cephalothorax is raised high with the front two pairs of legs tip-toeing, abdomen bent firmly to the ground and palps rounded (Fig. 9.9b).

Finally, the tips of their front legs touch and then, as they move even closer, are brought into contact along most of their length (Fig. 9.9c). At this stage the palps and fangs are stretched sideways and also come to grips. This position may be held for 30 seconds or more, each spider pushing against the other until one of them breaks away and moves off. Under normal circumstances, one or both would promptly leave the scene, but in captivity they may again come together in this way two or three times. However, the sequence gradually becomes less vigorous until it finally ceases. In the wild, the value of this behaviour may be to maintain a suitable space between males and to ensure that only the fittest males mate.

Mirror-image reactions

It is amusing, to say the least, to watch the male's reaction to his own image in a mirror. From a distance of some 7 to 8 cm he may catch sight of his own reflection, after which he treats the image as another male. Leg-raising and sidestepping, he approaches the mirror (Fig. 9.10), the directional sidestepping of his self-image being exactly the same as that of an opponent. Upon contact with the mirror, he strains against it but gets none of the signals that would emanate from another male. Sometimes, he climbs onto the mirror and gropes behind it. Now and again he turns and moves away, then swivels abruptly to face the mirror again as if trying to take the other by surprise. Initially standing stock still, he then embarks on another leg-raising and sidestepping display. After two or three futile efforts to 'touch his opposite number', however, he gives up and actually shows a strong tendency to ignore the mirror at later trials.

It seems rather a shame to destroy such a pretty little story and anyone who takes the time to repeat that experiment with one of these male spiders will find it hard not to let their imagination run away with them, too. It seems evident that the initial stimulus between two males is a visual one but the final consummatory act is a chemotactile one. This explains the efforts that are made to have contact in order to complete the chain of co-ordinated behaviour patterns. When he turns away, as if to deceive his opponent, and then turns back sharply, it is because the reflection of this movement is seen by one of his posterior eyes, inducing the swivelling behaviour that we have observed in other situations. It is

Fig. 9.10 This male *Trite auricoma* is displaying to his image in the mirror exactly as he would to another male. He approaches in the same manner until he touches the mirror and after groping at it several times, he turns away. This shows that his initial reaction to another male is visual but that contact with the mirror reveals his mistake.

so tempting to place our own human interpretations on what we see that it is hard to realise the part that instinct plays. On the other hand, it does seem that the spider has 'learned' that there is 'something fishy' about a mirror image when he refuses to be tricked at a later stage. Scientifically, this is usually refered to as *habituation*.

Courtship

Courtship displays (Figs 9.11a, b) have been the subject of much interest and analysis, particularly regarding birds whose attractive dances were once thought to be of value in the selection of a suitable mate. Indeed, the Peckhams, who first described such performances by salticids in 1892, also held this view. Now it is considered that they have also evolved as a means of species recognition, thus facilitating mating and preventing cross-breeding. Such ritualised behaviour is in fact highly differential and it is significant that the appropriate displays generally bring an immediate response from members of the same species. So reliable is this that we have frequently employed it as a method to determine whether two spiders with superficial variation really do belong to the same species.

Females are by nature stay-at-homes so we were fortunate to find several recently, thus enabling us to observe and photograph both courtship and mating behaviour. Males we had in plenty for they appear to wander in search of a mate; almost every day in the mating season we would catch one on the walls of our kitchen. The initiative is taken immediately by the male while the female, if receptive, remains in a crouching position. Although the female may not accept the male at once, her threat stance – if genuine – is enough to deter the stoutest male heart. As in the first stage of inter-male behaviour, the front legs are raised at an angle of 120° and waved up and down while the spider sidesteps – two to the left, two to the right. Even if the spiders are not facing each other to begin with, this action brings them smartly round face to face. The male continues his movements somewhat jerkily as if to emphasise them, all the while keeping his eyes fixed firmly and judiciously on the female. When he is within 5 to 6 cm of her, his leg movements change – providing of course, that the object of his intentions is still there.

Female responses

Sometimes the female moves away so that the male is obliged to follow her, still displaying vigorously. Once more frontally positioned, his front legs are pointed forwards, still waving up and down although the movements are now much

Fig. 9.11a In the Salticidae, courtship outside the female's nest involves a visual display with leg-signalling and sidestepping by the male. If the female is receptive, as shown here, she merely crouches, watches and waits.

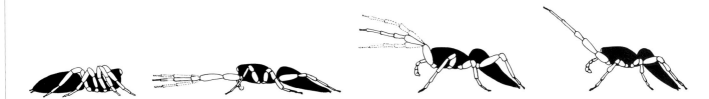

smaller and faster. The male's legs touch the ground in front of the female several times while he moves cautiously towards her (see Fig. 9.11b). She then raises her legs and touches his. Still using his legs, the male moves over on top of her. All show of resistance now ceases and she remains quiescent until mating finishes, a process which may take up to an hour. During mating, the male changes his position from the left side of the female using his left palp, to her right side using his right palp (Fig. 9.12). This change takes place after about 30 minutes. Since those first observations, we have witnessed courtship and mating many times, behaviour differing little from that first description, although copulation times varied from 30 to 45 minutes.

After mating, the females go into seclusion somewhere, but in the spring they can be found under stones and planks, or inside curled-up leaves, guarding their eggsacs within a retreat. At this time there is no sign of mature males, and it is not until late summer that they appear again. During the warmer months, juvenile spiders in many stages of immaturity are found, so we conclude that adults are mostly found in late summer and autumn, that females are inactive over the winter and usually lay their eggs in early spring. This pattern may vary, however, in different parts of the country.

More spectacular in both appearance and activity, but less widely distributed, is this jumping spider, *Trite planiceps*. Moving with great rapidity along the surface of the leaves of flax and cabbage trees, it can bridge a gap of up to 20 cm with both grace and precision. One of the largest of jumping spiders in New Zealand, it is distinguished by a shining ebony carapace and large black forelegs together with a slender green abdomen longitudinally marked by a central yellow stripe. Possessing the highly characteristic alert-and-swivel, stalk-crouch-and-jump style of hunting behaviour, this species demonstrates, by its accurate agility, the peak of arachnid visual acuity. Prior to jumping, both front legs are extended forwards in the direction of the anticipated leap (Fig. 9.13). This action

Fig. 9.11b *(top): As the male* (far right, moving left) *gets close, he shifts his front legs forward in the direction of the female* (left), *gradually lowering his body and legs as he inches towards her, at the same time vibrating these legs rapidly.*

Fig. 9.12 *(above): When the male* Trite auricoma *makes contact with a receptive female he vibrates his front legs on her carapace, then creeps on top of her and commences to mate. Her abdomen is twisted sideways, first in one direction and then the other, while the male's palps are applied to her epigynum.*

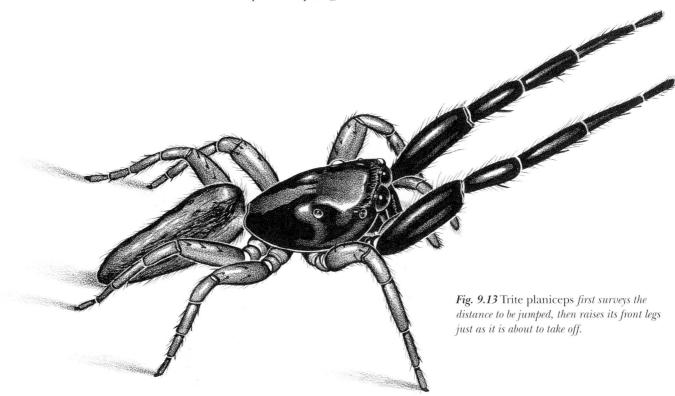

Fig. 9.13 Trite planiceps *first surveys the distance to be jumped, then raises its front legs just as it is about to take off.*

Fig. 9.14 a (below): Upon sighting a female Trite planiceps, the male immediately raises his first pair of legs sideways at about 120° to the ground. b (bottom left): This leg-raise posture is caught by the camera below. Notice the slightly extended fangs with spurs, the swollen palps and the dense bunches of hairs (scopulae) at the tip of the legs. It is these scopulae which ensure the firmness of the spider's grip on a surface.

Fig. 9.15 (bottom right): Trite planiceps guards her eggsacs within dense silk inside a rolled-up flax leaf. Four to five eggsacs are usually laid some two weeks apart and the first one hatches about the time that the third one is laid.

is seemingly associated with some form of distance estimation, a bit like sighting along the barrel of a gun – and almost as disconcerting too we found, for as soon as we came within about 20 cm, the subject of our observations would land on the camera, pencil, magnifying glass or even on our noses. Fortunately, perhaps, this spider is capable of an even more remarkable feat – that of returning almost instantaneously to the spot it has just left, should his objective prove unsuitable or undesirable. The legs-forward posture adopted by *Trite planiceps* prior to jumping is no doubt aimed at streamlining the body as it moves through the air and also positions the front legs so that they can either grasp the surface of the substrate on which the spider lands or be used to aid the spider in its seizure of prey.

At rest, this spider lies lengthwise with the first two pairs of legs stretched out in front and the third and fourth extended behind, a fact which should have led us to its hiding place sooner than it did. Our sweeping and beating of trees and shrubs disclosed not a trace but careful unrolling of the cabbage tree leaves and flax soon located them in all stages of growth. There, too, were the silken retreats, the moulting hideaways and the discarded skins. This spider is widely distributed throughout the country, but in the warmer North Island there is a greater tendency for them, especially the juveniles, to be found in low vegetation.

Courting and mating

There is no easy way to distinguish between males and females of *Trite planiceps* as there is in *Trite auricoma*. Perhaps the male is more slender than the female, and of course his swollen palps are further evidence. As well, a small spur develops on each fang (see Fig. 9.14b) although this is hard to see. However, their display activities vary considerably, with the female assuming the role of a leg-waving male under certain circumstances. During courtship she remains passive while the male performs a sideways leg-waving action (Figs 9.14a, b), side-stepping as he approaches. He displays only briefly to his own reflection and yet we have also obtained a partial reaction to a cardboard model which pictured the leg positions of a spider in frontal view. Several times, we have found an adult male closeted in a retreat with a nearly mature female which moulted within two days. With the male thus remaining in the vicinity of an about-to-moult female, successful mating is virtually assured. At two-weekly intervals four or five eggsacs are laid within a silken retreat and these are arranged in a row inside the rolled-up flax (Fig. 9.15) where the mother stays until all the spiderlings emerge. Guarding the eggsac at least provides a greater assurance that the early stages of their lives are protected against marauding creatures; but after hatching, spiderlings quickly disperse and it is not possible for the female to keep track of them.

Fig. 9.16a *(right): With his eyes on the camera, this little male* Euophrys *sp. (Salticidae) swivels around and lifts the front of his body up.* **b** *(below): Less conspicuous than the male, the female* Euophrys *keeps guard over her eggsac. This one was found inside a dried rolled-up leaf. Length 5–6 mm.*

An excellent example of the kind of jumping spiders we call 'hoppers' is a species found in large numbers on the outside walls of houses as well as amongst the garden plants in the North Island and as far south as Christchurch. But even though some sixty names have been bestowed upon New Zealand salticid species, we were not able to find a description which would exactly match this particular species. For the present we will refer to these spiders as *Euophrys parvula* (since this is the spider they most resemble) or 'house hoppers'. These delightful little spiders may be seen sunning themselves on almost every house within their ecological range but we noticed this little male (Fig. 9.16a) keeping his eyes on us from a leafy perch near our holiday retreat in Feilding.

Description and habits

These hairy, brownish jumping spiders are patterned attractively with dark and light areas which gives them a speckled or striped appearance. They often build their silken retreats in the cracks and crevices in the walls, and when the sun goes, even temporarily behind a cloud, they vanish within. When kept in captivity, they immediately begin to build a retreat. We found the only way to see them is to wait for the sun to come out or to bring them out under a bright artificial light. Many an hour we spent just watching them catch their food on these vertical walls, which they are able to traverse with the aid of special bunches of hairs (*scopulae*) beneath their feet, but always with the added assurance of a thin but strong dragline which they attach to the surface every now and again. To us, the most extraordinary feat seemed to be their ability, after stalking a fly to within about 5 cm, to defy gravity with a horizontal or vertical leap and land plumb on the doomed fly blithely grooming itself on the wall. It seems that the sheer momentum of the powerful jump projects the spider on to its objective before the pull of gravity takes effect. Yet they only occasionally jump upwards, usually circling sideways until they are at least horizontal with the target.

Display behaviour

Perhaps the most exciting period during our leisurely peering on walls was the observation of a pair doing their courtship dance, still out in the open on the vertical wall. All the twenty or so other salticid species that we have watched at one time or another use the front pair of legs to carry out their elaborate sequence of signals but the male *Euophrys* lifts his third legs high above his body at the same time drumming his palps on the surface and weaving his abdomen from side to side. To add to this conspicuous display, the male now sports a red forehead, acquired with the final moult (see Fig. 9.16a), thus signalling his maturity to a would-be mate. This spectacular courtship dance with the male circling and posturing in front of the female may often be witnessed during the summer months on the sunny side of your house in places as far south as Christchurch, if you keep your eyes open. Females lay their eggsacs in various places (Fig. 9.16 b) in nooks and crannies around buildings or amongst the plants in your garden.

Post-mating display by the female Euophrys

We found that a female too will display to males but only under certain circumstances. This happens when a recently fertilised female, or one known to have laid eggs, is courted by a male. When he is still some 7–8 cm away, the female performs a dance of her own. With her third pair of legs and abdomen raised high in the air, she sways from side to side. She does not come much closer to

Fig. 9.17 *Two male* Euophrys *dance in unison with their third legs raised and abdomens swaying from side to side, all the while circling from side to side.*

the male but turns first in one direction and then the other, as if to signal to any males that might be in the vicinity. Initially retreating at the sight of the female's display, the male may begin to court again, now approaching more quickly. This time the female crouches as she watches him closely and if he manages to come within 3 to 4 cm of her, she jumps at him. Fortunately, this is not a lethal jump and as soon as the male recovers from this attack he runs away. This 'rejection' dance or warning-away ritual is typical of *Euophrys* females once they have mated, but has not been seen in any other salticids that we describe in this chapter.

When two male *Euophrys* spiders meet they both begin a spectacular rhythmical dance, a pattern of movements which is more formally described as agonistic behaviour. Interestingly, this dance is closely similar to the display that the mated female uses to reject an unwanted male. In this two-male instance, however, one spider mirrors the other's performance, with third legs high and abdomens swaying from side to side, as they circle first in one direction, then the other (Fig. 9.17). It ends when spiders touch their legs and palps together, sometimes pressing their heads and pushing against each other as well, before one gives up and departs. Often, though, once the spiders make leg-contact they quickly separate and move away.

Other hoppers

Numerous small 'hoppers' are found on and beneath the bark of forest trees, but those more closely related to the house hopper live in rather exposed places. Along very pebbly beaches in New Zealand, well above the high tide mark, there are a number of species of large black or dark-grey salticids of which only one, *Marpissa marina* (Fig. 9.18), from along the Otago coasts, has been formally named. When hunting, this spider runs very briskly for a few centimetres, then stops, rubs its palps together, and looks around until the prey is within sight. Once prey is detected, however, *Marpissa* behaves like other salticids, creeping stealthily towards its unsuspecting quarry, then crouching and jumping at it. As long ago as 1891, Goyen realised that on these same cliffs and rocks there are three species of Diptera (flies) which these spiders resemble in colour and mode of progression and which may even be mistaken for the spiders themselves. Whether this mimicry benefits the spiders and/or the flies is not known but as *Marpissa* is the hunter, it is likely that it is the flies that are deceived. So perhaps as *Marpissa* runs in erratic 'stops and starts' along the face of a rock in search of them, flies fail to distinguish the spiders' movements from those of their own members. Once seized, the catch is carried away to a sheltered spot where the spider sucks up the pre-digested contents at its leisure. Sometimes, to escape a predator, *Marpissa* springs into space at the edge of a rock, then hangs head downwards by a thread, still holding its prey. In this position, it is eaten.

Similar habits can be observed in a whole series of related species (Fig. 9.19),

Fig. 9.18 *(below left):* Marpissa marina *(Salticidae) is found amongst the rocks along the Otago coasts. Known as the seashore jumper, the colour and patterning of this spider help to conceal it from predators. Length 10 mm.*

Fig. 9.19 *(below right): Riverbed salticids, like this one, are ideally camouflaged to escape detection. This undescribed species was found in Makarora Valley. Length 8 mm.*

Fig. 9.20a (above left): A colourful little male jumping spider, whose yellow clypeus in front suggests it may belong to the Trite group.
b (above right): Here, this spider adopts an alert posture as it poises with its front legs ready to be raised for a jump.

most of them dark coloured and mottled, which inhabit the shingly riverbeds and bare rock faces of the mountain sides high above the upper edge of the forests. All of these spiders take advantage of the sun to hunt. Amongst the many undescribed salticids in New Zealand is a small brightly coloured species which belongs to the *Trite* group (Fig. 9.20a). Distinguished by its large black chelicerae, pale palps, and red markings on the carapace and abdomen, this spider favours shrubs and low bushes as its habitat, and is widespread throughout the country. Like all jumping spiders, this spider raises itself up, looks ahead and assesses the distance before it jumps (Fig. 9.20b).

High-altitude salticids

A number of species of salticids are found on high mountain tops in Otago, in the South Island, where there is snow cover for about four to five months of the year. Although none of these have yet been formally described, it is believed that other similar mountain ranges in the South Island have their own distinctive groups of spider species (as well as associated flora and fauna), a situation brought about during the Pleistocene when such ranges were isolated by substantial ice formations. Mountains where salticids have been located include the Remarkables, Mt Pisa, and the Rock and Pillars – all in Otago. We have no doubt that surveys of other high-altitude regions would reveal similar speciation patterns to the ones we have found here.

Despite the lack of formal identification we know quite a bit about the life history and behaviour of two of these Otago salticids, one from the Remarkables which we will call Species A and the other from the Rock and Pillars which we will call Species B. It is sometimes asked why we do not give these spiders scientific names at once when we are confident they are different species. This situation demonstrates the explanation given in the Introduction to this book where we said that it was necessary to know about relationships with similar spiders – as we might expect to find amongst those on other mountain tops – before assigning them a generic name. As we have already seen, in the early days of colonisation, spider enthusiasts who worked in different regions often named spiders without knowing of an earlier classification or kindred species.

Although those salticids we describe as Species A and Species B are rather similar in size, general appearance and habits, the most important distinction lies in the colour of the hairy clypeal band below the eyes in the adult female. In Species A, all juveniles are a mottled white and black colour and very hairy, and at maturity the females develop a bushy white clypeal band (Fig. 9.21). On the other hand, Species B juveniles are much darker with very faint pale blotches, and adult females possess a bushy orange-yellow clypeal band (Fig. 9.22).

Fig. 9.21 (far left): The female mountain spider (Salticidae) shown here is found at about 1,300 m on the Rock and Pillar Range and is the one we call Species A. (See text for details.)

Fig. 9.22 (left): From the Remarkables (a range of mountains near Queenstown) at a height of about 2,000 m, comes this high altitude salticid. It is a mature female (Species B), identified as such by its luxurious orange clypeal band. Length 10 mm.

Winter survival

Although the habitats of Species A and B are covered with snow for some four to five months of the year, these jumping spiders make use of a range of strategies that enable them to survive. As winter approaches, they build very dense silk cocoons under rocks, and here they remain until the snow melts in the spring. In the laboratory, we found that these spiders still build thick cocoons at the same time of the year, despite the fact that temperatures were about 15° to 20°C higher than they would be on the mountain, so we conclude that changes in day/night length are important triggering factors. Nevertheless, unless spiders regularly experience low average temperatures of 1 to 2°C during the subsequent winter months, they do not survive. Kept in a refrigerator at appropriate temperatures, however, spiders crouch motionless in their cocoons, do not eat, and remain healthy, albeit sluggish, for four to five months. Although it is possible these spiders manufacture glycogen as an antifreeze mechanism, it is more likely that their winter survival is based on food stored in the body during the warmer summer months in conjunction with reduced respiration and metabolic rates.

Spring to autumn lifestyles

When the snow melts, these spiders emerge from their silk shelters and begin the search for food amongst the stones and sparse vegetation. A variety of other small invertebrates also have ways of enduring the winter snows and some of them will fall prey to hungry salticid predators. It is here too that these spiders will meet their respective conspecifics and, sooner or later, a mate. When a male mountain spider meets a newly moulted female, events follow quickly. Once he is alerted to her presence from some 10 cm away, he immediately raises his front legs laterally, slightly above the level of his body, sidestepping with his legs in this position. Sometimes the sidestepping is replaced by circling around and towards the female, with the front legs still raised but slowly brought to the front in the direction of the female. A receptive and hence unmated female simply watches him, swivelling as he circles. Then with his body low on the ground, first legs lowered frontally, vibrating and tapping on the ground, he inches towards the female. Slowly creeping on top of her, constantly tapping the carapace, he nudges at one side of her abdomen which twists around thus allowing him to apply his palp to her epigynum. Some fifteen minutes later, the male shifts his position and with a nudge, the female twists her abdomen in the opposite direction, so that he can insert the other palp. A non-receptive female (one which has already mated, for example) deters a male with a jerking threat-crouch posture. Other females are also treated with the same threat-crouch movements should they venture too close.

Fig. 9.23 *Mountain salticid males of Species A and B have black carapaces with mottled grey and black abdomens, losing their striking black-and-white juvenile patterning at maturity.* **a** *(above left): The hunch posture, exhibited here by one of the mountain salticids, is a sign of defence and antagonism. A similar stance is commonly employed by many salticids.* **b** *(above right): In this illustration, a male is shown responding to his mirror image with the characteristic agonistic behaviour that is found in both these species.*

Eggsacs are laid in late summer (March–April) in dense webs under big rocks and hatch in autumn (April–May). Spiderlings can be found moving about in these webs before the winter snows. They stay in these sheltered hideaways until after the snow melts, when they emerge to find their first food. It is likely they survive the winter snow by similar mechanisms to the adults and that they spend the following winter as half-grown juveniles, courting and mating in their second year. It is not known how long adults live in the wild, but it is probably two to three years.

Male interactions

When adult males meet, their reactions are very different to those of *Trite auricoma*. At first sight of each other, one or both will raise their front legs sideways, slightly higher than the body, but the metatarus will droop. One spider may jerk its body once or twice in the direction of the other, then the front legs are tapped sharply on the ground and this may result in the other spider taking flight. Sometimes the second spider stands his ground and the first combatant jerks again. If the second spider still persists, then both will adopt a threat-crouch posture (Fig. 9.23a). At this point the spiders are very close together and may grapple with each other, front legs entwined. However, it appears that each spider is attempting to tap the other on the carapace, behaviour which we have seen in male–female interactions prior to copulation. Grappling behaviour may persist for one to two minutes after which one of the contestants will break off and run away. Spiders are seldom injured in these encounters unless one of them fails to display or perform a threat-crouch. Males respond to their mirror image with typical agonistic behaviour (Fig. 9.23b), indicating that these spiders also recognise each other by sight.

Flattened jumping spiders

This account of the jumping spiders would not be complete without mentioning the oddest members of the family, the curiously flattened – variously coloured species of *Holoplatys* which look as if they have been pressed between the pages of a book as one might press flowers and leaves (Fig. 9.24a). This body form provides the clue to the habitat they prefer – small cracks and crevices in and under the barks of trees and between rocks, or even inside rolled-up flax leaves. Human constructions such as overlapping fence palings also provide an effective home and it is here that one species of these peculiar spiders is often seen today (Fig. 9.24b). Closely blending with the weather-beaten wood on which

they lurk, they exhibit a special behaviour known as thigmotaxis, a compulsion to have both dorsal and ventral surfaces of the body in contact with a surface which, when combined with a curious sideways gait, enables the spider to retreat immediately into any cracks and crevices available. In spite of their large

Fig. 9.24a (above): This mountain **Holoplatys** *is larger and blacker than the more familiar grey species and is rarely seen.* **b** *(left): The greyish, flattened* **Holoplatys salticid,** *seen here in its moulting retreat, is the one most likely to be found around human habitation, one of its favourite hideaways being between the overlapping boards of fence palings. The moult skin can be seen beside the spider.*

Fig. 9.24c *A different species of* Holoplatys, *often found within dried rolled-up flax leaves, is easily recognised by its brownish coloration.*

bulky forelegs and ungainly appearance, these spiders catch their food in the same manner and just as readily as other salticids.

There are a number of undescribed species of *Holoplatys*, most similar in size and shape but easily distinguished by colour and markings. One of these, usually found inside rolled-up flax leaves (Fig. 9.24c), may be the species whose fascinating courtship and mating activities have been described by Jackson and Harding. They found that these spiders possess a large number of displays as well as three different mating behaviours used in particular situations. If mature spiders meet out in the open, the male displays to the female with a series of acts consisting of front-leg-signalling, palp rotations, zigzag movements, as well as back-and-forth runs before mounting and mating. Females respond with hunched postures, short leaps and the occasional charge. However, females inside silk nests evoke a different response from males. Usually the male probes the nest, taps and strokes the silk, chews at the entrance and enters, brushing the female's face with his palps after which their faces touch. The female reacts to his first contact with the nest by heaving her body upwards, pulling at the silk and then thumping the carapace of the male as he enters. The most successful mating ploy is when the male comes across an about-to-moult sub-adult female in her nest. After probing and tapping, the male builds another cocoon next to the female's and waits therein until she moults. He then makes his way into the nest and mates with her.

CHAPTER TEN
SIX-EYED SPIDERS

Most spiders still possess the primitive number of eight eyes, but there are some families in which the anterior-median (AM) pair was lost very early in their evolution so that surviving groups now have only six eyes (Fig. 10.1a, b). With such an obvious feature in common, it is convenient to link these families together. It should be noted, however, that a number of species in traditional eight-eyed families have also lost one or more pairs of eyes, although such spiders are relatively rare. But these exceptions usually share other conspicuous characters, such as entelegyne genitalia and complex palpal organs, with their eight-eyed relatives, so they are not readily confused with true six-eyed families.

In addition to being six-eyed, these families possess primitive haplogyne genitalia (see chapter 1) but often have surprisingly complex internal features consisting of a wide range of sacs and spherical receptacula associated with extensive secretory glands. The male palpal organ is usually reduced to a simple bulb as in mygalomorphs and, as there is no external epigynal structure above the epigastric furrow, the embolus of the male bulb is inserted directly into the oviduct of the female during mating (Fig. 10.2).

Six families of six-eyed spiders are recorded from New Zealand – the Orsolobidae, Dysderidae, Periegopidae, Segestriidae, Scytodidae and Oonopidae. Dysderidae is represented in New Zealand by the introduced but now widespread *Dysdera crocota*, while Oonopidae and Periegopidae each have a single

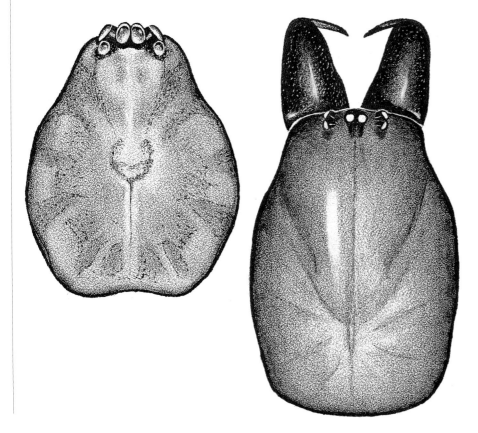

Fig. 10.1 Dorsal view of the carapace of two of the families of six-eyed spiders showing differences in the arrangement of the eyes. ***a*** *(far left): Orsolobidae (*Pounamuella australis*). **b** (left): Segestriidae (*Ariadna sp.*). For two other families discussed in this chapter, Periegopidae (*Periegops suterii*) and Dysderidae (*Dysdera crocota*), see Figs 1.14i and j.*

native species. Segestriidae and Orsolobidae are both widespread throughout New Zealand and are the dominant six-eyed spiders. Only once has a member of the Scytodidae been seen in this country. Segestriidae is the only family whose members sit and wait for prey, the rest being active hunters.

Orsolobidae

A new family, Orsolobidae, was established by Forster and Platnick in 1985 when it was realised that certain six-eyed spiders from Chile, Australia and New Zealand (previously recorded as unusual members of two well established cosmopolitan families, Oonopidae or Dysderidae) actually represented an entirely different lineage found only in the Southern Hemisphere. Because the Chilean members had been earlier designated as a subgroup Orsolobini of the family Dysderidae, this name, under the rules for scientific names, became the basis for the new family name.

Distinguishing features

Although orsolobids have superficial resemblances to oonopids and can also be confused with dysderids, they are quite different in many ways. Oonopids are typically small – in fact, usually minute – whereas most orsolobids are medium-sized although some, notably the Chilean *Orsolobus* and the New Zealand *Subantarctia*, are relatively large. Moreover, orsolobids all possess cheliceral teeth, (missing in oonopids) and many, but not all, have a prominent chevron pattern on the upper surface of the abdomen (a feature very rarely seen in the Oonopidae). While in size and superficial appearance these spiders look rather like dysderids, the peculiar flattened structure of the claws and other difficult-to-see structures on the tarsi of the legs serve to distinguish them. These obscure but significant tarsal structures were not noticed when the spiders were first recorded from Australia and Chile.

The uniqueness of these spiders initiated a further search for them in several southern countries and this produced 170 species, most previously unknown. Subsequently, Charles Griswold of the Californian Academy of Sciences discovered a number of African species. This means that their distribution encompasses all the southern lands, as might be expected for a group of animals once widely distributed on Gondwana before it broke up and drifted apart.

The first New Zealand orsolobid

It was while Graham Turbott, a well-known ornithologist, was coastwatching in the Auckland Subantarctic Islands during World War II that he collected the first New Zealand orsolobid. Later, in 1955, it was described as *Subantarctia turbotti* after this locality and its collector. *Subantarctia* species are relatively large, often more than 5 mm in body length, and are recognised by the handsome reddish cephalothorax and cream abdomen. As they look just like Mediterranean dysderids, this is the group where these new spiders were first placed. Soon after, a large collection of orsolobid-like spiders from all over New Zealand was examined and 29 new species, many strikingly different from *Subantarctia* and some originally placed in the Oonopidae, were added to this family. Even at that time, two unusual structures near the tip of each tarsus set these spiders apart from Oonopidae and Dysderidae.

Tarsal organs

At first these two new structures were difficult to see with a light microscope but thirty years later, when the family status was validated, a detailed study was made possible by the development of Scanning and Transmission Electron Microscopes (SEM and TEM). Initially called *tarsal tubercles*, these structures varied from mounds ringed by decorative lobes to ones armed with prominent spines and were always situated on the upper surface of the tarsus near the claws. Further forward from each tubercle, a pair of bristles arising from a circular pit and usually branching near the tip were discovered. Named *tarsal thorns*, they were positioned just behind the flexible lobe, which bore the two claws. Moreover, the surface on the back of the claws was covered with fine denticles as well as the usual two rows of teeth on each side, all of which were clearly visible under a light microscope. However, with the advent of SEM to magnify the surface structures and TEM to investigate internal tissues it became clear that the tarsal tubercle was a development of the usually inconspicuous but well known arachnid sense organ identified in spiders as the *tarsal organ* (see chapter 1).

Previously, the only structure known was a shallow blister with an apical slit leading into a shallow chamber provided with sensory lobes, and believed to function as a taste receptor. When the orsolobid structure was examined using TEM techniques, typical bundles of neurones leading to knobs and spines were revealed. These were soon identified as sensory receptors and it was concluded that this structure was an elaboration of the well known tarsal organ. 'Tarsal thorns', however, were undoubtedly new structures and it was assumed that they were innervated receptors which respond when the rough surface on the back of the claws is pressed onto them while the spider walks (see Fig. 1.8c). However, it was also interesting to see that when magnified, these structures varied from species to species and so proved to be important defining characters.

Abundance of orsolobids in New Zealand

Although 106 species of Orsolobidae from New Zealand are now described, this represents only part of the local fauna because this country, despite its smallness, is home to the most abundant and varied representatives of this family. These species are grouped into fifteen genera whereas there are only eleven from all other countries combined. Most genera are defined by the structure of the male palp and female internal genitalia and thus are difficult to identify in the field. However, although many are characterised by the pronounced chevron pattern on the abdomen, others are relatively easy to identify because of the purplish pigment covering the abdomen. Examples are *Tautukua*, *Pounamuella* and *Bealeyia*.

By comparison with most species, members of the genus *Subantarctia* are 'giants' which may be up to 5 mm long and, with their reddish-brown carapace and legs as well as pale cream or grey abdomens, look very much like baby dysderids. *Subantarctia* (Fig. 10.3) is the only New Zealand genus without chevrons or patterning on the abdomen and so is easily recognised. Like all orsolobids, its species are ground-dwelling hunting spiders which, although favouring forest conditions, are sometimes found in tussock country, or in litter amongst subalpine plants well above the bushline. Although the first species was found in the Auckland Islands, *Subantarctia* is widely distributed in mainland New Zealand and nine species have already been named. These spiders are often found resting inside thin silken tubes under stones or logs during the day, and they hunt for small invertebrates on the forest floor at night.

***Fig. 10.2** (far left): During mating, haplogyne spiders such as* Subantarctia trina *(Orsolobidae) adopt a venter to venter position (see Fig 2.16b). In this species, the male inserts both palps at once. The smaller spider (male) is seen here mating with the much larger female.*

***Fig. 10.3** (left): This orsolobid spider from Mt Taranaki,* Subantarctia penara, *is an active hunting spider. Like all species in this genus, the dark mottled abdomen of this spider has no particular markings.*

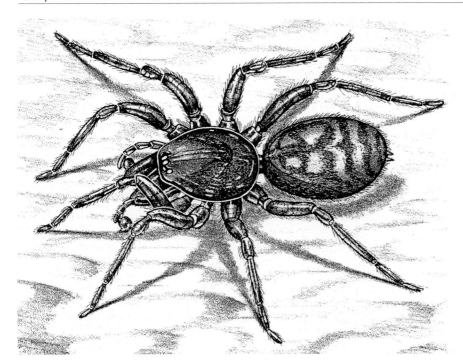

Fig. 10.4 Duripelta *species (Orsolobidae) can be recognised by the chevron pattern on the abdomen. In this male spider, the hard ventral plate can be seen extending up onto the front and sides of the abdomen.*

Those chevron-patterned orsolobids which can be more readily identified belong to *Duripelta* (Fig. 10.4). Moreover, all its species are characterised by hard plates on the abdomen. In females, such abdominal scutes are usually restricted to the ventral surface behind the epigastric groove but for some reason they are more extensive in males, plates sometimes extending as far back as the spinnerets. In some species, males have another plate on the dorsal surface. Whereas the domed eggsacs of *Subantarctia* are laid within the female's silken retreats, the flattened eggsacs of most other species are attached to the undersurface of a log or a fallen leaf on the forest floor, the female remaining for some time to guard it.

A spider without eyes

Perhaps the most unusual orsolobid is *Anopsolobus*, known only from a single female specimen found living underground at Brightwater in Nelson in 1972. Despite further searches, no more have ever been found. This spider is totally blind, having lost all its eyes and, as it also lacks pigment, is a very pale cream colour. It was quite by chance that this single specimen, about 2 mm long, was collected with other terrestrial arthropods, some also blind, from below the surface of the earth. Originally, the purpose had been to collect aquatic invertebrates living in subterranean water by means of a special trap lowered some 4 m down a bore but when the water table receded during an extended dry spell the trap was left high and dry. Instead of the desired sample, a completely unexpected haul of terrestial animals was discovered in the trap. They were all invertebrates belonging to various groups such as mites and false scorpions, and included this one spider. Known as interstitial animals, they were part of a community living in interstices among rocks and shingle well below the earth's surface and were modified for life in this deep dark habitat in much the same way as the spider. Details of the tarsal organ and the internal genitalia of this spider led to the conclusion that it was related to the widespread forest-dwelling genus *Tangata* (Fig. 10.5) which has distinctive abdominal chevrons and six functional eyes. The period of time that must have elapsed since the ancestors of *Anopsolobus* descended below the earth's surface can only be guessed at, but it would need to be measured in the millions of years in order to account for the evolutionary changes which have taken place.

Fig. 10.5 Tangata orepukiensis *(Orsolobidae) belongs to a widespread forest-dwelling genus. Here, the female is seen with her eggsac. The eyeless interstitial spider,* Anopsolobus, *is related to this group.*

Dysderidae

Northern Hemisphere natives

As a group, the dysderids are native to the Northern Hemisphere where they are centred around Mediterranean countries. Curiously, one species, *Dysdera crocota*, is able to live away from its original home and, in the wake of human settlement, is now a familiar spider in many parts of the world including New Zealand. This conspicuous spider, some 12 to 14 mm long, with its bright orange-reddish carapace and legs contrasting sharply with its pale cream abdomen, attracts immediate attention whenever it is seen (Fig. 10.6a). In addition to its striking appearance, the chelicerae are very large, and because they project forwards they give the spider a most ferocious mien. This, however, is no cause for concern as these spiders rarely show any tendency to bite.

Strictly nocturnal in its habits, this spider secretes itself within a tightly woven silken nest during the day, only emerging occasionally to hunt at night. After feeding, *Dysdera* may remain in its nest for lengthy periods and this may be one reason why it is so rarely seen (Fig. 10.6b). Jerome Rovner demonstrated that in accidental flooding, *Dysdera* is able to maintain itself from the bubble of air in this nest, which thus acts like a physical gill. Furthermore, it was shown experimentally that an air bubble can last for many days in the nest, up to ten times longer than an air bubble held by a submerged spider without a nest. This shows that, in addition to serving as a safe haven and a moulting, oviposition and brooding chamber, the nest itself represents an important means of survival for the spider should it be covered with water.

Specialist feeder

In his book *The World of Spiders* Bristowe ponders about the sort of food the slow-moving *Dysdera crocota* might catch. To find out, he provided them with a variety of prey that they might be likely to encounter. But ants and earwigs were avoided

Fig. 10.6a (below left): This distinctive spider, Dysdera crocota *(Dysderidae), is often found under stones in the garden where it hides during the day.*

Fig. 10.6b (below right): Dysdera crocota *has built its daytime retreat amongst grass. The silken nest has been pulled aside to show the spider. John Cooke reported that the female may remain in its nest for up to a month while guarding its eggsac and newly emerged spiderlings.*

and items such as moths and flies flew quickly out of reach. Eventually, finding that *Dysdera* would attack and eat slaters (woodlice), he described how the spider twisted the front of its body and, because of its elongated chelicerae, was able to pierce the slater below with one fang, and above with the other. With this technique the spider avoids contact with the slater's lateral glands responsible for exuding unpleasant-tasting substances which make it noxious to most other spiders. Perhaps another explanation for its choice of food is that *Dysdera* is able to acquire this unpalatability itself and so use it as a weapon against parasites and other enemies. Such a possibility has not yet been tested.

The supposed predilection of *Dysdera* for slaters has been challenged by Simon Pollard and his colleagues. In a series of experiments, they provided *Dysdera* with prey options and, although these tests were inconclusive, they did show that *Dysdera* would consume a variety of other captive arthropods. However, as the insects offered were effectively reduced to the same level of activity as the spider, perhaps the apparent preference for slaters is because *Dysdera* finds them easy to catch in the wild. Our own nocturnal observations tend to bear this out. On stone and concrete retaining walls, we have regularly seen *Dysdera crocota* leisurely prowling amongst the hundreds of slaters moving slowly about in the dark.

Mate recognition

Courtship in *Dysdera* is not a very auspicious affair and mating takes place in an almost casual fashion. If a male should happen to touch a female, they pivot around until both spiders are facing, the female at the same time spreading her chelicerae and reaching forward with her front pair of legs. When contact is made the front legs of both spiders become entwined and they move towards each other with fangs extended. The male then places his chelicerae inside those of the female's thus ensuring that both are inoperable. Tapping the female with his front pair of legs, the male uses the second pair to stroke the underside of her abdomen simultaneously rubbing his palps alongside those of the female. This pacifies the female who allows the male to crawl beneath her. Now, assured of his safety, the male inserts both palpal organs at once into the female aperture at the genital furrow and so mating is accomplished (Fig. 10.6c). At the peak of summer, a bundle of eggs held together with only a few strands of silk is found in a retreat with the mother spider.

John Cooke concluded that mate recognition by these spiders was achieved mainly by tactile and olfactory sense organs and perhaps also by the exchange of chemical substances, there being apparently little reliance on sight. To test this notion, he occluded the eyes of a number of spiders and observed their courtship to see if they could still mate successfully. He found that the loss of

Fig. 10.6c *To mate, the male Dysdera crocota must first subdue the female before creeping under her to insert his palps. This posture is shown in Fig 2.16b.*

vision had not the slightest effect on their ability to find each other and mate, nor did it prevent them catching food. Moreover, some spiders were still thriving even after two years of artificial 'blindness'.

Segestriidae

Tunnel dwellers

Segestriids are all moderately large spiders and their six eyes are arranged in three pairs with a clearly visible space between each pair. While these spiders have not yet been officially revised we already know there are many different kinds in this country. At present they are placed in two genera: *Ariadna* which is a worldwide group: and *Gippsicola*, recorded from Victoria, Australia. All of them look very much alike. Both genera are well fitted for life in narrow tunnels, having long tubular bodies and the front three pairs of legs directed forward (Fig. 10.7 a). Fortunately, it is not necessary to delve deeply into the minor structural differences which separate these two genera because in New Zealand the colour patterns of the abdomen are enough to tell them apart.

Widespread genus

Ariadna, the most widespread and common genus, has been recorded from the seashore to the mountains. Along the coastline, the narrow silken tunnels are occasionally found under stones but more often in the nooks and crannies on the rock faces above the high tide level. In open country, their retreats may be found beneath rocks but in the forest they prefer to use holes vacated by wood-boring insects (Fig. 10.7b) sometimes found in the smaller branches and twigs but also on the trunks of trees. Tunnels are always generously lined with tough silk which extends to the outer rim so that occupied burrows are easily seen (Fig. 10.7c). The dorsal abdomen of all but one of the species is brown or purplish, while the central pigment is usually darker thus forming a median streak.

***Fig. 10.7a** (above):* Ariadna *sp (Segestriidae) is shown here inside its silk-lined tunnel. The first three pairs of legs, oriented towards the front, enable the spider to move easily within the confined space of its home.* **b** *(right): At night* Ariadna *can be seen waiting for prey at the entrance of its burrow. The three pairs of frontally positioned legs, some in contact with radiating silk threads, are poised to seize an unwary insect. (Photograph Jerome Rovner)*
c *(far right): The entrance to* Ariadna's *tunnel is often heavily lined with silk. A network of silk extends beyond the rim and serves to alert this sit-and-wait spider to the presence of prey.*

Fig. 10.7d Ariadna septemcincta *is the most distinctive species in the genus* Ariadna *because of the reddish-brown chevrons on its abdomen.*

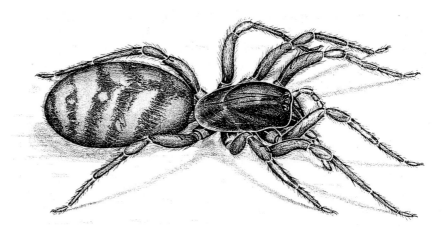

The one exception is *Ariadna septemcincta* which has reddish-brown chevrons down the dorsal surface of the abdomen (Fig. 10.7d). This species lives only in the forest where its burrow is usually found in rotting logs on the forest floor but it has also been recorded living in burrows in the ground.

In this country, *Gippsicola* apparently lives only in the forest where it occupies old insect burrows in the trunks and branches of trees but there is little difficulty in identifying their homes, because, unlike any native species of *Ariadna* (except for *septemcincta*) ten or more stout threads radiate out from the mouth of the burrow (Fig. 10.8a). Perhaps these simple traplines demonstrate the first steps taken by primitive ground-dwelling spiders towards the evolution of the multitude of aerial webs used by more recent spiders. *Gippsicola* rests at the mouth of the burrow with its front legs on some of these threads, which are raised above the surface by small supporting pillars thereby acting as highly efficient trigger lines. Here it sits and waits for something edible to stumble across the 'warning' lines. The spider itself is distinctive because its pale abdomen is adorned

Fig. 10.8a (left): The tunnel home of Gippsicola *sp. (Segestriidae) can be readily differentiated from most* Ariadna *species by the conspicuous silk lines radiating out from the tunnel entrance.*

Fig. 10.8b (below): The long cylindrical body of Gippsicola *sp. and the double row of dark patches which show up against the pale abdomen make this spider easy to recognise.*

with a double row of conspicuous dark patches (Fig. 10.8b). The eggs are laid within the burrow by both groups of spiders. When the young emerge they do not necessarily move away at once, for often a number of large but still immature juveniles will be found sharing the burrow with their mother.

Periegopidae

Periegops suterii

It is very curious that one of our rarest spiders should have been collected by chance at an early stage of this country's faunal investigations. This spider, found on Banks Peninsula, Christchurch and, to date, the only named *Periegops* species in New Zealand, was initially described in 1891 by A.T. Urquhart as *Segestria suterii*. Two years later, however, Eugène Simon was sent another specimen to which he gave the name *Periegops hirsutus*, unaware that a similar spider had already been recorded. Although these separate discoveries may mean that this spider was more common in the last century, this is not necessarily so, since it is highly likely that the specimen studied by Simon also came from Banks Peninsula. Moreover, until very recently we would have been confident that this species existed only in a few patches of relict bush on Banks Peninsula. In 1995, by a strange coincidence, further specimens were recognised from three other sites, four in the Riccarton forest reserve in Christchurch by David Blest; another from a trapdoor spider burrow in the East Cape Region, of the North Island by Grace Hall, a Landcare arachnologist; and yet another from the Alderman Islands. The latter two spiders, both females, are remarkably similar to the Canterbury species but until a male is found it cannot be decided whether or not they represent a second species of *Periegops*.

A new family

Despite their similarity, these spiders may indeed be separate species, a supposition boosted by the discovery of a further species living in forest near Brisbane, Australia by Valerie Todd-Davies. Moreover, as in many rare species, the original designation of *Segestria suterii* was not compatible with Northern Hemisphere spiders and even last century, Simon, noting its unusual features, suggested the need for a separate family. Despite agreement among scientists no action was taken because for so long only a single species was known. Subsequently, with the addition of the Australian spider, *Periegops australia*, a new family Periegopidae was established in 1995.

In fact, the Australian and New Zealand species are quite difficult to tell apart except for slightly different abdominal patterning and variations in the structure of the palp. However, the character that differentiates *Periegops suterii* from other New Zealand six-eyed spiders is the obvious separation of the three pairs of eyes (Fig. 10.9). Although the rarity of these spiders means that little is known

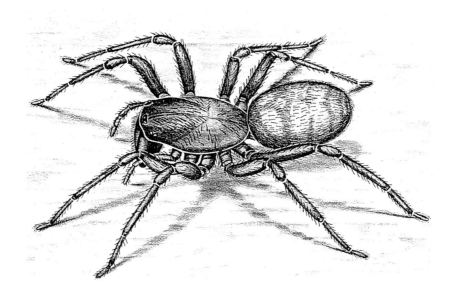

Fig. 10.9 Now officially known as Periegops suterii (Periegopidae), this spider can be recognised by the wide spacing between the three pairs of eyes and its abdominal markings.

of their habits, a certain amount can be deduced. For example, a reduction in their silk glands supposes that they do not rely greatly on silk and are unlikely to construct a snare. Collecting records suggest that, like *Subantarctia,* they construct tubular retreats under logs or stones and the fact that both males and females have been trapped in pitfalls points to an active predatory lifestyle. The chance discovery of three adult males associated with a single female under a log seems to indicate that females release a pheromone to attract males when they are receptive. Collecting records all reveal that these spiders are primarily forest species living on the forest floor.

Scytodidae

Spitting spiders

Known as spitting spiders, members of this family are found in tropical and some temperate climates. They do not occur naturally in New Zealand and only once has a scytodid been found living here. This spider, first seen crawling along the wall of a North Island school by David Court, an observant school teacher with a good knowledge of spiders, proved to be *Scytodes thoracica.* No further sightings have been reported, which is rather surprising as these spiders could readily survive in this country.

Originally from Europe, *Scytodes thoracica* is now quite widely distributed, being common in the northern states of the USA, for example. It is a domesticated species, recognised by its pale yellowish tint patterned with black spots, the high rounded carapace and long spindly legs. When, in its wanderings over the walls and ceilings of houses, it detects prey about 10–12 mm in front, it gives a jerk, and a viscous gum and venom mixture spurts from the oscillating chelicerae. Two sets of sticky, transverse, zigzag threads pin the victim to the ground and the spider bites it, pulls it free and feeds on it at its leisure. The two huge bilobed glands which produce the toxic spitting glue are housed in the rounded carapace. Overseas, this spider is known to prey on silverfish, moths and small cockroaches.

Another *Scytodes* species found in the southern states of USA reveals some interesting variations. First, it apparently spins an aerial web although there is no mention of its use in prey capture. According to Cole Gilbert and Linda Rayor, however, this *Scytodes,* first taps its prey, gives a shudder then spits, after which it cuts and pulls its prey away from the dried spit. Holding the prey with its third legs, the spider then begins to wrap it in silk. This is accomplished by the alternate use of the fourth legs, drawing loops of silk from its spinnerets and wrapping them around the prey. The trussed victim is held by the chelicerae as the spider feeds on it.

Oonopidae

A spider puzzle

The rarity of oonopids in New Zealand is puzzling because it is the most widely distributed of the six-eyed families and in some situations is one of the commonest leaf litter spiders. The single species, *Kapitia obscura,* has only been found a few times and appears to be restricted to the southern half of the North Island. This little spider, less than 1.5 mm long, is pale brown with a cream abdomen and, unlike many overseas representatives which have hard plates, the abdomen is soft. Although this spider might seem, at a glance, to be a minute orsolobid, its chelicerae have no teeth, its claws are not modified as in orsolobids and the tarsal organ is a simple pit. The absence of these particular features means oonopids are, in fact, very different from the orsolobids.

CHAPTER ELEVEN
ORBWEB SPIDERS

As spiders evolved to take advantage of the ever-changing environment and consequent habitat diversity so, of course, did other invertebrates. Although many of these creatures still found their living on the ground, others acquired new features and habits and migrated upwards to the grasses, shrubs and trees that now grew in warmer, drier climates. Equipped with the silk-spinning skills that marked their emergence, spiders did not need to take to flight as insects had done, but developed aerial snares to trap this new elusive prey.

It was once thought that the perfect symmetry of the orbweb was the peak of spider development, that it marked the summit of achievement in snare-building skills. There is no doubt that orbwebs have been highly successful as prey-capture traps, the five families found in New Zealand and elsewhere being testimony to this. These are the Araneidae, Tetragnathidae, Nanometidae, Theridiosomatidae, and Uloboridae. A further family Deinopidae, is mentioned because an Australian species has twice been found here. Although, in these groups, the basic orb design is very similar, there are many ways in which the web has been modified to suit not only the spider's special needs but also its changing environment. But arachnologists are now beginning to think that orbwebs arrived early on the evolutionary scene and that many other kinds of silk snares were subsequently modified to trap the greater range of prey as it became available.

In this, the first of two chapters about web-building spiders, we begin with the orbweb families, not just in deference to this particular line of thought about their evolution, but also because this is the web with which most people are familiar. It is the one which decorates our shrublands, forests and gardens, and is also the web of myths and legends and romantic verse.

Fig. 11.1 This large, vertically oriented orbweb, some 30 cm wide and 45 cm in height, was built by Cryptaranea venustula *(Araneidae), a spider widely distributed in forest habitats throughout New Zealand. A web of this size takes the spider from two-and-a-half to three hours to build. Twenty-three evenly spaced radials have been put in place, and the spider then circled more than fifty times to lay down first the non-sticky scaffolding (now removed) and then the sticky spiral. These tasks required the extrusion of over twenty metres of silk, pulled from the spinnerets and carefully positioned by the one or other of the fourth legs.*

Orbweb snares

The conspicuous orbweb snares spun by many of these spiders are well known but although they are widely accepted as part of the landscape there is always a sense of wonder that this perfection in symmetry and structure is the work of such lowly animals (Fig. 11.1). Could they be the outcome of some form of intelligent reasoning, or at least the end result of learning passed down from parent spiders? Such notions are quickly dispelled, however, as the parent spider often dies before the spiderling is ready to construct its first snare. Moreover, if we remove an eggsac and isolate the spiderlings as they emerge, they are still capable of executing all the steps required to produce this characteristic snare. We find, too, after studying the actions employed in building this web, that the spider is restricted to a rigid routine in which one particular behaviour must always precede the next. If, for example, the earlier stages are partly destroyed, the spider is quite unable to work out that it is useless to continue without repeating these first steps. Instead, it continues with a pre-determined set of actions even though the final result is a travesty of the structure that the spider set out to create. Such stereotyped behaviour is said to be instinctive, resulting from a programme precoded in the spider's genes. While the stimuli triggering this chain of reactions and culminating in the finished web are by no means fully understood, by observing the spider in action it becomes clear that the end result is achieved by a series of relatively simple actions.

Orbweb Construction

Exploring the site

Most orbweb spiders construct their webs after dark. But sometimes they begin during daylight hours if the weather is dull and overcast, and this is the best time to observe them. To begin with, however, spiders spend some time – even two or three nights – exploring their surroundings, laying a dragline as they go. They walk along various branches, twigs and leaves, now and again dropping on draglines, perhaps returning via one of them to a previous spot. These meanderings provide the spider with information about available attachment points and the area's suitability for a web, while at the same time establishing boundary threads from which the orbweb can later be suspended (Fig. 11.2). Some preliminary threads may be gathered up and eaten if these rovings reveal no appropriate sites. It is clear, however, that this preliminary information may

Fig. 11.2 *Araneid spiders travel considerable distances as they move around looking for suitable web sites. This orbweb spider,* Colaranea viriditas, *deposits silk as it walks about and, at the same time, controls the position of the thread with a hind leg.*

differ from site to site and that the spider is somehow able to incorporate these variations into its web-building. No doubt there are some key requirements, and if these are not met the spider may move to another place.

Bridging the gap

Having completed these preliminaries, the spider begins the process of building the orbweb. First it must establish a horizontal line from which the web is hung. This 'bridging' thread is usually laid down in one of two ways. The spider first attaches one end of a silk thread to the twig where it is resting and then walks down, across and up (or around) other twigs or branches while holding this thread away from its body with a fourth leg. When it reaches another twig a suitable distance away it pulls this 'loose' line across the gap and attaches it at this point. If, however, there is no easy route by which to do this, the spider waits for the wind to carry a floating thread across the gap until it becomes firmly anchored to a plant or tree on the other side. Once the gap is bridged, the spider runs across this line, letting out a thicker line and at the same time removing the original line. This new, stouter line, which we will call the bridge, becomes the basis for the ensuing orbweb.

Establishing the framework

The next set of actions establishes the outer boundary threads and the means by which silk spokes radiate out from the centre of the finished web. First, an additional line is constructed by the spider as it crosses the bridge again, this new line also being held free with a fourth leg. This line, however, sags below the original bridge thread but is fixed at the same end point. The spider moves to the middle of this sagging thread, apparently judging this position by a comparable line tension on either side. After attaching another thread at this spot, the spider lets itself down on a silk thread (Fig. 11.3a) until it reaches a twig somewhere below. The vertical line is then pulled tight and firmly fixed. This neat trick stretches the transverse line to which it is attached, downwards, with the result that the threads take up a Y-shape and the future hub of the web is thus established at the junction of the three arms (Fig. 11.3b). Moving along one of the upper arms of the Y the spider carries another line as before, attaches a further thread to the centre of this, then carries this up to the bridge to be fixed and pulled tight (Fig. 11.3c). Now, there is not only another radial in place, there is also a transverse line below the bridge line which will form one of the upper limits of the orb itself (Fig. 11.3d).

The spider continues to construct further radials using preliminary threads to anchor the principal attachment points. Because gravity is no longer so useful, the spider must adopt the technique of doubling up on a radial and carrying this loose line to a new attachment point. At the same time it adds marginal threads to the lower sectors until, supported by a few widely spaced radials, the outline of the web takes shape (Fig. 11.3e). The rest of the intermediate radial threads are then put in place. By moving down an existing radial to the hub and attaching a further thread to the centre, the spider carries it free to the boundary line then moves along this line for a short distance before attaching the thread it is carrying, so forming a new radial. By continually moving from the hub to the outer margin the remaining radials are quickly filled in so that there is an array of evenly spaced radials, just like spokes in a wheel. At intervals, the spider revolves around the centre hub laying down threads that secure the radials until the temporary spiral scaffolding is put in place. When this stage is reached, the spider's behaviour changes abruptly as it begins to circle out from the centre, attaching a widely spaced spiral thread of dry silk to the radials as it goes (Fig. 11.3f). This 'scaffold' not only keeps the radials in place until the final sticky spiral is laid but also provides a 'footing' and a guiding line as the spider moves around.

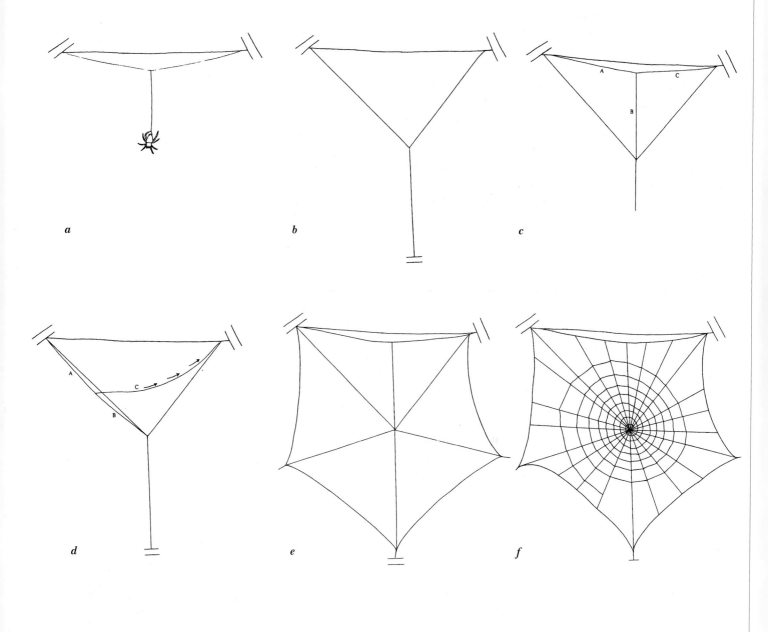

Figs 11.3 Orbweb construction.
a After a silk bridge-line has been established the spider drops on a thread from the middle of it.
b This vertical thread is attached to a twig below and pulled tight, thus forming the first three radials.
c A loose thread (A, B) is laid along one of the upper radials and connected to the opposite corner by thread C.
d Thread C is pulled tight, and the first stage is completed.
e Other sectors are completed similarly by making use of boundary threads (not shown) laid down earlier. The inner radials are filled in, one by one, by the spider carrying a free thread with a hind leg and attaching it to an outer margin thread.
f Dry silk is used for the scaffolding, which consists of a widely spaced spiral winding out from the centre.
g Using the scaffold thread as a guide and the radials for a footing as well as attachment points, the permanent sticky spiral is laid down from the outer edge to the centre, the spider gathering up the dry spiral as it goes. As each part of the new spiral is fastened to a radial, it is stretched so that the original sticky coating is broken up into globules (shown in Fig 11.4).

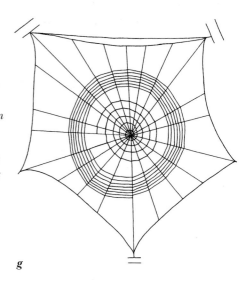

Laying down the sticky spiral

The sticky spiral starts at the outer edge of the web and, as the spider proceeds with laying this down, it removes the temporary scaffold. Here is how the spider carries this out. It lays the permanent spiral by crossing between the radials on the dry scaffold thread at the same time pulling out a new line of silk, now coated with a sticky substance. But these sticky spirals are laid much closer together than the dry scaffold spirals, which are gathered up at the same time, and so the web gradually assumes its familiar shape (Fig. 11.3g). The sticky material is at first evenly coated over the inner thread but as the spider attaches each thread between the radials the line is stretched so that the surface tension of the fluid silk layer is broken. As a result, the sticky coating gathers like a string of beads fairly evenly spaced along the length of the basic silk core (see Fig. 11.4). The spider stops laying down the sticky spiral when it is some distance from the hub so that there is always a clear area near the centre where it normally waits.

Accessories

Although the web is now ready for its main function of trapping food, many spiders make some additional modifications. Some species remove the threads from the hub when they have finished the construction to leave it empty of silk. Other species spin conspicuous zigzag mini-sheets of silk radiating out from the hub to form what is commonly known as the stabilimentum (Fig. 11.4). This eye-catching feature may take many shapes and sizes. Its purpose, however, has been the subject of much speculation, ranging from claims that it provides additional stability for the web to suggestions that it makes the spider itself look bigger, or more difficult for predators to see, or even that it enables birds to avoid an otherwise almost invisible web. Because these stabilimenta vary from species to species it is not easy to describe their construction, but if you are lucky enough to see one being built, you will realise that the whole process is, like the web itself, a smooth progression of actions with one step leading to the next until the whole thing is complete.

Questions often asked

What happens to silk after the spider removes it? The answer is that the spider usually eats it – not as dry silk, but after liquefying it with digestive enzymes which ooze from its mouth. Indeed, the same thing happens later to the silk which the spider uses to wrap up prey and, in some species, even to the web itself.

Why is the spider not trapped in its own sticky threads? It is generally believed that the spider is protected by a coating of fluid which is continually renewed from glands associated with the mouth and chelicerae. The legs and other parts

Fig. 11.4 Colaranea viriditas *hangs head downwards from the hub of its completed web. Above the spider is the stabilimentum, a long narrow structure strengthened with zigzag bands of silk put in place after the orbweb is completed. One possible function of this accessory is to alert birds to the presence of the web, which would otherwise be destroyed if they flew through it. Visible, too, in this web are the globules of sticky silk scattered along the main spiral threads.*

of the body become coated with this layer when spiders groom after feeding, this being accomplished by drawing the legs, one by one, through fluid flowing from around the mouthparts and slowly wiping the rest of the body with them. An interesting experiment was carried out by dipping the legs of one of these spiders in a solvent, and then putting it back on the web. At once it could be seen that the spider experienced difficulty in walking on the web until the fluid coating was renewed.

Do spiders really need gravity to be able to build orb webs? The answer is no, and this is how we found out. Because gravity was always thought to play an important part in orbweb construction, particularly as spiders started to build, it was often supposed that they would not be able to make webs in outer space. In 1972, Judith Miles, a high school student in Massachusetts, USA, put forward a proposal to include a spider web-building experiment in Skylab 2, a spacecraft being prepared for a two-month outer space flight by NASA[1] in 1973. Two adult female spiders, *Araneus diadematus*, and two flies were carried aloft in small vials. The spiders were released onto frames suitable for web-building, one after a week in space, and the second after four weeks in space. Four days after its release each spider built an orbweb and duly made several more in Skylab under zero gravity conditions. All webs were photographed. Early webs were generally normal but later webs became smaller and somewhat irregular. However, these results, analysed by Peter Witt and his colleagues, showed that spiders could manage the essential requirements of web-building under such conditions. The frames provided for the spiders probably enabled them to carry out the preliminary stages without gravity by using techniques that we described above. Other factors, such as an inadequate supply of food as well as the long-term effects of weightlessness on relatively short-lived, low mass animals may have contributed to the decline in web regularity.

What happens when an insect flies into a web? After a web is complete, the spider adopts a head-downwards position at the hub, clinging motionless with its claws until the struggles of a trapped insect bring it rapidly to life. Its first action is to tweak some of the radial threads, which often sets the insect into motion again. However, the spider's main purpose is to determine which threads reveal the greatest tensions or vibratory signals, for this will indicate the route that leads unerringly to the trapped insect. On reaching its struggling victim, the spider usually first feels it with its legs or palps and, if the size and taste are acceptable, it is rapidly wrapped with a swathing band of silk and then bitten (Fig. 11.5). Wrapping is accomplished as the spider twirls it round and round with its hind legs, at the same time pulling silk from its spinnerets by the alternate use of these legs, winding the silk around the victim like thread on a bobbin. Several quick bites are now administered before the wrapped-up bundle is cut free from the web and transported to the hub where it will be eaten.

Do spiders repair their webs? After a good night's catch, the web will be badly torn but few spiders are able to make effective repairs. Some spiders use the web for several days with gaps bridged by draglines as the spider moves about but many of them tear down the mutilated structure and construct a new web on the framework of the earlier one. We found that the beautiful green *Coloranea viriditas* (Fig. 11.6), which we photographed in our garden, belonged to this house-proud group as it regularly rebuilt its webs every day for more than the week it took us to get the photographs we wanted. During the daytime many webs remain empty, but this does not mean that the owners have gone. Invariably a strong silk cable leads from the hub of the web to some secure retreat nearby where the spider hides, at the same time keeping the claws of one leg hooked to the cable so as to receive signals transmitted by vibrations from prey caught in the mesh.

[1] National Aeronautics and Space Administration.

Fig. 11.5 (top): *After wrapping the prey with silk, this araneid spider* Eriophora pustulosa *bites its victim several times. Now the spider will leave this tattered segment of its web for the security of the hub where the catch will be eaten.*

Fig. 11.6 (bottom): *This attractive spider,* Coloranea viriditas, *comes in a variety of different colours and abdominal patterns (see Fig 11.2). The green hues and leaf-like 'shield' of the one shown here help to conceal it among the foliage where it lives. This native araneid species is common in gardens all over the country.*

How do male orbweb spiders court their mates? Strangely enough, no one seems to have recorded the mating behaviour of any of New Zealand's orbweb spiders. We outline here what one might expect to find and hope that some of our readers are prompted to fill in this gap. As in many spiders, the males are much smaller than the females and a few, such as the araneids *Arachnura* and *Celaenia*, are truly midgets. Quite possibly, some very minute males are able to get close to the female almost without her knowledge but most males need to approach cautiously to ensure that they do not become her next meal. Preliminary recognition is normally achieved by a series of vibratory signals which the male transmits by plucking and tweaking at the threads of her web as he slowly approaches. At first he may be repulsed, whereupon he rapidly retreats or drops out of harm's way on a dragline. But the urge to mate is strong and he keeps repeating his attempts until her aggressiveness fades, and this encourages him to come quite near. However, his troubles are not yet over because he must now spin a special mating thread and coax the female on to it. This he does with a series of jerks and tugs so that eventually she ventures onto this thread and hangs onto it upside down. Once the male succeeds in getting the female into this position, mating proceeds quite smoothly. As she hangs motionless upside down he moves towards her, clasps her with his legs and inserts first one palp and then the other while both are suspended downwards (see Fig. 2.16f). No doubt species-specific pheromones or scents emitted and detected by one or both spiders also play a role in male and female attraction.

Worldwide Groups

Orbweb families such as Araneidae and Tetragnathidae belong to a higher worldwide group known as the Araneoidea. However, the related out-group, Deinopoidea, contains *Deinopis*, the net-casting spider (not found in New Zealand) and Uloboridae, this family having representatives in New Zealand, (e.g., *Waitkera*). The Theridiosomatidae have some derived features which set them further apart from the mainstream orbweb spiders.

The orbweb spiders most people see in New Zealand belong to the worldwide subfamily Araneinae (Araneidae) and these, in general, are recognised by the almost triangular shape of the abdomen when viewed from above (Fig. 11.7). As can be imagined, these widespread and easily collected spiders attracted the attention of spider enthusiasts in the late nineteenth century and some sixty species were described from New Zealand. Unfortunately, the emphasis at that time was on their colour patterns, which we now know may differ radically from one specimen to another. As a result, many were named a number of times and it was very difficult to be sure about which species the names belonged to. Fortunately some of the original specimens named by Urquhart at that time were deposited in the Canterbury Museum, Christchurch, and eventually it was shown that only fifteen of the earlier names were valid. To these were added five new species, bringing

Fig. 11.7 *The triangular abdomen of* Cryptaranea albolineata *makes this and other orbweb spiders of the subfamily Araneinae (Araneidae) easy to identify. Seen here with an eggsac,* Cryptaranea albolineata *is a species with a very broad ecological tolerance, being found in forests and gardens as well as coastal habitats.*

***Fig. 11.8** (above)*: Eriophora pustulosa, *an Australian immigrant, is now one of the commonest orbweb spiders in New Zealand. It is seen in gardens, and on and around houses and schools everywhere. The patterns and colours of this species vary enormously, ranging from shades of whites, blacks and browns to greens and yellows, even orange hues, and often mottled with patches of white. This spider's successful colonisation of many habitats may lie in the different colours and patterns that it is able to assume.*

***Fig. 11.9** (below left): Although its colours are deceptive,* Eriophora pustulosa *can be recognised by the knobs at the end of the abdomen. Seen here are three in a transverse row while behind a further two are arranged longitudinally, although only one is visible. Many* Eriophora *species also have two humps at the upper corners of the triangular abdomen, clearly visible in this illustration of* pustulosa.

***Fig. 11.10** (below right): This* Eriophora decorosa *spider is quite unlike* Eriophora pustulosa *in appearance, and the triangular abdomen is less obvious. The end of the abdomen has been drawn out into a short, stout tail, obscuring the two knobs below. The yellowish abdomen with its dark band extending to the tail is a characteristic feature of this species.*

the total New Zealand fauna to twenty. Most of these species are found throughout New Zealand and include some also found in Australia and now thought to have been introduced from there. Indeed, the orbweaving spiders most often seen around gardens and houses in this country and which are well established here include those that accidentally crossed the Tasman Sea, which lies between Australia and New Zealand. We will discuss some of these introductions first.

Australian Colonisers

Perhaps it is not surprising that the commonest orbweb species found throughout New Zealand is the Australian *Eriophora pustulosa* (Fig. 11.8), which belongs to the Araneidae. It is not certain, though, whether this spider ballooned across the Tasman Sea or was brought in by chance during early colonisation but the fact that it is present on all outlying islands, including the remote New Zealand Subantarctic Islands, supports the former view. Initially this spider was named by Walckenaer in his Histoire Naturelle Insectes Aptères as *pustulosa* from Tasmania although in the same publication it was called *verrucosa* from New Zealand. Moreover, because this spider assumes an endless array of colour patterns on the dorsal abdomen, ranging from pure white to various greys and black as well as all shades of yellow and brown and even green, it eventually acquired at least ten different names. However, *pustulosa* was the first name, and so *pustulosa* it remains.

Fortunately, despite these guises, *Eriophora pustulosa* is easily recognised by five small knobs at the end of the abdomen. Three of these knobs are arranged in a transverse row with two more in a longitudinal row behind (Fig. 11.9). These ubiquitous spiders and their webs are not only found in the garden but also in any moderately open spaces although they seldom penetrate far into dense forest. Nevertheless, from time to time they appear on some remote tramping hut deep in the bush but never far from that hut. A second species, *Eriophora decorosa*, which could be mistaken for a young *pustulosa*, has fewer knobs on the posterior of the abdomen (Fig. 11.10). This is also an Australian spider, rarely found but nevertheless widely dispersed, raising the suspicion that it is not actually a permanent resident but merely balloons over now and again.

The large web is built in any suitable place, one *Eriophora pustulosa* even finding the frame of our kitchen window an accessible – and profitable – spot. Here, all sorts of moths and other flying insects, attracted by light at night, came to flutter against the window and the spider benefited accordingly. Resting below the eaves during the day this female remained linked by a thread to the centre of the web thus enabling it to move quickly back and forth. One afternoon, when a grasshopper landed in the web, the spider quickly abseiled to the hub then swung down again to the struggling insect, wrapping it vigorously with silk, then biting. Moths, caught at night, were always bitten before being wrapped, a habit previously recorded in overseas araneids. This female spider, light grey

Fig. 11.11 (right): A sub-adult male, Eriophora pustulosa, *hangs in the remnants of his web as he chews and sucks on a fly. Although his palps are enlarged, he will undergo another moult before becoming an adult. Like the females of this species, his abdomen is a typical triangular shape with the characteristic knobs at the rear.*

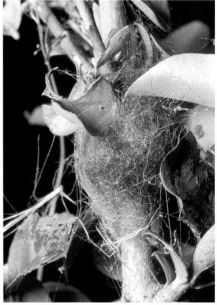

when it first arrived, stayed for six months but by the time it departed it had turned a dark chocolate brown, the same colour as our eaves were painted and where it rested during the day. Can we surmise that these spiders are able change their colours to match their surroundings?

Male *Eriophora pustulosa* are not midgets, as in many other orbweb spiders, but are about two-thirds the size of females (Fig. 11.11). In their search for females, they often dangle on long threads for considerable periods, sometimes for more than 24 hours. Mating has not been observed. Several eggsacs are laid sequentially in foliage (Fig. 11.12a, b) or in a corner on buildings or under the eaves. Each egg-mass is embedded in loose, wiry silk, at first brown but later turning to olive green, and as there is usually a number they are near the female's orbweb. After about three weeks, the spiderlings hatch but it may be a week or more before they emerge. Eggsacs are often littered with cast-off skins resulting from their first shedding but another moult will take place before they leave the security of the eggsac.

Wasp attack

Like many other spiders, araneids have their enemies. We thank Denis Gibbs of Hamilton for the following story after he witnessed an attack on a juvenile orbweb spider, *Eriophora pustulosa*,[2] by *Vespula germanica*, the german wasp. It was a mild winter's day and the spider was resting motionless in its orbweb which was strung in an exposed position between the branches of a lemon tree. The wasp alighted nearby, then flew up and over the web landing on a leaf on the other side (i.e., the side where the spider was suspended). From this position it stared at the spider for several minutes, almost as if it was assessing the best way to approach its potential target. Suddenly it rose up in the air, then dived straight at the spider. Seizing the hapless creature in its forelegs, the wasp crushed it with its strong mandibles and began to devour the victim. Disturbed at some stage, the wasp broke away from the web with ease and flew away with the remains of the spider in its jaws.

Fig. 11.12a (top right): Two eggsacs belonging to Eriophora pustulosa *have been laid together in foliage. These eggsacs, olive-black in colour, are held together and fastened to a branch with coarse wiry silk.*

Fig. 11.12b (above): An eggsac with the silk pulled aside shows the newly hatched spiderlings and their discarded, crumpled white skins. Pale and colourless at this stage, the spiderlings will moult again before leaving this shelter.

Other people have reported seeing similar attacks by german wasps on *Eriophora pustulosa* and concern was expressed that some of our native orbweb spiders might also be under threat from this accidentally introduced pest. Experiments were undertaken by Richard Toft. His results from four study sites in beech forests in the Nelson Lakes National Park strongly suggested that native orbweb spiders were indeed vulnerable to predation by these wasps. It was recommended that efforts to control wasp numbers by poisoning be undertaken early in the spring to summer seasons in order to reduce this possibility.

More Australian immigrants

In addition to *Eriophora pustulosa* there are other Australian orbweb spiders which, like *Eriophora decorosa*, are found only rarely but are continually reinforced with

[2] Previously known as *Araneus pustulosus*.

Fig. 11.13 This plump and colourful araneid, Neoscona orientalis, *has been seen a number of times in this country. These spiders are well known in Australia and the Pacific Islands and the bright and colourful patterns on the abdomen are evidence of their tropical origin.*

further immigrants from across the Tasman. Two of these araneids are large and very conspicuous, but neither has been found in large numbers despite being recorded in New Zealand many times over the last hundred years. The first arrival, *Eriophora heroine*, which is almost 20 mm in length, has been regularly sighted from all parts of the country ever since Urquhart described it under the name *Epeira brounii* in 1884. The second, the equally large *Neoscona orientalis* (Fig. 11.13), was also described by Urquhart from Te Karaka as *Epeira orientalis*. That these two spiders are native is doubtful since they are only seen occasionally in New Zealand. In Australia and the Pacific Islands, however, these spiders are common and widespread and, although their origin is elsewhere, Urquhart's name will have 'date' priority over any other subsequent name.

Bird-catching spider

An Australian spider belonging to the Tetragnathidae has also made its way to New Zealand. This is our most spectacular immigrant, the big *Nephila* spider (Figs 11.14a) which is common in north-east Australia; and known as the Bird-catching spider in a number of tropical countries. Reports of these spiders with their huge orbwebs of up to two metres in diameter (Fig. 11.14b) regularly occur from northern parts of New Zealand but it is not yet known if they breed here. Without a doubt, adults are able to survive in the warmer regions of this country but they have never been recorded in sufficient numbers to indicate that colonies are established. It is reported from Australia that *Nephila edulis*, a diurnal feeder, catches a variety of flying insects, particularly leafhoppers and that the commensal spider genus, *Argyrodes* (Theridiidae, see chapter 12), is a regular inhabitant of its web, often living there in considerable numbers. The tiny *Nephila* males, several of which are regularly seen on the fringes of adult female webs, remain at a distance for some time. When one of them eventually approaches and mates with the female, he usually does so when the female is eating, otherwise he is likely to be attacked.

An odd newcomer

Poecilopachys australasia (Araneidae), known as the two-spined spider, (Fig. 11.15a) is a comparatively recent immigrant from Australia, having arrived in the Auckland area some thirty years ago. Apparently able to breed in this climate, these spiders have slowly spread over most of the North Island although they have not been successful in colonising the far south. The female *Poecilopachys* is a medium-sized spider with a plump oval abdomen marked with yellow and cream bands and is distinguished by two large dorsal horns projecting from its abdomen; the male is quite small and is without horns. Writing in *Tane* in 1974, David Court says that the mature female is mostly found resting on the lower surface

Fig. 11.14a (below left): This large and elegant spider, Nephila edulis *(Tetragnathidae), is a common sight in many parts of Australia where it is indigenous and is now quite frequently seen in the North Island of New Zealand. Reports suggest it is slowly spreading.*

Fig. 11.14b (below right): The huge vertical orbwebs made by Nephila *spiders are supported by a network of threads at the back and front. The spider, dwarfed by the web it has made, hangs head downwards at the hub. (Photograph David Court)*

of the leaves of citrus trees (Rutaceae) and that when it is disturbed, waves of colour pulse along the bands of cream and yellow. They are also one of the very few orbweb spiders which lay sticky spirals without first putting a non-sticky spiral in place. Moreover, Densey Clyne notes that the web differs from the typical araneid orb by being built in an inverted triangular framework within which the rather small orblike portion becomes extended in the lower sector with additional cross threads. This structure may be set at an angle or is sometimes horizontal. Adult males are small, being less than a third of the body length of the female and do not spin a web. The golden, spindle-shaped eggsacs are very distinctive and are always hung near the web (Fig. 11.15b).

Another really striking tetragnathid spider is *Leucage dromedaria* (Fig. 11.16) with its greenish legs, pale carapace and prominent hump on the anterior of its silvery abdomen. Originally described as a native by Urquhart under the name of *Nephila argentata*, it was subsequently found to be identical to a common Australian species by which name it is now known. Considering the large number of *Leucage* species in Australia and other countries, it is rather surprising that only

Fig. 11.15a (above left): One of the strangest spiders to make its home here is Poecilopachys australasia *(Araneidae). The purpose of the two horns or spikes which project from the end of its wide and oval abdomen is unknown.*

Fig. 11.15b (above right): Seen here putting the finishing touches to its large spindle-shaped eggsac is the two-spined spider, Peocilopachys australasia. *As in most araneids, the sac is hung near its web which, in this instance, has been built in a kowhai tree* (Sophora *sp.*). *(Photographs David Court)*

Fig. 11.16 (below): The silvery Leucage dromedaria *(Tetragnathidae) builds its large orbweb in a horizontal position close to the ground and hangs beneath it. An Australian spider, it seems to be quite well established in the North Island.*

one species has arrived here. *Leucage dromedaria* builds a typical orbweb except that it is strung horizontally amongst the undergrowth on the floor of more open types of forest and has a maze of threads reaching to the ground below. The spider hangs onto the underside of the orb so that, as you look down upon it, its sombre ventral surface merges with the dark background of the forest floor. This means that it is barely visible from above and its cryptic dorsal colouring renders it less easy to perceive from below

Araneidae

Native forest dwellers

In New Zealand, a number of prettily coloured orbweavers with pale green abdominal surfaces are found near the margins of forests. The species illustrated in this book, the green orbweb spider *Colaranea viriditas* (see Fig. 11.4), is found from one end of the country to the other as well as on the Chatham Islands. Once it was restricted to lowland tussock plains and forest margins but today it is to be found in gorse, scrub and long grass. The vertical web is provided with a conspicuous stabilimentum above the hub, and below it the spider, hanging head downwards, is often seen during daylight hours awaiting the entrapment of prey. The eggsac looks similar to that of *Eriophora pustulosa* and, as in that species, is often bound to a twig near the web.

Zealaranea crassa, another widely distributed orbweb spider generally found on hedges or amongst gorse and manuka scrub, is greyish brown with a prominent white band across the front of the abdomen. Although this white marking is usually the most distinctive feature, it can be quite variable. Sometimes it is divided into two white spots or markings (Figs 11.17a, b) and occasionally a second white band runs posteriorly from the centre of the transverse band to form a T-shaped mark. At other times white pigment may cover much of the upper surface. This spider is a true native species which, in its natural state, prefers manuka-type scrub and bush margins although it is not common higher up the mountains. Nevertheless, it has taken advantage of exotic hedges in areas cleared of native forest and so has become the second-most common spider in our pastures and gardens.

Fig. 11.17a (below left): Zealaranea crassa *is another araneid species with variable colours and patterns on the abdomen. Shown here is a female with a green and white mottled pattern on the abdomen highlighted by two white circles at the front. This species is common in pastures and garden.*

Fig. 11.17b (below right): Another variation can be seen in this Zealaranea crassa, *which has two black crescent-shaped patches marked with smallish white circles on the front of its abdomen.*

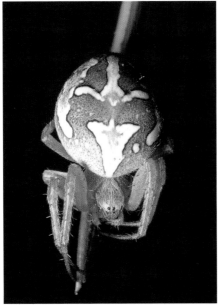

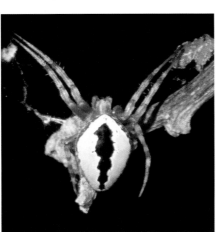

Fig. 11.18a (top left): This large gravid female, Colaranea verutum *(Araneidae), with its mottled green base colour, is distinctively marked on the back and sides of the abdomen. Its orbwebs can be seen amongst ferns and shrubs as well as in tussock country.*

Fig. 11.18b (top right): The Colaranea verutum *spider with its typical araneid eggsac has a reddish base colour with central yellow markings.*

Fig. 11.18c (left): A North Island forest dweller, this colourful spider known as Colaranea melanoviridis *is easily recognised by the black diamond-shaped markings down the back of the abdomen.*

Fig. 11.19 (above): This attractive spider, its abdomen patterned in yellow, olive-green and brick-red, is Novaranea laevigata *(Araneidae). Although commonly found in native grasslands and tussock habitats, it is widely distributed throughout the country.*

Zealaranea crassa builds a typical orbweb but with closely spaced spirals. The centre is bitten out at once but the gap gradually becomes covered with irregular threads as the spider moves about. A silk cable leads directly from the hub to vegetation nearby. When prey is trapped, the spider first shakes the web before moving in to administer a 'long bite'. The prey is then wrapped with a band of silk but with very little rotation of the victim. After a further series of 'short bites' the prey is taken to the hub if it has been captured at night. Prey caught during the day is usually transported to the resting place at the end of the cable unless it is very large, when it may be eaten in the web.

Those native orbweb spiders which live primarily in the bush and subalpine scrub are often brilliantly coloured. For instance, a number of species of *Colaranea* are conspicuously patterned with green, brown and yellow set against white and black areas. These spiders construct typical orbwebs amongst ferns and scrub both in the bush and higher up in the subalpine scrub and tussock country. The most variable species is *Colaranea verutum*, no two of which seem to be the same (Figs 11.18a, b). Amongst the most striking species is *Colaranea melanoviridis* (Fig. 11.18c), seen mainly in the forests of the North Island, and *Novaranea laevigata* (Fig. 11.19), typically found throughout the country in undisturbed grassland and tussock. All these more colourful spiders construct stabilimenta above the hub of their webs, accessories which suggest that, in their cases at least, they function as deterrents to predators.

***Fig. 11.20a** (above left): Suspended upside down from a silk line, this golden spider is* Arachnura feredayi. *It belongs to the Araneidae and is known as the 'tailed' spider because the back of the abdomen is endowed with a long thin tail-like appendage.*

***Fig. 11.20b** (above right): Although at present also known as* Arachnura feredayi, *this brick-red spider may eventually be described as a different species. The spinnerets are on the underside of the abdomen, seen here near the hind leg which is holding the thread.*

***Fig.11.20c** (below): The web of* Arachnura feredayi *shows the V-shaped sector which lacks silk. This sector, however, is where the eggsacs are positioned with the spider sitting at the hub below them. The whole effect is of a dead twig caught in the web and, indeed, we have seen several webs where a twig is actually incorporated above the row of eggsacs, making it look even more realistic.*

Spiders with 'tails'

The tailed *Arachnura feredayi* is an extraordinary spider (Fig. 11.20a). The yellow-brown female illustrated has an 18-mm-long body, one-third of which is 'tail'. This elongated 'tail' is really an extension of the upper abdomen because the spinnerets are located about halfway along the underside of the body. The tail portion is slender and can actually be wriggled, but it is difficult to imagine any particular use for it as the spider mostly rests motionless at the hub of her web. Perhaps an occasional wriggle diverts a flying insect which then becomes caught in the web. While most spiders encountered are golden brown occasionally one is found with a completely different coloration (Fig. 11.20b) and it is not certain yet that these are the same species. The web, usually strung close to the ground, is unusual in a number of ways – for example, an upper V-shaped sector is never filled in with the sticky spiral so that the web appears unfinished, and it is always surrounded by a network of irregular threads.

So far we have been talking about the female *Arachnura*. But what of the male? At first sight, it is hard to believe that a spider little more than one millimetre long and perhaps one-fiftieth of the bulk of its potential mate could belong to the same species. Quite apart from its tiny size, the male is very different in appearance because it lacks the incongruous tail of the female. Only a slight peak at the back of the hard abdomen is testimony to its relationship. But how can such a tiny creature effectively consummate such a union? Since mating behaviour in this species has not been observed we need to refer to observations of the closely related *Arachnura higginsi* in Tasmania to anticipate what happens in *Arachnura feredayi*. The minute male lurks in the irregular threads on the outskirts of the orbweb and approaches a female by vibrating a radial thread in the usual way. A special mating thread is constructed near the hub where the female is resting and after the male entices her on to it, mating takes place in the way described for other orbweb spiders.

The eggsacs of *Arachnura feredayi* are pale brown but, instead of being hidden away, they are placed in full view in a vertical string in the open upper sector of the orb with the lowest eggsac at the hub where the spider rests head downwards (Fig. 11.20c). Some six or seven sacs may be found all bound together and each sac contains up to fifty eggs. The most recently laid eggsac is always near the centre of the web and we were puzzled by the spider's ability to judge this placement so accurately until we read more of the habits of *Arachnura higginsi*. He observed that the spider constructed its first eggsac at the centre of a newly built web and then rested below it. Next day she demolished the old web and her lower resting position became the centre for the new web. She then laid another eggsac which, of course, was below the first. By continuing in this way, it is apparent that after six or seven eggsacs have been laid in a vertical line, the position of each new web will become very much lower than the centre of the original web. This means that the bottom end of the string of eggsacs where the spider rests will thus encompass the more recently laid eggsac.

In Tasmania, *Arachnura higginsi* is found in large communities where each spider has its own orb but these are often interconnected. This habit has not been seen in *Arachnura feredayi* although similar associations are frequently found in the next orbweb weaver we mention.

Communal associations

Cyclosa trilobata females are easily recognised by their stout but short tails, flanked on each side by a prominent swelling (Fig. 11.21a). Their overall colour is glossy black with silvery patches down the upper surface of the abdomen but some specimens also have reddish markings. Males, less than 7 mm in length, are only half the size of the females and their three lobes at the end of the abdomen are much smaller. Although single isolated webs are occasionally found, ten or more webs are frequently linked together, usually in a shady spot and never more than about a metre from the ground. Each web is a typical orb but all possess a stabilimentum which is woven horizontally across each side of the hub and into which debris is incorporated. Spiders rest at the hub so that their bodies are in line with this structure, and are then very difficult to see. Unlike *Arachnura*, eggsacs are not constructed in the snare itself but are deposited on a nearby twig. Each sac has the shape of a small cone of fine brown silk (Fig.11.21b) and, as several are usually placed in a row, the finished product may be mistaken for the seed pod of the kowhai (*Sophora* sp.) tree.

With its carapace gleaming with a thick covering of silver hairs and its silver abdomen suffused with yellow and red along the median surface and sides, *Argiope protensa* is a striking spider (Fig. 11.22a). Its body may be 20 mm in length with about a third of it forming a pointed tail; as a result the spinnerets are on the underside of the abdomen. Because the yellow shading on the abdomen becomes more pronounced as the spider reaches maturity, young spiders appear almost entirely silver. More common in the North Island, *Argiope protensa* is to be seen at the hub of its large orbweb during summer. Orbwebs are often built at an angle to the ground and many have some form of stabilimentum. Prey consists of flying insects (Fig. 11.22b) and since many of them are quite small, the spider needs a plentiful supply. The male is very much smaller than the female and, although the mating of *Argiope protensa* has not been observed, it probably follows the pattern of its relatives overseas where the female binds the male with silk from her spinnerets as he inserts his palps. Nevertheless, it appears from

Fig. 11.21a *(above):* Cyclosa trilobata *(Araneidae) is a shiny black spider with a silvery band down its back. Three prominent humps at the end of its abdomen help to identify it. Uneven stabilimenta extend above and below the dense hub and are joined here by a semicircle of thick silk.*

Fig. 11.21b *(below): The cone-shaped eggsacs of* Cyclosa trilobata *are fastened securely with silk to twigs a short distance from the orbweb. They are not guarded by the female but are protected by a hard outer covering and their likeness to seed capsules.*

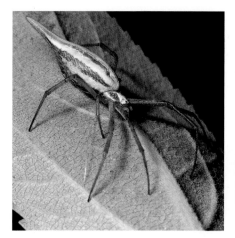

Fig. 11.22a (right): The tailed spider shown here is Argiope protensa, *an araneid which is easily recognised not only because of its elongated abdomen but also because of its red, yellow and silver stripes.*

Fig. 11.22b (far right): Photographed in the act of wrapping prey, Argiope protensa *demonstrates its skills. The hind legs are involved in pulling and throwing silk, the first and third pairs handle the prey while the second pair support the spider in its web. The spinnerets are clearly visible.*

Fig. 11.23a (below): The spider that builds this greatly extended orbweb or ladderweb is known as Cryptaranea atrihastula *(Araneidae). Most of these spiders build their webs against a tree trunk but because they are constructed of such fine silk they are almost invisible. We were lucky enough to find one that stood out from the trunk, but even so we had to dust it with cornstarch to obtain this photograph.*

Fig. 11.23b (bottom centre): The female Cryptaranea atrihastula *looks as if it is coated with bits of lichen but these are natural patterns on its body. They serve to protect this spider in the lichen-encrusted trees where it hides during the day. There is considerable variability, in such cryptic camouflage, which depends on the spider's daytime retreat.*

Fig. 11.23c (bottom right): This male Cryptaranea atrihastula *has a light-green body colour so that the lichenicolous effect is enhanced (cf. Fig 11.23b). The front two pairs of legs in the male have more spines than the female.*

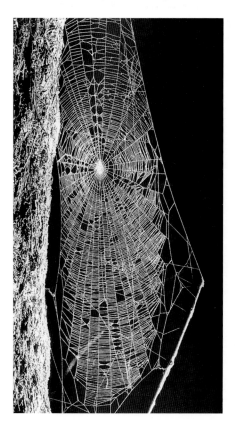

these accounts that this action does not necessarily indicate any sinister designs on the female's part as, at least in the first mating, the male eventually manages to escape his bonds. The smooth greenish-coloured eggsacs shaped like high domes are attached to a twig somewhere close to the web.

One of the most astonishing things about orbweb spiders is that not all of them make orbwebs. In some species, for example, the orb has been extended to form what is generally known as a ladderweb. In New Guinea, *Tylorida ventralis* (Tetragnathidae) makes a web about a metre long with a ladder reaching from the orb at the top almost to the ground. In other species, the orb has been reduced to a triangle, such as in *Hyptiotes* (Uloboridae), or even omitted altogether. Representing the extended orbweb in New Zealand is *Cryptaranea atrihastula* (Araneidae), the ladderweb spider. On the other hand *Celaenia*, better known as the bird-dropping spider, no longer builds a snare to catch prey but makes use of a totally different predatory technique.

The ladderweb spider

The web built by *Cryptaranea atrihastula* consists of a central orb and two ladder-like extensions above and below (Fig. 11.23a). Made of the finest silk, this snare is generally built against (but not touching) the trunk of a tree where it is all but invisible. The longitudinal axis of the web always corresponds to the vertical axis of the tree – this means that its ladder-like proportions are neatly adapted to the shape of the tree. The spider itself (Figs 11.23b, c) is said to be lichenicolous because its abdominal pattern mimics the colour and appearance of lichen, cryptically concealing it during the day when it retreats via a silk cable to a nearby lichen-encrusted tree trunk.

Construction of the orb is similar to that described above except that between eight and sixteen extra-long vertical radials are put in place above and below the hub. Upon completion of the non-sticky spiral, the spider loops back

and forth over the longer radials, completing first the upper ladder then the lower one which tends to be the longer of the two. Finally the spider lays down the sticky ladders, then the spiral section, and ends up at the hub as do other orbweb spiders. The spider waits here at night but hides during the day.

Webs serve as 24-hour traps and for prey storage. If, for example, craneflies or day-flying moths become entangled in the web during the day, the spider leaves its retreat briefly to wrap and bite its catch, leaving the wrapped bundle in situ until nightfall. Several items may be caught during the day, but at night the spider returns first to the hub, then tweaks the web to verify the position of its earlier catches and retrieves them one by one.

A specialised moth trap?

Our observations suggest that the ladderweb design functions basically as a moth trap. The fineness of the sticky mesh does not allow even the smallest moths to fly through it, so these are trapped. But large ones, however, often manage to bounce off. Although 'bouncing off' allows some moths to escape, others slide and slither their way down the web and are sometimes caught near the hub by the spider. Moths that pass the hub and fall all the way are chased down the web by the spider and are frequently caught before they can reach the bottom and escape. Now it so happens that when a moth strikes a web, its scales detach with ease, leaving behind scale-laden patches or 'moth scars'. The extra vertical length of this web design provides the spider with more opportunity to seize those slipping, sliding moths. Moreover, the more scales they lose, the more likely they are to stick to the web. 'Moth scars' seen by torchlight provide evidence of the spider's catch rate for the night. With the approach of daylight, however, the spider removes its tattered web, eating the remains as it does so, and builds another web. If no prey is caught, webs are usually left in place, sometimes for several days.

Closely related and very similar in appearance to *Cryptaranea atrihastula* is another lichenicolous araneid, *Cryptaranea subcompta* (Fig. 11.24) which builds not a ladderweb, but a typical orbweb. Its web stands out at right angles from a tree or sapling and is much more exposed. The advantage of this position is that it allows the spider to catch a greater variety of prey. Like *Cryptaranea atrihastula*, however, it rests inconspicuously during the day on a lichen-encrusted branch or trunk. Surprisingly, despite its cryptic retreat, this spider is found in large numbers in the nests of *Pison*, a parasitic wasp, whereas *Cryptaranea atrihastula* is not targeted in this way. Does this mean that the finely meshed ladderweb behind which this spider hides during the day can be seen by a sharp-eyed wasp and thus prevents its capture?

Bird-dropping spiders

The most extraordinary orbweb spiders in New Zealand are the bird-dropping spiders of the genus *Celaenia* (Fig. 11.25a). These spiders do not make an orbweb at all, as we will later explain. In the female, the carapace is always drawn into a narrow peak in front where the eyes are grouped together but it is the abdomen which seems so grotesque. Unlike most spiders, this is bunched out at the sides so that it is much wider than it is long. The dorsal surface has peculiar bumps and projections so that when the spider is at rest with the legs drawn up to the side it is difficult to distinguish it from the debris or bark where it lies (Fig. 11.25b). Some species are a dirty blotchy-white colour, and this is how the name 'bird-dropping' spider was derived. They are usually found some 2 or 3 cm below a leaf or twig dangling upside down from a silk thread which is still attached to the spinnerets and held by the front pair of legs. Quite apart from their unusual appearance, these spiders merit another claim to fame because they have completely abandoned the use of silk to capture prey. It is on the basis of morphological features alone that they are classified as orbweb spiders.

For many years, the way these spiders captured prey was a mystery, but this

Fig. 11.24 Another related spider, Cryptaranea subcompta, *very similar in appearance to* Cryptaranea atrihastula *but which builds a typical orbweb at right angles to the tree to which it is attached.*

has now become clearer after some interesting observations in Australia. Similar species in that country are often called orchard spiders and occasionally death-head spiders because of their outward resemblance to a skull, and we hoped they might provide some clues as to the habits of our own species. We soon heard that these spiders were often found eating moths, which they grasped with the heavily spined two front pairs of legs while hanging from a silk thread. Since this seemed a rather haphazard way to obtain food, the question arose as to how often a moth would come close enough for the spider to catch it?

Soon afterwards, it was discovered that the moths caught by *Celaenia* were all males. Further observations revealed that moths actually flew towards the dangling spider as if they were attracted to it in some way, acting rather as if they were approaching female moths of their own kind. The idea arose that perhaps these spiders exude a scent mimicking the female moth's own attractant pheromones, which are designed to entice prospective mates. If so, the spiders' scent would act as a decoy for male moths and this would account for these observations. Further support for this idea came from highly magnified SEM pictures of the segment of the first legs of the female *Celaenia*, which revealed that they are peppered with minute pores leading to secretory glands, the possible source of such a scent.

Do our own species behave in the same way? To find out how New Zealand species behave we put some females and eggsacs on shrubs in the garden and kept them under observation throughout the summer. Their behaviour was exactly the same as their Australian counterparts (Fig. 11.25c) so we carefully removed some captured moths before they were eaten and, under the microscope, confirmed that they were all males. Most of them proved to be tortrix moths. On the other hand, male *Celaenia* are always extremely small and, since we had never seen them catching prey in this manner, we wondered whether, as happens in some related spiders, they might be nearly mature by the time they left the eggsac. To test this theory, we had to watch many spiderlings after they had hatched.

Celaenia's round eggsacs, each 5 to 6 mm wide, are covered with dense dark-brown silk and have four or five knobs around the sides (Fig. 11.25 d). Each eggsac has a stout stalk but, as the female continues to construct her full quota of seven or eight sacs, they are held together in a bunch with a light envelope of threads rather than being suspended by the stalks. After waiting some weeks for the spiderlings to emerge, we decided to expedite matters by regularly spraying the sacs with water, and within two days the spiderlings began to emerge. At intervals of seven to ten days, further sacs released their young, all of which emerged from neat round holes on the upper surface of the cocoons. Our first hopes of solving this puzzle were dashed when we saw their size because all of them were much smaller than a fully grown male. A closer examination showed that these pale little spiders with strong bristles on the abdomen looked all the same and would clearly require several moults before males could be distinguished from females. These barely visible spiderlings moved to and fro over the twig, laying down their draglines as they went, until all the leaves were clothed with fine silk (Fig. 11.25e).

Our next question was whether young *Celaenia* actually constructed orbwebs to trap food until they were big enough to catch moths, but at that time our observations failed to answer this. It was not until some years later that a letter from Stewart Lauder of Greymouth, a keen observer of spiders, provided us with the answer. He had watched a colony of *Celaenia* in his garden and found that spiderlings, like the adults, also hung from a thread but instead captured tiny midge-like flies in much the same way as adults captured moths. He sent a number of these flies to us for identification. They proved to be psychodid flies which, by an odd coincidence, are known as moth flies. We found that these

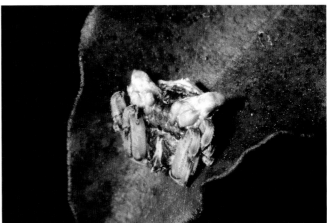

Fig. 11.25a *(top left):* This curious-looking spider is Celaenia sp. which, although belonging to the Araneidae, does not build an orbweb. It is seen here suspended from a rather sparse tangle of threads but it often hangs from just a single one.

Fig. 11.25b *(top right):* Celaenia is a genus of spiders generally known as bird-dropping spiders because they resemble this faecal matter. The Celaenia olivacea shown here, is one of the species which merits this nickname.

Fig. 11.25c *(above left):* Photographed in our garden, this Celaenia sp. is eating a moth much larger than itself. It was seized as it hovered close to the spider, which was hanging from a single thread in typical fashion.

Fig. 11.25d *(above right):* The female Celaenia usually lays several brown parchment-like eggsacs which are tied together with silk. She hangs beneath them until they hatch.

Fig. 11.25e *(right):* These newly emerged Celaenia spiderlings have laid down a network of fine silk as they move about on the plant nearest to their eggsacs.

flies too were males, which suggests that spiderlings also produce a scent to attract these tiny flies until they are big enough to attract and seize moths. The chemical nature of these pheromones or scents is not known but we hope that one of these days someone will be curious enough to find out.

Tetragnathidae

Big-jawed spiders

Another family group of orbweb spiders, the tetragnathids, sometimes called big-jawed spiders (Fig. 11.26a), are primarily associated with water or wet situations in all the parts of the world where they are found. In New Zealand, tetragnathids construct orbwebs with widely spaced spiral threads and most are also found near water, particularly open swamps. Here, they catch mainly the flying adults of various aquatic insects. This spider can be recognised by its long, thin body, elongate legs and strong, heavily-toothed chelicerae, features which easily separate it from other spiders which might be encountered in a similar habitat. Webs are occupied mainly during the evening but spiders are seen there during the day if the weather is overcast. Even in captivity, at low light levels, you can easily watch a spider make its snare for it requires only ten to fifteen minutes to fashion a simple orbweb and then take up a position at the hub. When the spider is at rest, the two front pairs of legs are stretched out in front and the two hind pairs out behind so that it looks just like a twig caught up in the threads (Fig. 11.26b). Since most species are dull in colour, mainly various shades of yellow or brown, this adds to the deception.

Native species

Although a number of these spiders were named by early arachnologists, some of them have since been found in Australia and others are virtually worldwide in their distribution. However, there are a few small, undoubtedly native, tetragnathids which live in forested habitats (Fig. 11.26c). Females are characterised by an unusual, humpbacked abdomen and many are brightly coloured with green and red, often matched by similarly coloured streaks on the carapace. These attractive little forest tetragnathids string their webs horizontally across low foliage and do not always live near free-standing water, although it must be said that they are only found in very moist forested areas.

Neither the habits of our own forest species nor their Australian relatives have been studied, so we can report little of their life histories. An important characteristic of tetragnathids, however, is the simple palpal bulb in the male and the absence of an epigynum in the female, so this means they are haplogyne spiders in contrast to most other orbweb spiders. Overseas, it has been noted that there is no obvious courtship in tetragnathids as a male approaches his mate (Fig. 11.26d). Once their front legs touch, they move closer so that in both spiders these legs are stretched out sideways. As the chelicerae touch, the fangs

Fig. 11.26a (below left): This male Tetragnatha *(Tetragnathidae) has a whitish abdomen with faint black markings. Here you can see the large chelicerae with their fangs projecting in front of the spider; hence the common name, big-jawed spiders.*

Fig. 11.26b (below right): With a body length of just 6 mm, these pale greenish, slender tetragnathids are hard to see when at rest on a twig or even in their webs. Typically, the long front legs are stretched out in front with the four back legs to the rear.*

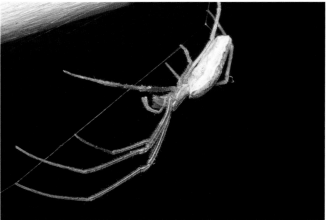

of the male clasp the chelicerae of the female, firmly wedging them with specialised uppermost spines. This first approach is made while both spiders hang upside down from a leaf or twig but, once the chelicerae are locked, both spiders drop down on threads so that they are hanging vertically, venter to venter, while the male applies his palp. Because the male chelicerae of our native species have those same exceptionally strong teeth near the tip as found in overseas *Tetragnatha* species, it is likely that mating behaviour is similar. Eggsacs are invariably laid within a curled leaf or a similarly protected niche and the female remains with the eggsac for some time.

Fig. 11.26c (above left): Many native tetragnathids are brightly coloured with red, white and black markings as seen here but others have hues of green, red and brown. The female has a short and rounded abdomen although some species have a more prominent hump.

Fig. 11.26d (above right): A female tetragnathid climbs cautiously along a silk thread as she awaits a mate. She maintains contact with this thread with four legs as well as a silk line from the spinnerets.

Vertical webs

Most of the conservatively coloured orbweb spiders which construct vertical webs in forests throughout New Zealand belong to the genus Meta of the subfamily Metidae (Tetragnathidae). These native spiders are characterised by long front legs and a simple epigynum, little more than a small lobe. One of these species, *Meta lautiuscula* (Fig. 11.27), is commonly found around Nelson where it constructs a normal vertical orbweb in the wetter parts of the forest. The colour of its abdomen is varied, like most of these spiders, but consists generally of subdued blues and reddish browns with distinctive pale anterior patches. The carapace is brown with a prominent black triangular patch extending from behind

Fig. 11.27 Long front legs are a feature of this spider, Meta lautiuscula *(Tetragnathidae) found around the wetter forests of Nelson. Abdominal colours vary but blues and reddish browns are common.*

Fig. 11.28a (above left): Larger than Meta lautiuscula, Meta arborea *has a pale abdomen with easily recognised blue and black patches. It is found in native forest in the North Island.*

Fig. 11.28b (above right): This large web, about a metre in diameter and angled over a stream, was built by Meta arborea. *The spiral mesh is widely spaced, suggesting the female owner was expecting large prey to be caught.*

the eyes. Another species, *Meta arborea* (Fig. 11.28a), which is found all over the North Island, is larger and has a pale abdomen with a few small black spots some of them being in the form of broken chevrons. These spiders build very large webs, up to a metre wide, with widely spaced sticky threads. They are also found within the forest, and their webs may be tilted at a slight angle over running water (Fig. 11.28b).

The European Zygiella x-notata

This spider was first collected from Taradale, near Napier, in 1967 by an observant school boy, Russell Hutton. The appearance and genitalia of the female are very similar to another European spider, *Meta segmentata* and both species were placed in the Metidae at that time. Today, however, they have been assigned to the Tetragnathidae.

Zygiella orbwebs are made of extremely fine silk and are often hard to see. They are built in about 30 minutes and are generally renewed daily. Juvenile *Zygiella* spiders build typical all-round spiral webs and sit in the middle to wait for prey. As they grow bigger, and depending to some extent on the available framework, they begin to build webs with a missing segment in the upper region. This is bridged by a signal thread leading from the hub to the spider's retreat above the web. Beacause of their tendency to build in association with houses and sheds, their distribution is almost cosmopolitan.

Zygiella x-notata came to everyone's attention in the 1960s, being one of the spiders used by Peter Witt and his colleagues to test the effects of a variety of drugs on their ability to build a typical adult web. A drop containing a selected drug was placed on the spider's mouth as it hung suspended from a thread. Changes in web geometry were measured. Although a variety of web distortions in relation to specific drugs were recorded, it was not possible to determine which sensory perceptions were responsible.

Nanometidae

Three distinct groups

Nanometidae is proposed as the family name for an interesting group of orbweb species long known from New Zealand and Australia. Although in New Zealand these spiders are divided into three distinct groups, they all share some interesting characters. For example, they possess branched tracheal systems in the abdomen rather than single tubes, as found in most other orbweb spiders, as well as distinctive male and female reproductive organs. However, the most surprising feature is the presence of a unique sound or vibratory organ in the males. This consists of a set of small teeth on the coxa of both hind legs which are rubbed against a raised ridge on the plates that cover each of the booklungs. Moreover, all these spiders construct horizontal webs rather than the vertical ones found in most araneids. This horizontal orientation restricts the position-

Fig. 11.29 Nanometa *species (Nanometidae) all favour humid habitats where they build their webs in low shrubs and foliage. Small silvery-backed spiders, their front legs are characteristically long. This photograph also shows that the lateral eyes are touching each other.*

ing of the snare and so they are always found near the ground.

The most numerous and widely distributed genus of these spiders is *Nanometa*, related to the Australian spider *Nanometa gentilis*. All construct webs in low shrubs and foliage in habitats which are constantly humid. These relatively small silvery spiders have elongate abdomens, and the anterior two pairs of legs are much longer than the other two pairs. When at rest, these front legs are positioned forward in line with the body while the posterior pair are directed to the rear (Fig. 11.29). Spherical eggsacs are hung near the web.

The second group is quite different in many ways. The New Zealand *Orsiella lagenifera* (Urquhart) is typical of the various species found in Australia and New Caledonia. These spiders appear to be restricted to shady streams where they construct horizontal snares above the water surface, often anchored to stones or debris sticking out of the water (Fig. 11.30a). Their overall colour is pale brown with a dark brown folium on the dorsal abdomen and a pair of silvery bands on the ventral surface (Figs 11.30b, c). In appearance they closely resemble some aquatic species of the big-jawed *Tetragnatha* spiders, which also construct similar horizontal webs above streams.

Fig. 11.30a (top): *Restricted to shady streams,* Orsiella lagenifera *(Nanometidae) build their horizontal webs above running water. They are pale brown spiders with darker patterns on the abdomen, and have long legs. In many ways they are similar to tetragnathids for which they are sometimes mistaken.*

Fig. 11.30b (above left): *Female* Orsiella lagenifera.

Fig. 11.30c (above right): *Male* Orsiella lagenifera.

Fig. 11.31 An Eryciniolia purpurapunctata *male can be distinguished from* Nanometa *by the clear separation of the two lateral eyes on each side of the carapace (in* Nanometa, *they are touching.)*

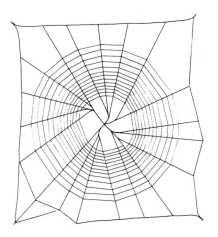

Fig. 11.32 Tiny theridiosomatid spiders (Theridiosomatidae), 2–3 mm in length, build orbwebs about 10 cm in diameter. Once the orb has been completed, the radials are grouped together as shown in the illustration. The spider attaches a thread to the centre, carries it away to a nearby twig and then pulls it taut to form a cone. The resulting shape has been described as being like a miniature inside-out umbrella. Holding the web in this position, the spider stretches its legs out onto the central radials. The web is released when prey strikes it, thus causing further entanglement.

The third group consists of a genus initially named by Urquhart as *Erycina* although it is now known as *Eryciniolia* because '*Erycina*' was already in use for another group of animals. The only known species, *Eryciniolia purpurapunctata* (Fig. 11.31), differs from *Nanometa* partly because the two lateral eyes on each side of the carapace are widely separated (in *Nanometa* they touch) and partly because the abdomen does not have any silvery pigment (as in *Nanometa*).

Ray spiders: Theridiosomatidae

Tiny spiders

These tiny ray spiders belong to the wider orbweb group (Araneoidea) and are found worldwide although they are much more abundant in tropical countries than in cooler climates. Although not yet officially recorded in New Zealand, a number of species have been found here, but as they are so very small (never much more than 2 mm) they are rarely seen unless specifically searched for. It was not until 1976 that they were recorded from Australia when Jörg Wunderlich described three species from the east coast forests of New South Wales and Queensland.

Although tiny, these spiders are easily recognised in the hand with a simple magnifying lens. They have a relatively large globular abdomen which overhangs the cephalothorax and are mostly inconspicuous greys or brown. If the specimen is a male then identification can be readily verified because the palpal bulb is relatively large and seems out of all proportion compared to that of other spiders. These spiders are so small that they could well have been part of our chapter on midget spiders, but because their use of the orbweb is so unusual, we decided to include them here.

Why they are called ray spiders

The name ray spider comes from the appearance of the orbweb after it has been modified by these spiders. Apparently designed to catch smaller nematocerous flies such as midges, which abound in the moist forested regions where they live, the orbweb functions in a very different way in this spider family. To begin with, the spider builds a fairly typical, generally vertical, orbweb in low foliage either in the forest or on a bank or cliff face where it is shady and moist. It is the next stage which is so unusual. The spider cuts out the central hub and some of the associated spiral thread. Then it pulls the radial threads together in groups of two or three all around the web so that each group ends in a single thread attached to the centre of the web, giving the impression of many sectors radiating out from the hub (Fig. 11.32). The spider proceeds to attach a thread to the centre of the web, moves away from the hub to a nearby twig and then, with the first two pairs of legs, tightens this thread so that the orbweb is pulled into the shape of a hollow dome.

Held under stress by this retaining thread, the spider waits for prey, but once a fly hits any of the web sectors, the retaining thread is released and the fly is entangled. The spider moves in quickly to bite its victim. Because the web has been subdivided into all these sectors, the remainder of it is seldom damaged and the spider continues capturing prey with the other portions of the web. It is assumed that the ray design adds strength to the central portion of the web, allowing it to be held under tension, and that its rapid release once an insect strikes it not only enhances entanglement but also helps to prevent damage to other sectors.

The eggsacs of these spiders are distinctive. They are usually globular and suspended by a long thread near the web. Indeed, when searching for ray spiders in the United States, we were told by more experienced local arachnologists to search first along a sheltered forest bank for the eggsacs and, having found them, to then look for the spiders. This tip is useful in New Zealand, too.

Uloboridae

Cribellate spiders

The orbweb spiders mentioned above all complete their snares with a spiral of sticky thread, beaded with globules of adhesive fluid which serve to entangle prey. Uloborids, on the other hand, are small cribellate spiders which produce an extremely fine entangling silk from minute spigots on the cribellum (see chapter 1). This bundle of slender threads is drawn out by the row of bristles on the metatarsus of the fourth legs (calamistrum) and twisted around a pair of the stronger standard threads produced by other spigots on the posterior-lateral spinnerets. This crinkly compound thread is then used for the entangling spiral of the orbweb. Uloborids are widespread and, particularly in the tropics, you will find a great variety of them, many constructing very unusual webs.

In New Zealand there is but a single species, the primitive *Waitkera waitakeriensis* (Fig. 11.33), which is restricted to the North Island. Not only is *Waitkera* similar in appearance to *Eryciniolia* but their habits are somewhat alike so it is not surprising that they are often found together. *Waitkera*'s abdomen is conspicuously white with a few black patches along the sides but the carapace has a pale olive-greenish tinge apart from a large black area behind the eyes. The legs are also olive-green with dark bands. Webs are constructed in low bushes and ferns in damp forests and are often horizontal or distinctly inclined with numerous threads surrounding them. The owner is always found on the underside of the hub. Sometimes these spiders make use of knockdown threads above *Cambridgea* sheetwebs as attachment points for their orb and so, particularly in the far north, there is a close association between these two spiders. The eggsacs of *Waitkera* are typical for the family and are often built in an angle in the frame threads above the orb.

Deinopidae

Two genera of these cribellate spiders, *Deinopis* and *Avella*,[3] are found in Australia. They belong to the wider group (or out-group) of the Araneoidea, one reason being that their rectangular snares are considered to have been derived from orbwebs. These sticky traps, held in readiness by the four long, stick-like front legs are then stretched and flung at passing prey. Because of this, they are

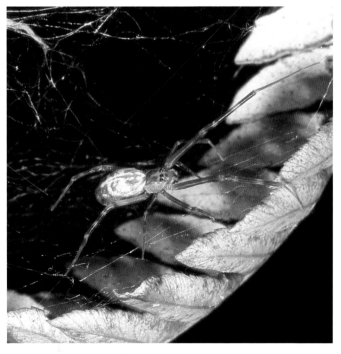

Fig. 11.33 (below left): The primitive Waitkera waitakeriensis *(Uloboridae) is found only in the North Island.*

[3] Transferred from *Menneus*.

widely known as net-casting spiders although they are also occasionally called 'stick' spiders as they resemble dead twigs when at rest during the day.

Two of these spiders, a male and female *Deinopis subrufa*, have been found in New Zealand. A preference for dry, sparsely forested habitats means that some areas in northern parts of this country would suit their lifestyles. Deinopids are nocturnal spiders and this genus has large posterior-median eyes specialised for night vision. This distinguishes them from *Avella* whose posterior-median eyes are small and unspecialised and whose net-casting technique, according to Austin and Blest, is probably triggered by vibration.

CHAPTER TWELVE
SPACEWEB SPIDERS

Two main trends are evident as spiders evolved. One of these encompassed their use of silk to build aerial snares, the other led them to abandon silk for this purpose and to become cursorial hunters. Of course, such trends are not always straightforward and many variations occur within these two divisions. Aerial snares can be broadly separated into two groups. The first are described as two-dimensional – orbwebs and their derivatives, for example. The second group are those three-dimensional structures which we call spacewebs, such as the sheetwebs built by *Cambridgea foliata* (Stiphidiidae). This spider's web may be up to a metre in diameter and is buttressed by a network of threads above and below the sheetweb, all of which are necessary to its particular prey-catching tactics (Fig. 12.1). It occupies a space of up to a cubic metre and is called a spaceweb. In this chapter we discuss those spiders which build spacewebs of various dimensions: Theridiidae, Stiphidiidae, Pholcidae, Linyphiidae, Synotaxidae, Cyatholipidae, and Neolanidae.

Cobweb spiders: Theridiidae

The theridiid web

Characteristically, theridiid webs consist of an irregular network of strong threads only a few of which are sticky and, because of their untidy nature, are commonly called cobwebs. The builders of such webs belong to a large group of small to medium-sized spiders sometimes known as comb-footed spiders. This is because many of them have a prominent row of curved, toothed bristles on the tarsal segment of the hind legs (Fig. 12.2) used to comb out the swathing silk with which they immobilise their prey. As we have seen in other families, some species of theridiids have also developed specialised methods of catching prey which do not include swathing silk and in these species the comb is reduced or even completely absent.

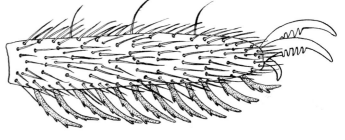

Fig 12.1 *(below left): This spaceweb built by* Cambridgea sp. *(Stiphidiidae) shows a dense horizontal platform with a network of silk above and below it. The extensive upper section is used to intercept flying insects which are knocked down onto the central sheetweb. The spider waits underneath and seizes insects as they struggle to escape. The remains of an old sheetweb can be seen below it.*

Fig 12.2 *(below): One of the characteristic features of the Theridiidae or comb-footed spiders is the presence of a row of curved bristles on the tarsi of the hind pair of legs. This is used to comb out silk from the spinnerets to throw over prey. As a result, the struggling insect is immobilised.*

Fig. 12.3a (top left): This introduced Achaearanea tepidariorum *(Theridiidae) frequents the outside of houses in the North Island of New Zealand.*

Fig. 12.3b (top right): Nesticodes rufipes *is another introduced species usually found around buildings in more tropical countries. (Photographs Frances Murphy.)*

Fig. 12.4a (bottom left): Achaearanea veruculata *is a small native theridiid which is generally found in its web under stones and logs as well as on the walls of houses. The small round abdomen and reddish-brown markings make it easy to recognise. The light-brown eggsac is larger than the spider's abdomen.*

Fig. 12.4b (bottom right): The colour variation of Achaearanea veruculata *shown here has a body with greenish hues and darker green markings on the abdomen.*

Although theridiids come in a wide range of shapes and sizes, most are readily recognised by features such as the globose abdomen, the tarsal comb on the fourth legs and the relatively small eyes grouped together near the front of the head, in addition to the habit of hanging upside down in their webs. In New Zealand it is not always possible to provide up-to-date scientific names for members of this group because there have been no recent revisions or detailed studies, even though large collections are now available. In general, theridiids are not only widely distributed throughout the world, they are also remarkably uniform in structure and habits so that many generic names are universal. As far as southern regions are concerned, however, some generic names may not be retained when the fauna is revised.

Early classification

Most theridiids named in the early days of spider classification in New Zealand were placed in the long-established genus *Theridion* but today we realise that there are actually no native representatives of this group found here. Indeed, the two species of introduced theridiids commonly found around houses in the North Island, *Achaearanea tepidariorum* (Fig. 12.3a) and *Nesticodes rufipes* (Fig. 12.3b), have recently been transferred from *Theridion*. Furthermore, of the remaining twenty or so species which A.T. Urquhart described as *Theridion*, most are undoubtedly more closely related to *Achaearanea* than to *Theridion*, but so far only a few with this name have been linked to the many theridiid spiders that are actually known in this country.

House and garden theridiids

Most native theridiids found about the house and garden are *Achaearanea* species, those small reddish-brown or sometimes greenish spiders associated with typical theridiid webs. The commonest one is *Achaearanea veruculata* (Figs 12.4a, b) and you will find its small cobweb on the outside walls of houses and sheds all

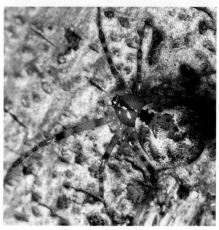

over the country – one of the few native species to achieve such a wide distribution. Indeed, this reverses the common trend for it is usually the introduced species that spread and the native ones that become restricted. Somehow *Achaearanea veruculata* managed to reach the Scilly Islands off the coast of England, where it is now successfully established as one of the more common spiders. Perhaps it was imported with native New Zealand plants, many of which also thrive in this northerly habitat. Other members of this genus vary in their colour and pattern although their habits are similar.

Forest theridiids

The majority of the native theridiid species inhabit forests or scrubland where they are found mainly above the forest floor or the ground. The commonest species living on the forest floor are the rather dull-coloured spiders with cream or grey abdomens, one of which was named by Urquhart as *Theridion ampliatum* but which we now know as *Achaearanea ampliata* (Fig. 12.4c). This rather plump little spider constructs a small cobweb amongst the litter and under fallen logs. In the same habitat are to be found a number of species of minute black spiders, in appearance rather like money spiders. These have yet to be revised and, in the meantime, are placed in the northern genus *Pholcomma*. A few species have hard plates on the abdomen just like some of the anapids but the carapace is not elevated as in those spiders and the shape of the abdomen resembles that of a typical theridiid. Above the forest floor many species of *Achaearanea* live in ferns and low shrubs, including a dazzling array of brightly coloured species (Fig. 12.4d, e) which construct their webs well above the ground where they trap mainly the smaller long-legged flies.

The katipo spider

Best known of all theridiids in New Zealand is undoubtedly the katipo, but we should point out that there are actually two native 'katipo' species, both of which are quite small. In each species, the abdomen of adult females is about the size of a garden pea and the overall colour is black. However, the best-known species, *Latrodectus katipo*, has, in addition to the partial red hourglass patch on the underside of the abdomen, a prominent red stripe on the upper dorsal surface (Fig. 12.5a). The second species, *Latrodectus atritus*, loosely labelled the black katipo, also has an indistinct hourglass beneath its abdomen but lacks the red dorsal stripe above, so is a uniform shiny black (Fig. 12.5b). In both species

Fig. 12.4c *(below):* Achaearanea ampliata *is an amber-coloured theridiid with dark and light markings on the abdomen. It was formerly known as* Theridion ampliatum.

Fig. 12.4d *(bottom left): This* Achaearanea sp. *is a pale yellowish-green but is distinguished by two large white patches edged with black on the dorsal surface of its abdomen.*

Fig. 12.4e *(bottom right): A jagged red stripe on the abdomen is a distinguishing feature of yet another* Achaearanea *species. Notice the small eyes at the front of the carapace.*

Fig. 12.5a *(left):* Latrodectus katipo *belongs to the family Theridiidae and is commonly known as the katipo. The body of this adult female is about 8–9 mm long and the red stripe down the back of its abdomen is distinctive. In some spiders the red stripe is framed by a white border.*

Fig. 12.5b *(above): Except for the absence of the red stripe, the black katipo (*Latrodectus atritus*) is very similar to the true katipo (*Latrodectus katipo*).*

juveniles and males are quite different in size and colour from the females. Amongst young spiderlings, males and females look alike and cannot be distinguished until about the fourth instar. Up to this point, they are whitish with diagonal black markings on the side of the abdomen and perhaps a trace of red down the back. This juvenile appearance is retained by the male which reaches maturity in the fourth or fifth instar (Fig. 12.6a). However, females continue to grow (Fig. 12.6b) and go through two to three more moults during which the black pigment increases and red coloration develops. This leads to their familiar black appearance and distinctive red markings.

Only the female katipo is capable of biting humans and then only when it is adult and most conspicuous with its black colour relieved by the prominent red stripe above. The male is much smaller than his mate and is completely different in appearance. The carapace is brown with a central black band and the abdomen glistens white with three black streaks on the side and two irregular black bands down the top enclosing a string of small orange diamond-shaped patches. Indeed, the adult male does not differ much from the immature stage of both sexes; one way of looking at it would be to say that the male never really grows up. By the time the female is one moult away from maturity, she is mainly black, the dorsal stripe has intensified to red, and the last lingering traces of white mostly disappear with the final moult.

Although the development of *Latrodectus atritus* is very similar to the katipo, the red stripe is missing so that from above, all you see is a black spider. Rearing studies reveal that the eggs and young of this species are dependent on much

Fig. 12.6a *(above left): The male* Latrodectus katipo *is only about one-sixth the size of the adult female so is very small compared to its mate. Its red markings are easily seen and the swollen palps are just visible in this illustration.*

Fig. 12.6b *(left): This juvenile female (*Latrodectus katipo*) is very similar to the adult female except that the abdomen is smaller and has more white markings on its back.*

higher temperatures than those required by *Latrodectus katipo*. Other observations suggest that when these two kinds of katipo live on the same strip of beach they do not interbreed as one would expect if their differences were merely colour variations. Subsequently, laboratory studies show that they do not generally crossmate but when sometimes they do, the eggs are infertile.

The katipo snare

All over the world, theridiid webs are usually much the same and in New Zealand, the katipo (*Latrodectus katipo*) web is one example. In a clump of marram grass the katipo first constructs a broad silk base with many supporting threads above and below, including a number of sticky guy-lines anchored to debris in the sand. A cone-shaped retreat is built among the marram stalks usually in the lower part of the web. Moreover, if the katipo is kept in a large cage, the web pattern is much the same, the spider being able to proceed without needing to work amongst the usual vegetation. In the wild, the marram grass provides support, shelter and, as these plants harbour a variety of invertebrates, food as well. The husks of wandering ground insects, largely beetles as well as moths, flies and the occasional spider, are generally found near or below their nests and can be used to pinpoint the whereabouts of the spider itself.

Because the web is always near the ground, the main catching threads are just above the ground so that ground insects are most likely to fall victim to this snare. Many of the beetles are much larger than the spider (Fig. 12.7) but once entangled they seldom escape, not because the poison is so deadly but because of the methods the spider uses to make sure they are completely immobilised. At the first movements of the trapped insect, the spider jerks and pulls at the threads which, by resisting in a similar way to a fishing line, indicate the direction of potential prey. As the insect struggles the anchor lines break away from the ground and their elasticity means that the trapped prey becomes suspended one or two centimetres above the surface. The spider then runs down to the victim but, instead of plunging in her fangs, she stops, touches it first and then turns around so that the spinnerets are directed towards it. As in most theridiids, the tarsi of the hind legs have a row of curved bristles which are arranged like a comb and the spider uses these to scoop sticky silk from her spinnerets and flings it over the insect with a series of rapid movements. After the insect is firmly trussed she bites it several times, usually at the joints, after which she runs up and down strengthening the web before administering the long bites which will end the last struggles. Once this happens, the spider hauls the prey well above the ground to hang in the web until she is ready to eat.

When food is readily available the female spider continues to trap and often her larder will contain five or six insects awaiting her pleasure. Actually, it is difficult to know when an insect has been fed upon because theridiids feed differently from some other spiders. Instead of crushing the prey as they exude digestive juices, they force fluid enzymes through the small bite holes and after

Fig. 12.7 The katipo often catches prey much larger than itself. Seen here is a female throwing silk over a large staphylinid beetle desperately trying to escape the entangling threads.

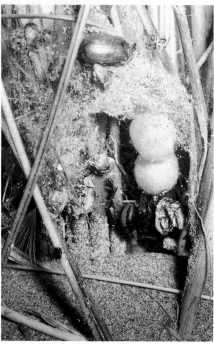

Fig. 12.8a (above top): A male and female Latrodectus katipo *are about to mate. In this* Latrodectus *species, the male is generally not at risk from the female.*

Fig. 12.8b (above): The female katipo usually hangs her eggsac near the base of marram grass and builds a silk shelter above and around them. This becomes covered with sand, effectively hiding and sheltering the eggsacs in a little sand cave. Here, the marram grass has been pulled aside to reveal the spider, eggsacs and the remains of prey.

predigestion has taken place, they suck up the resulting broth. When the meal is completed, the empty husk of the insect is left. Most times, these carcases are cut out of the web and dropped to the ground and are often a clue as to the presence of the spider. Because the growth period of the male is much shorter and his ultimate size so much smaller, his feeding activities are not as vigorous. However, his hunting behaviour is very similar.

Courtship, mating and egg-laying

Female katipos remain in the same web for long periods of time unless food is scarce. It is here that the male, which wanders as an adult, will find her. He enters her web and vibrates it at intervals as he slowly approaches the mature female. Usually she is rather aggressive at first and often chases the male, which then turns tail and runs, but eventually she becomes docile and allows him to approach. He will reach forward with his front pair of legs and tap her as she hangs upside down in her web. After the male performs a courtship consisting of bobbing, plucking and tweaking actions on the web, interspersed with periods of cautious advances and being chased by the female, mutual agreement is reached. The female accepts him as a mate and remains quietly suspended in the web. The male then moves onto her ventral abdomen, tapping rapidly as he does so until his abdomen is above hers and he is facing in the same direction (Fig. 12.8a). He inserts his palps one at a time, leaving the female briefly between the two insertions; each copulation generally lasts from ten to fifteen minutes. After completion the male moves away and begins to groom by running his palps and legs through his fangs and wiping them over his body. Although it is often said he falls victim to the female, this is a myth – but there is one notable exception in this genus as we shall see later.

Eggsacs appear during November and December. They are round, slightly less than a centimetre in diameter and cream in colour. Each female constructs five or six sacs in the course of three to four weeks. She hangs them up in a bunch in the centre of the web and then spins more silk above them. This invariably becomes covered with sand so that her eggs appear to be hidden within a little sand cave (Fig. 12.8b).

Distribution in New Zealand

Latrodectus katipo (sometimes referred to as the true katipo) is found along sandy beaches in both the North and South Island but does not extend further south than Karitane on the east coast of the South Island and Greymouth on the West Coast. The black katipo is not found at all in the South Island but is restricted to the northern coasts of the North Island and, in some localities (e.g., Bell Block, north of New Plymouth) may occur in the same area as the true katipo. Eddie McCutcheon of New Plymouth, who has made a study of the distribution of these spiders, reports that the black katipo extends no further south than the Stony River dunes on the west coast and Waipatiki Beach on the east coast of the North Island.

Both species are found mostly on sandy beaches where they normally spin typical theridiid webs at the base of clumps of grasses and sedges growing on the sand. Introduced marram grass, which has been extensively planted on many beaches to stabilise the sand dunes, has largely replaced the native pingao and spinifex so this type of vegetation forms the main habitat today. Once, favourite hiding places for katipos were old rusty four-gallon tins discarded by motorists in the days when it was necessary to carry extra petrol in the car. However, it is not unusual today to find katipos associated with driftwood and debris, drink cans, even corrugated iron and other refuse lying above the high tide mark. We remember finding katipos in old empty cordial cans lying here and there amongst sunbathers on a Canterbury beach, who were seemingly unaware of these spiders lurking nearby. The increased daytime temperatures as well as the attraction for insects resulting from such casual 'habitats' suit the breeding requirements of the katipo and this is often where they tend to aggregate in preference to their usual habitat.

These simple field observations show that higher temperatures benefit the development rates of *Latrodectus katipo*, a finding which can be linked to the tolerance limits of their natural southern distribution in New Zealand. Rearing experiments show that temperatures higher than about 17°C need to be maintained during the development of eggs and young even though, as adults, they are capable of withstanding much lower temperatures. *Latrodectus atritus*, however, requires temperatures of about 22°C or more to flourish.

The Australian redback spider

One of the more recent immigrants to this country is *Latrodectus hasselti*, commonly known as the redback spider. A chance discovery of this spider near Wanaka in the South Island in 1980 led to studies that indicated significant differences between the katipo of New Zealand and the redback of Australia. Although the redback's web is built on a similar pattern to the katipo's, it is larger and the silk is stronger, often yellowish in colour. The female redback too is larger than the katipo; its bright red dorsal stripe is usually wider and extends nearly all the way down the back. The most definitive character, however, and one which can only been seen with a 10x hand lens or a microscope, is that the katipo's body has a dense covering of short, very fine hairs, whereas the redback has two easily distinguishable kinds of hairs – one clearly long and fine whereas the other is shorter and stouter.

Its method of prey capture is similar to that of the katipo to which it is closely related. The spider sets sticky trap lines, running quickly to the site of the disturbance, and after touching the struggling creature, turns its back and rapidly enswathes it in silk (Fig. 12.9a). However, its range of food is greater. Reports that it has been seen with small lizards and mice and even a small bird are well documented in Australia. In New Zealand, the options are more limited, it seems, as the bulk of its prey in Wanaka (about 63 per cent) consists of grassgrub beetles (*Odontria* spp.) as well as other beetles, blowflies and the occasional spider (Fig. 12.9b).

Another factor which characterises the redback is its propensity for producing large numbers of offspring. An adult female lives for up to two years and during this time may lay from ten to twenty eggsacs, each of which may contain up to two hundred eggs. Given temperatures of about 25–27°C, a young spiderling may become an adult in from six to eight weeks. Although this suggests that huge populations can arise very quickly, a limiting factor is that these spiders have strong cannibalistic instincts. This means that quite large numbers of them may fall prey to members of their own species, particularly when food is scarce.

*Fig. 12.9a (below left): The Australian redback spider (*Latrodectus hasselti*), a close relative of the katipo, catches prey in a similar manner. Often it remains in a head-up position and wraps its prey from above.*

Fig. 12.9b (below): First found living in the wild near Wanaka, the redback generally makes its nest in amongst the rubble under stones and occasionally at the base of scrubby vegetation. When this redback's nest was uncovered, it revealed two eggsacs and the remains of prey hidden amongst debris, all bound together with silk. (Photograph Frances Murphy)

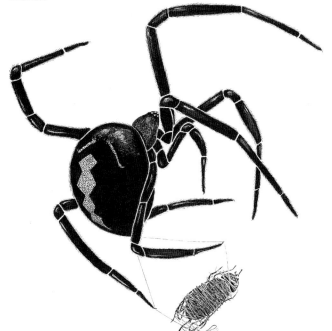

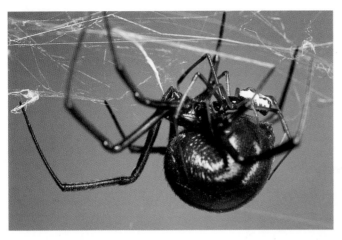

Fig. 12.10 *The mating sequence of the Australian redback spider* (Latrodectus hasselti). *(See text for details.)*
Fig. 12.10a *(above left): After courtship the female remains quietly suspended in her web and the male climbs on her abdomen to mate.*
Fig. 12.10b *(above right): The male launches himself into the air as he begins the somersault.*

The female redback cannibal

As we have explained above, when the katipo mates, he climbs onto the ventral surface of the female's abdomen as she hangs upsidedown in her web and, with both spiders facing in the same direction, he inserts a palp, first one and later the other (Fig. 12.10a). But the redback male behaves in a very different way after he inserts the first palp. As soon as the hematodocha inflates, he first stands on his head (Fig. 12.10b) and then somersaults so that his abdomen is lying against the female's mouthparts; he is now facing in the opposite direction and the dorsal surface of his body rests against the ventral surface of the female's cephalothorax (Fig. 12.10c). This seems to be a very dangerous manoeuvre and so it proves to be.

As soon as the male lands against the female's mouthparts, juices flow from her mouth and envelop the rear of the male abdomen. Her fangs can be seen

Fig. 12.10c *(above left): After somersaulting, the male lies against the female's mouthparts. She begins the process of digestion.*
Fig. 12.10d *(above): When the first mating is complete, the male pulls away. The coiled embolus can be seen as he does this.*
Fig. 12.10e *(left): Before the second mating takes place, the male leaves the female for a few minutes. The injury to the back of his abdomen is clearly visible. The male will groom himself and briefly court the female again before mounting her again. Note the red ventral hourglass in both male and female.*

masticating this area and during this time the hematodocha of the male pulsates as sperm is transferred. After ten minutes or so, the male starts to wriggle and within seconds he is endeavouring to free the embolus (of his palp) from her epigynum (Fig. 12.10d). As she tries to restrain him with her palps and fangs he gives a tug and escapes. He retreats a short distance from the scene and begins to groom. The male's rear abdomen is pinched in as a consequence of the female mastication (Fig. 12.10e) but otherwise he appears to be unharmed. You might think that this would deter him from mating again but this is not so. After a few minutes of grooming and a brief courtship, he returns to the scene and goes though the same procedure, this time inserting his other palp before turning a somersault. Now the female is more decisive and plunges her fangs in so that she can begin sucking out his living juices, by then presumably predigested from her earlier enzymatic injections. After a further ten minutes or so, as the male stirs a little, she hastily wraps him thoroughly with silk and completes her meal.

This is not just an occasional happening. We have watched over sixty matings by *Latrodectus hasselti* in the laboratory, and this procedure is followed almost every time. Moreover, sexual cannibalism, as it is called, has been observed in the wild in Australia by Maydianne Andrade, who carried out further studies on this spider at the University of Toronto in Canada. Her experiments showed that the male gains several advantages by this behaviour. The duration of courtship is longer so that he can transfer more sperm and hence fertilise more eggs, other males may be hindered from mating, and it is likely that the female gains some useful nutrients from his body which help to nourish her developing embryos. This behaviour by redback spiders is one of the most stereotyped examples of sexual cannibalism known in nature, more particularly because of the male's apparent complicity in the procedure.

Worldwide distribution of Latrodectus

Many different species of *Latrodectus* are found throughout the world and most have very similar features and habits to the katipo, although colour and patterns vary. This commonality led one spider specialist to suggest that, rather than thiry-five or so species, there were only five or six species which had managed to become widely distributed. This conclusion led to the notion that the two kinds of *Latrodectus* in New Zealand were merely colour variations of the same species, and that this 'species' was the same as the Australian redback, and then perhaps that they were varieties of the well-known black widow, (*Latrodectus mactans*) of the United States of America. Moreover, this proposal had the effect of limiting studies on these spiders in this part of the world since a great deal was then known about the black widow and it was thereby assumed that further studies were not required. A further consequence was the reduced awareness by port interception authorities of the harmful status of these overseas spiders. This view is not accepted today and most of the original species have been reinstated.

False katipos

Although the katipo is by far the best known spider in this country we should mention that some closely linked, although relatively harmless, relatives often appear inside houses and cause needless panic. One of these is *Steatoda grossa*, sometimes known as the house cobweb spider (Fig. 12.11), which actually comes from Europe. The dark-brown to blackish female sometimes has a distinctive pale crescent-shaped patch near the front of the plump abdomen with a few more pale patches behind this. In the slender male the pale patches are more extensive so that the dark pigment on the upper abdominal surface is broken into transverse bands. Their untidy tangled space webs may collect dust in the upper corners of rooms but the spiders themselves are rarely seen. Indeed, most spiders which live in our houses do so in the ceiling under the roof and between

Fig 12.11 Steatoda *is another widespread genus belonging to the family Theridiidae. Shown here is* Steatoda grossa, *a spider which probably came to New Zealand from Europe. Although this spider is for the most part black, the abdomen may vary from all black to black with white spots or to reddish-brown.*

Fig. 12.12 An African species, Steatoda capensis *is frequently mistaken for the katipo and as a result is often called the false katipo. It is the black body plus the occasional splash of red or orange colour at the back of the abdomen that leads to this mistaken identity.*
Fig. 12.12a (top left): A female Steatoda capensis *with a reddish-brown abdomen.*
Fig. 12.12b (top right): An immature female, Steatoda capensis.
Fig. 12.12c (above): A male Steatoda capensis

the walls rather than in the rooms, only emerging at night when food can be trapped. When mature, however, males leave their webs and retreats and roam about looking for females.

Another species, *Steatoda capensis*, is to be found in more natural surroundings. This species was described, amongst others, by Sean Hann, and cited as originating from South Africa. It is now found along many beaches where it may share a similar habitat to the katipo. It is this spider which is also believed to be responsible for displacing the katipo along the Wellington to Wanganui coastline. *Steatoda capensis* is usually an all-over shiny black with few markings but often has a small bright red, orange or yellow patch near the tip of the abdomen and a crescent-shaped band near the front and for these reasons it is often mistaken for a katipo (Fig. 12.12a, b). Males are most unlike adult females but are similar in coloration (Fig. 12.12c) and they also possess a stridulatory system. This consists of a hard plate on the upper abdomen near the carapace, which in turn has a series of ridges on the posterior margin against which the hard plate rubs. Two other species, *Steatoda lepida* and *Steatoda truncata* are endemic and are found in both mountainous and lowland regions.

Biological control potential

Some years ago Bruce Given, a well-known New Zealand entomologist, passed onto us these interesting observations he had made of *Steatoda*. While surveying an area for signs of grass damage, he noticed numerous cowpats in a one-and-a-half hectare paddock not being grazed at that time, on a dairy farm north of Whangarei. These cowpats were all fairly dry and often angled above the ground by grass growth. Upon closer inspection he noticed that most of these cowpats sheltered one to three *Steatoda* spiders, their webs being built between the upper cowpat and the ground. More significantly, however, these webs contained numerous remains of adult white-fringed weevils (*Graphognatus lencola*), the larvae of which cause considerable damage to grass roots.[1] The weevils themselves sought refuge during the day under the cowpats and were then caught by *Steatoda*. It was thought at the time that the spiders were contributing to the control of these pests.

Unusual theridiids

So far we have described typical cobweb spiders which are alike in body shape and form, but there are many others, usually smaller species, which have distinctive modifications, and many have quite distinctive behaviours. The grotesque *Phoroncidia* has an unusual single-line snare (discussed in midget spiders, chapter 13) but they are not alone in the use of this simple snare. *Rhomphaea* is easily recognised by the increased height of the dorsal abdomen to form a cone so that the abdomen is much taller than it is long (Figs 12.13a, b). In males, the abdomen is noticeably thinner and the palps, which bear the mating organs,

---

[1] The white-fringed weevil came into New Zealand from South America.

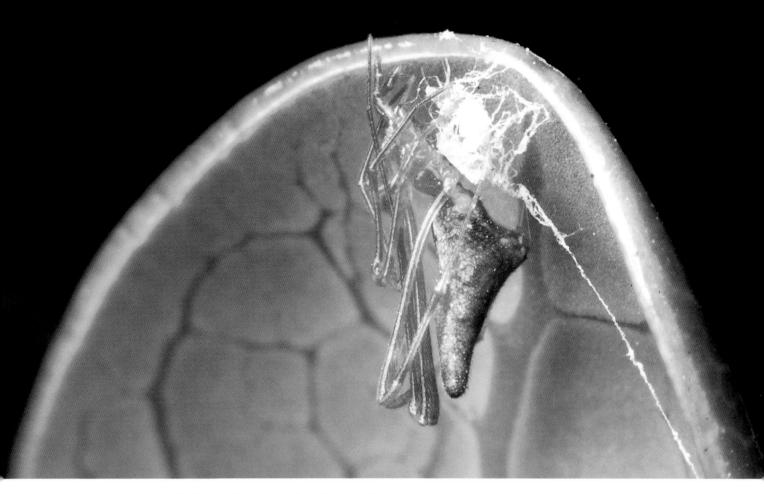

are relatively long (Fig. 12.13c). It is usual to find these spiders suspended upside down from a single line strung between twigs and often supported by vertical threads, particularly at night. These lines are often used by other spiders to move from twig to twig. When they move *Rhomphaea* intercepts them and, by this means, captures its prey. Observations by Mary Whitehouse reveal that this spider often entices 'prey' onto its web by its movements and then immobilises its spider victim by enveloping it in a sticky triangle of silk drawn from the spinnerets by its hind legs.

A common inhabitant of scrub such as manuka and tussock in the North Island, *Dipoena* is rather an unusual genus. Although there is nothing particu-

Fig. 12.13a (above): This female Romphaea *sp. guards her small bundle of eggs as she hangs from a thread strung inside a rolled-up leaf. See how the abdomen is extended dorsally to form a cone, thus giving this spider its unusual appearance.*

Fig. 12.13b (far left): The abdomen of this female Romphaea *sp. is so greatly extended so that it lies beyond the folded legs. As it clings by the front legs to its single-thread 'snare' it is supported by a dragline attached to the spinnerets and held by a hind leg.*

Fig. 12.13c (left): Male Romphaea *species are shaped somewhat differently to females. In this species, just the end of the abdomen is cone-shaped while the palps are very long. In this illustration they are folded up closely near the head.*

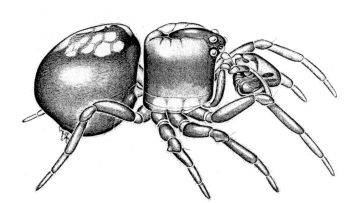

Fig. 12.14 The male Dipoena *(Theridiidae) has a strangely elevated abdomen and carapace with eyes on a mound at the top as shown here. However, the female does not have these strange modifications. Body length 2 mm.*

Fig. 12.15 *(below): This odd little spider (*Euryopis *sp.) is often found in the same habitat as* Dipoena *although the male of this species is not modified in the same way.*

Fig. 12.16 *(bottom left): The male* Moneta *sp. (Theridiidae) has very long legs and palps. Unlike the female (see Fig 1.9) its abdomen is not drawn up into a stout rod but has an elongated shape.*

Fig 12.17 *(bottom right):* Episinus *is one of the most widespread theridiids inhabiting low-growing ferns and shrubs. It clings upside down from a few threads of silk and catches its prey from this position.*

larly striking about the female, the male is quite extraordinary because the carapace is monstrously enlarged, extending above into a relatively huge, flat-topped 'cylindrical box' with the top surface creased and wrinkled (Fig. 12.14). These strange spiders, which are found all over the world, are purported to prey specifically on ants as does its close relative *Euryopis* (Fig. 12.15) which is often found in the same habitat. However, the male *Euryopis* is not modified in this way and is little different from the female.

Moneta is another theridiid which is at home on low shrubs. The female's disguise is extraordinary. Its abdomen is drawn up into a slender rod and, as the spider normally rests along a twig with its two front legs forward, the hind pair to the rear and the third pair pulled in against the body, its pose presents an imitation of a bud on a plant (see Fig. 1.9b). The abdomen of the male, however, is not modified in this way (Fig. 12.16) and, of course, his life as an adult is very short. Once he has mated, the male's usefulness has ended. The most widespread theridiids in the lower parts of forest shrubs and ferns are the species of *Episinus* (Fig. 12.17). These brown spiders may be seen at night hanging upside down just above the ground, clinging to two or three threads of silk. Like their European relatives, they capture their prey from this position.

A kleptoparasitic spider

In many countries, both tropical and temperate, orbwebs are occupied not only by their rightful owner but also by very small and often colourful spiders which, at first sight, appear to be spiderlings but which, on closer examination, are found to be adults. These are the small commensal theridiids of the genus *Argyrodes*, which spend their lives in the webs of a number of the orbweb spiders. They make their tiny webs at the margin of the orbweb and venture onto it to feed on smaller prey trapped in the web or more often on the remains of larger prey left by the host spider. Such spiders are said to be kleptoparasites.

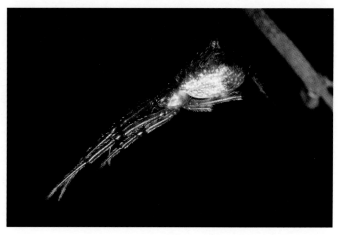

There are no native *Argyrodes* species but the silvery *Argyrodes antipodiana* (Fig. 12.18a) is frequently found in the web of the introduced Australian orbweb spider *Eriophora pustulosa*. The eye is first caught by the gleaming silvery sheen of the conical abdomen, glistening like a drop of quicksilver in the sun as the spider moves about the web of its huge host with seeming impunity. The male is similar to the female except for the carapace which is strongly modified in this species, the eyes being borne on two large lobes. Eggsacs, shaped like miniature toadstools, are attached to the web in clusters of three or four (Fig. 12.18b). Although *Argyrodes* is most often seen in the web of its Australian compatriot it is also occasionally found in the webs of some of our native orbweb species. Although the usual host of *Argyrodes* is the hardy *Eriophora pustulosa* which is found all over New Zealand and even in the Subantarctic Islands, the restriction of these little kleptoparasites to warmer climes is demonstrated by their absence from the southern half of the South Island.

Fig. 12.18a (left): Agyrodes antipodiana *is a kleptoparasitic theridiid spider most likely to be found in the orbweb of the introduced Australian spider* Eriophora pustulosa. *The eyes of this male are borne on two lobes protruding from the front of the head.*

Fig. 12.18b (right): The female Argyrodes antipodiana *guards her little eggsacs, which are shaped like miniature toadstools.*

New discoveries

Detailed observations and experiments on *Argyrodes antipodiana* were carried out by Mary Whitehouse while at the University of Canterbury, and these have added considerably to our knowledge of these fascinating spiders. She notes that *Argyrodes*' own web is not only sticky in comparison to the host's web but that it is an escape route, a refuge and a snare as well as a larder. Insects trapped in it are first wrapped, then bitten and items stolen from the host are stored there. Moreover, *Argyrodes* was able to pillage much larger prey from its *Eriophora* host than it could capture itself, and as an added advantage this prey was often already partially digested. This kleptoparasite moves between webs with ease, providing itself with a bridge line by hanging from a thread which it holds with a fourth leg and releasing silk which is carried away by the wind until it is snagged on a twig or another web. Such 'ballooning' behaviour is common to most spiders but *Argyrodes* can use this technique to move from host to host.

Stiphidiidae

A recent immigrant

The genus *Stiphidion*, from which the name of the family is derived, is one of a number of Australian cribellate genera, one species of which was undoubtedly introduced into the northern part of the North Island some years ago. Because at that time it was not thought to be related to Australian cribellate spiders, it was deemed to be native and named *Amarara fera*. Recognised eventually as *Stiphidion facetum* from Australia, a spider known there since the early part of the century, this name now has priority. Although this spider seems to be actively extending its range to the south it has been recorded only once or twice from Wellington, for example, and is not yet established in the South Island.

In the north, *Stiphidion facetum* (Fig. 12.19) may be found on clay banks, tree

Fig. 12.19 This distinctive cribellate spider with its long, banded legs and mottled abdomen is Stiphidion facetum *of the family Stiphidiidae. An Australian immigrant, it generally builds its snare beneath an overhanging surface.*

trunks and quite commonly, beneath bridges, while in Auckland it is very often found under the eaves of houses. This spider prefers to construct its snare beneath an overhanging surface if this is available but will settle for any less-than-vertical slope. The web takes the form of an inverted cone held firmly in place by guy lines anchored to the margins of the sheet. The spider sits on the substrate below the cone waiting for insects to blunder into this trap.

Native stiphidiids

The three genera of stiphidiids native to New Zealand are all ecribellate, a marked contrast to most Australian stiphidiid species including *Stiphidion* which are cribellate. The commonest species group in mainland New Zealand and the Chatham Islands is *Cambridgea*, its sheetwebs being found on buildings, or in gardens, hedges, scrublands and forests all over the country. Indeed, *Cambridgea* is one of the few groups of native spiders to have occupied human-modified habitats so very readily.

The sheetweb

Ten or more *Cambridgea* species are distributed throughout the country and their webs are all very similar, differing mainly in size. The basic design consists of a horizontal sheet or platform, held taut by numerous lateral lines, and drawn down into dimples underneath by a number of near-vertical ventral threads. Above the sheet, which may be up to a metre square in the largest of these species, a complex network of threads acts as an device to intercept and knock down flying insects, which then drop onto the catching sheet. A tunnel-shaped retreat opens below this sheetweb and gives the spider easy access to its undersurface. Rarely seen out in the web during the day, the spider shelters in this retreat but emerges in the evening to cling beneath it and wait for prey.

Predatory behaviour

The snare has no sticky threads but they are obviously unnecessary because after an insect has flown into the upper knockdown threads and fallen onto the densely woven sheetweb it has little chance of escape. Detecting a disturbance, the spider rushes rapidly to the scene on the underside of the web, and thrusts first its front legs, then its fangs, through the silk to hold and bite the victim several times (Fig. 12.20). The victim is then dragged through the silk and wrapped with silk. If something else hits the web while the spider is thus engaged, it rushes to the next disturbance and repeats the sequence. Several meals may be wrapped and hung in the web in this way, the spider eating the last catch first and then returning to devour the ones caught earlier. Small prey such as flies, for instance, are not wrapped but larger prey are always quickly swathed in silk. When a moth is knocked down onto the web it bounces up and down several times, as if on a trampoline. This is because the scales on its wings detach as the moth hits the silk so that it is able fly off but is then knocked down by the threads above, so it falls to the web again and again. The spider rushes franti-

Fig. 12.20 Cambridgea *bites its prey, in this case a drone fly, several times before wrapping it in silk.*

cally about trying to seize the moth each time it lands but finally goes to the upper surface and catches it there. The venom of these spiders is extremely effective as even large insects succumb in a few seconds.

Garden sheetweb spiders

Cambridgea species are among the most widespread stiphidiids, some having spread beyond their native forest habitats and adapted not only to garden environments but also the underside of houses and sheds. All species of *Cambridgea* look very much alike (except for size) with conspicuous black bands down the middle and sides of the carapace and some degree of black mottling on the abdomen. Very small species whose body length is about 5 to 6 mm or less are still restricted to forest habitats, and it is species of up to four or five times this size that are commonly seen in gardens.

Although the webs of the largest species, *Cambridgea foliata* (Fig. 12.21), from the North Island may be up to a metre wide, the ones most likely to be seen in the garden are perhaps half this size. This species too always constructs its webs from a tubular silk retreat made by lining a hole or gap in a branch or tree trunk whereas other *Cambridgea* species do not make use of such specific bases. Interestingly the related Australian *Corasoides australis*, which constructs an almost identical web close to the ground, does so from a silk-lined tunnel which may be up to 20 cm deep in the ground. The eggsacs of *Cambridgea arboricola*, usually numbering three or four, are hung in the web near the retreat but are not always easy to see because the conspicuous white silken cover enclosing the eggs is hidden under a thick coating of debris (Fig. 12.22a).

Stridulatory system

The males of many species possess an ingenious stridulating device associated with the pedicel and abdomen (see Fig. 1.13). This consists of a number of hard ridges, differing in configuration for each species. These ridges are present on the upper rear surface of the abdomen, as well as a stiff lobe projecting from the pedicel. As the spider vibrates its abdomen up and down, these two surfaces rub against each other and sounds are produced. Because the ridges are fixed in size and number for each species, either a distinctive species-specific sound and/or a particular vibration of the sheetweb could be produced whenever this stridulatory device is activated. However, these possibilities have not been tested. Because this stridulating organ is found only in male spiders, it is likely to have a role in transmitting messages to the female during courtship and/or the mating ritual. Note that not all species possess this organ. For example, the largest of these species, *Cambridgea foliata*, which also differs from most of its congeners in several other ways, does not. The way these spiders mate has never been observed but as male chelicerae are usually very much larger than those of the female, perhaps they are used to grasp the female in much the same way as the male *Tetragnatha* immobilises its mate.

Eggsac construction and camouflage

While little is known of the mating habits of *Cambridgea*, or indeed of any stiphidiids, the construction of their large round eggsacs is easily observed if spiders are kept in captivity. We should mention first that eggsacs are actually difficult to find in the wild unless one knows what to expect. This is because the original conspicuously white sac is somehow coated with a thick layer of debris which appears to be loosely bound with silk so that it looks like a roundish ball of debris. We were curious to see how this happened, as suitable material is often some distance away.

We kept a gravid female in a cage and she soon constructed a sheetweb in the upper region. At some stage, a small patch of silk was spun below the sheetweb and first a large number of eggs was laid in a pile on top of it. These were then enclosed in a coating of flocculent silk before being covered evenly with a smooth

Fig. 12.21 One of the largest species in the family Stiphidiidae, Cambridgea foliata *is found mainly in North Island native forests. An ecribellate spider, its large spaceweb often measures up to a metre in width. Body length 20–22 mm.*

Fig. 12.22a The eggsacs of Cambridgea arboricola *are camouflaged by a thick layer of debris and are placed in or near the extensive webs.*

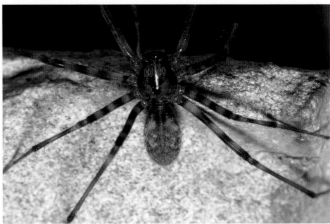

Fig. 12.22b (left): After the female Cambridgea *has fashioned her eggsac, she gathers debris (here, wood shavings) from below (details in text). She hauls a small bundle of shavings up and binds it to the eggsac with silk, repeating this process until the eggsac is covered.*

Fig. 12.23 (right): Nanocambridgea gracilipes *is found in North Island, Nelson and upper West Coast forests. Here, this male exhibits the long slender legs which are a feature of this group.*

white layer of closely woven silk. Prior to eggsac construction we had placed a layer of debris on the cage floor so the next stage proved to be most interesting. After completing the final white silk coating, the spider dropped on a thread, attached silk to a bit of debris and, after climbing back up to the eggsac, hauled this tiny load up and bound it to the sac with silk. This action was repeated many times until a thick layer of debris completely concealed the white eggsac (Fig. 12.22b). But if we did not provide any debris, the spider would descend as usual and search the floor several times before finally leaving the eggsac in its pristine state. Even when a spider makes its web on a rock face in the wild it incorporates small pieces of rock and dried grass into the outer silk covering. Clearly *Cambridgea* spiders deliberately conceal their eggsacs – the camouflage is not merely the accidental incorporation of nearby material.

A close relation

The slender-legged *Nanocambridgea gracilipes* (Fig 12.23) has a similar distribution to *Cambridgea foliata* being found in the North Island, in Nelson and partway down the West Coast of the South Island. Both the web and the spider itself show a close relationship with *Cambridgea*, even possessing the same kind of abdominal stridulatory organ as found in some species of *Cambridgea*. However, in *Nanocambridgea*, the stridulatory organ is underneath the pedicel and consists of a transverse projection behind the sternum which rubs against ridges on the ventral surface of the abdomen. Its web is most often constructed on a damp sloping bank in shady areas of the forest but sometimes it is found amongst low shrubs and ferns. Although similar to a typical *Cambridgea* web it is not as robust. The eggsacs are similar to those of *Cambridgea*.

A stiphidiid without a web

Ischalea spinipes, named *spinipes* because of the strong bristles on its legs, is another native stiphidiid which at first does not seem to have any close relationship with either *Cambridgea* or *Nanocambridgea*, one reason being that it does not make a snare. Its strangely flattened and elongate body (Fig. 12.24) and slender spiny legs once led to suggestions that it might belong to the Tetragnathidae (cf. Fig. 11.26a). However, further examination revealed anatomical similarities to the Stiphidiidae and this is where it is now placed.

Most of these spiders are a pale yellowish-brown, rather like the colour of straw, but sometimes pale green specimens are found. They have a habit of lying along the midrib of a leaf with their front two pairs of legs bent slightly to the side and the hind ones to the rear, and in this position they are very difficult to see. Moreover, no sign of a snare has ever been found so we must conclude that these spiders hunt their prey on ferns and shrubs close to the forest floor, having presumably abandoned their snares at some stage in the distant past to become cursorial hunters. They hunt at night and as their movements are slow and deliberate, it is difficult to imagine them catching an active insect but, as we

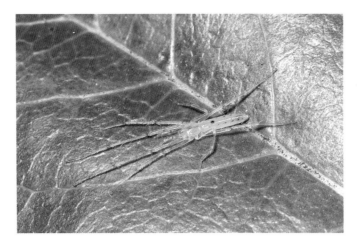

Fig. 12.24 The elongated abdomen and legs of Ischalea spinipes *are characteristic of this species. However, the pale green colour of this spider from Lake Hauroko appears to be a feature of those members that live in the moss-covered forests of Fiordland, in the South Island.*

have failed to keep them alive in captivity, we do not yet know how they go about it. The marked change in the shape of these spiders from that of other space-web builders to the elongate body and slender spiny legs is undoubtedly linked to its vagrant habits.

The pale green *Ischalea spinipes* (Fig. 12.24) comes from Lake Hauroko in Southland and shows a common colour variety of this spider which lives in the moss covered forests of Fiordland. However, in drier forests in Westland and the North Island where the spider tends to dwell amongst the dead fronds of ferns and similar habitats, the general coloration is a pale cream or yellowish brown, often with a darker band down the middle of the abdomen. There appears to be a definite correlation between the colour of these spiders and their habitat although they do not seem able to change pigmentation to suit their surroundings.

Pholcidae

Daddy-long-legs spider

Although most people in the North Island know a 'daddy-long-legs' spider when they see one, this is not necessarily so in southern regions. This is because *Pholcus phalangioides* is susceptible to the winter cold and is rarely seen south of Christchurch even though it generally enjoys the protection of our homes and outbuildings. Even central heating has offered little encouragement to its southward spread because, according to W. S. Bristowe who studied *Pholcus* in Great Britain, the dryness of centrally heated air inhibits the hatching of its eggs.

As you might guess, *Pholcus phalangioides* was accidently introduced into New Zealand and, indeed, to many countries around the world. Although this species is the one most people see in New Zealand, two smaller species with typically long slender legs but shorter, almost oval, abdomens have also arrived here. One of these, generally seen under rubbish in gardens and backyards in the North Island, has an attractive, greenish abdomen broken up by two pairs of white patches, while the other, found in the South Island, is a uniform brown. So far, these more recently introduced pholcids have never been found away from their urban haunts and their eventual identification may provide a clue to their country of origin.

The pholcid snare

The extensive maze of fine threads which makes up the pholcid snare is usually hung near the ceiling and there is a definite preference for corners. Somewhere amongst this silken network, the spider can be seen suspended upside down from the tips of enormously long, slender legs which are attached to an elongate body (Fig. 12.25a). The broad carapace has a slight mound and this bears the eight eyes, whose arrangement is a characteristic feature of this spider (Fig. 12.25b). Few signs of life are apparent until a small insect blunders into the web, and then, as if with ten-league boots, the spider strides over to the

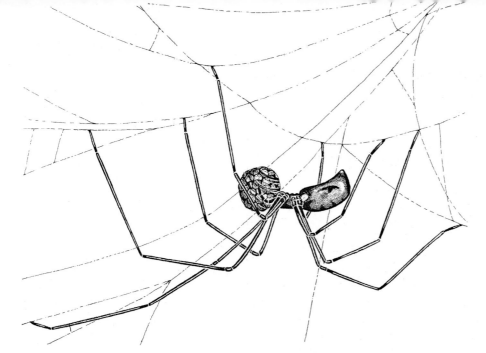

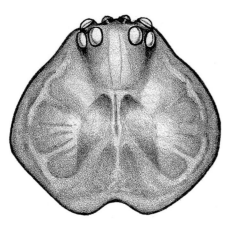

Fig. 12.25a *Commonly known as the daddy-long-legs spider,* Pholcus phalangioides *is a cosmopolitan species which is also found in New Zealand. Seen here suspended in a web, the female carries the bundle of eggs in her fangs during the breeding season. Belonging to the Pholcidae, its body length is about 6–7 mm.*
Fig. 12.25b *The carapace of* Pholcus phalangioides *showing the anterior-median (AM) eyes in front while the anterior-lateral (AL), posterior-lateral (PL) and posterior-median (PM) eyes are arranged in two groups of three.*
Fig. 12.25c *The female* Pholcus phalangioides, *here seen from below, continues to hold her eggsac as the young hatch and make their way into her web.*

scene of the disturbance. If, during this approach, the insect ceases to struggle, the spider pauses, turns in various directions, tugging the silk with its legs until the insect's struggles are reactivated. But *Pholcus* always halts well out of harm's way, remaining about a leg's length from the struggling victim. A few strands of silk are then thrown around the insect to immobilise it before the spider moves closer and trusses it up so that further struggles are impossible. If too large an object becomes entangled a very different reaction is observed. The spider at once begins to shake its whole body so that all that can be seen is a blur as the entire web vibrates. It is not necessary to wait for something large to be entangled, because even a finger poked into the web will start the spider vibrating. Perhaps the purpose of this behaviour is either to shake free an object too large for the spider to handle or to entangle it further.

How Pholcus mates

Mating in *Pholcus* is of special interest, not because it is in any way spectacular but because the male inserts both of his palpal organs simultaneously into the reproductive openings in the same way as seen in many six-eyed spiders such as *Dysdera*. This is how it happens. The female hangs head downwards in the midst of her extensive silk web, her legs grasping different threads. The male enters the web and at once moves towards the female who now begins to flail her front legs. Although only about half the size of the female, the male does not pause but continues to approach, his front legs waving until their legs touch. Initially, as the male pushes hard against the female's two front legs, she resists but eventually bends them sideways and backwards. To all intents and purposes, the male first appears to be head-butting the female but then, as his abdomen falls back into a horizontal position and starts to vibrate, a close look reveals that he has inserted both palps into her genital openings and is now mating. This position

is held for about one to two minutes, after which the spiders part. The entire sequence of events takes less than six minutes.

Care and development of eggs

Pholcids take care of their eggs in a most unusual way. After the eggs have been laid, they are held together with a few strands of silk and this lightly silked bundle is then carried about by the chelicerae and palps of the female until the spiderlings hatch (Fig. 12.25c). Egg-carrying females are common in the warmer months from summer until autumn, many producing successive batches right up until then. Spiderlings leave the egg bundle as they hatch and hang suspended, looking rather like a line of drying clothes, from some of the threads of their mother's web. Each batch of eggs takes about three weeks to hatch and, according to Pièrre Bonnet, the spiderlings pass through five moults before they reach maturity. Bonnet found that at the height of the season this might take about twelve weeks but as the weather becomes colder and food less abundant, growth slows and those spiders which do not complete their fifth moult before winter do not reach maturity until the following spring. No one knows if these observations hold good for spiders established in this country

Linyphiidae

Spiders belonging to this widespread family are generally known as sheetweb spiders and although the webs of some species are not very prominent, they usually follow the same three-dimensional design. The slender-legged members are inconspicuously patterned in grey, black and brown. They are sometimes confused with theridiids but can be separated from them by the prominent spines on their legs. As well, the presence of a distinctive process on the cymbium of the linyphiid palp serves to distinguish them. Some of these species spend rather secluded lives amongst debris on the ground or in low shrubs but others are the predominant spiders in grasslands and pastures.

Most linyphiids found here are native and, apart from the money spiders, just two species are introduced. One of these is the common *Lepthyphantes tenuis* (Fig. 12.26) which abounds in grassland throughout New Zealand, while the other is the puzzling but distinctive *Ostearius melanopygius*, immediately recognised by the unusual reddish or orange abdomen and the black patch around the spinnerets (Fig. 12.27). Originally believed to be native, this spider has now been found in many parts of the world including England and, because there are no other closely related linyphiids, its original homeland is uncertain. Perhaps it is a relict of the southern Gondwana that spread from that land but it is

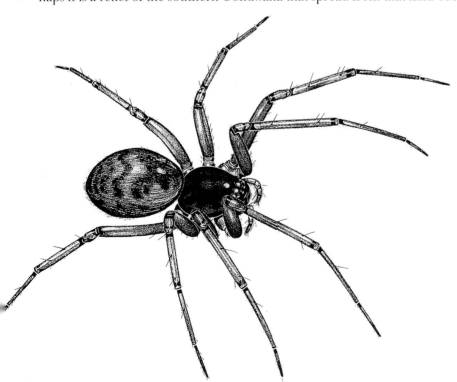

Fig. 12.26 *Lepthyphantes tenuis* *is an introduced species belonging to the Linyphiidae. Found in pasture and grassland, it is common all over the country but particularly in the South Island. It may also occur in disturbed native forest and pine plantations. This spider is grey to black with white markings.*

Fig. 12.27 *The only species in its genus,* Ostearius melanopygius *is a small spider widely found in the Southern Hemisphere. In the field it is easily recognised by its reddish-orange abdomen and dark carapace. It has been found in debris, litter and even in poultry manure.*

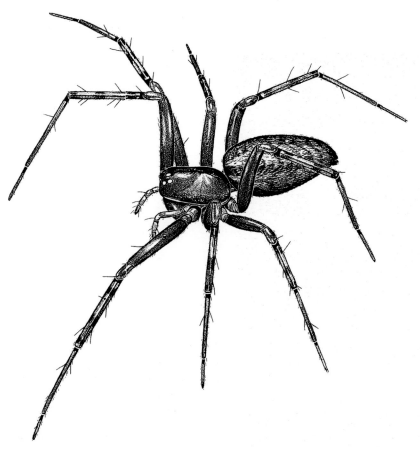

just as likely to have originated from some northern land. The slender-legged *Laetesia* (Fig. 12.28), characterised in the female by a peculiar three-lobed epigynum, is one of the more familiar groups of grassland linyphiids, indigenous species of which are found in both Australia and New Zealand.

Until recently only about twenty linyphiid species were thought to exist in New Zealand and few of these were named. Now, resulting from studies of New Zealand collections by David Blest and Frank Millidge, 95 species are known and more are yet to be described. All are ecribellate spiders which have tarsi with three claws as well as four tracheal trunks arising from a single spiracular opening. Although several subfamilies have been proposed, only the Mynogleninae need concern us here.

The Mynogleninae group

By far the most abundant group of linyphiids is the Mynogleninae, mostly small to medium-sized spiders some 2–10 mm in length and rather similar in appearance (Fig. 12.29). They are mostly found in marginal or transitional habitats such as manuka scrub bordering forests, pastures and marshes. Some species, such as *Mynoglenes major* and *Mynoglenes titan* for example, are associated with creek beds and often build their sheetwebs over running water.

Fig. 12.28 *(above):* Laetesia *is a genus of small spiders about 1–2 mm in length and in most species the abdomen is patterned in black and cream.* Laetesia trispathulata, *shown here, is a native grassland spider found all over the country.*

Fig. 12.29 *(left): This male* Paralinyphia *species is one of a group of linyphiids belonging to the subfamily Mynogleninae. Small to medium-sized spiders, they are mostly found in forest margins, pasture and marshland.*

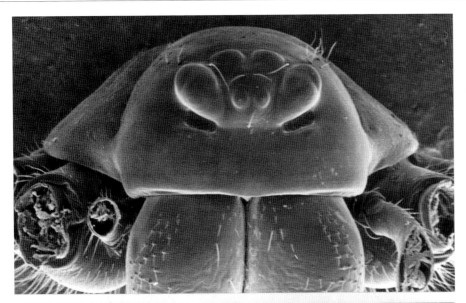
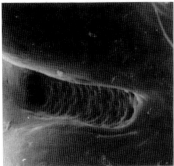
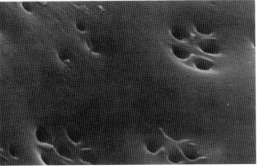

Fig. 12.30a (top): Scanning electron micrographs of a frontal view of **Mynoglenes titan**. *This spider is adapted to creek-bed habitats, building its flimsy sheetweb above the surface of flowing streams in heavily shaded areas. Below the lateral eyes are two depressions called sulci.*

Fig. 12.30b *(below left): A close-up view of a sulcus.*

Fig. 12.30c *(below right): The floor of the sulcus is perforated with pores. (SEMs courtesy David Blest)*

The uniqueness of this group arises from two depressions known as sulci, which are set below the lateral eyes. The floors of these sulci are perforated by numerous minute pores from which a secretion is discharged (Fig. 12.30a, b, c). Similar pits are found in male Erigoninae (see chapter 13) where they are known to have a function during mating but in the mynoglenines these sulci are present in both sexes and also in juveniles so a similar function is unlikely. The only other reasonable suggestion is that they have a defensive role but there is, as yet, no evidence to support this theory. Interest in this group of linyphiids, initially known only from New Zealand and the Subantarctic Islands, grew when they were also found to be relatively abundant in East Africa. This discovery was not the end of the story because their range has now been extended to other parts of Africa and Australia although not yet to South America. In New Zealand they are found in most habitats and make rather unstructured webs. The eggsac is usually a planoconvex structure attached to the underside of a log or stone where these spiders generally live (Fig. 12.31).

Fig. 12.31 *(right): This* **Paralinyphia** *species has built its eggsac on the underside of a log. Typically for this group, the eggs are contained in a flattened circular disc of silk firmly attached to the log with many threads.*

Synotaxidae

A recent family

Like several of other families discussed in this book, Synotaxidae was established only recently. Prior to this, a number of species of long-legged spiders from South America were given the name of *Synotaxus* and placed in the Theridiidae. When further relatives were found in Chile, Australia and New Zealand, it was decided to separate them from the Theridiidae and establish a new family – Synotaxidae – for them.

All synotaxids have long, slender legs and males often have stridulating structures associated with the abdomen and the carapace or the pedicel. South American *Synotaxus* spiders are unique in that they weave a peculiar lacelike web, often above running water, while the webs of other groups in this family are more typical of structures made by some of the theridiids and linyphiids.

Sub-family Pahorinae

The most widely distributed New Zealand synotaxids belong to the subfamily Pahorinae, which have not so far been found in any other country. Pahorines are all small, slender, long-legged spiders which build sheetwebs on the trunks of trees or in low shrubs and ferns. In two genera, *Pahora* and *Pahoroides*, the sheetweb is cone-shaped with many upper threads which not only support the cone but also act as a device to intercept and 'knock down' flying insects (Fig. 12.32a, b). The spider waits below the cone onto which prey drops. The webs of three other genera – *Wairua*, *Nomaua* and *Runga* – are somewhat similar except that the sheet is flat instead of domed. All males have a pair of pits in the eye region, the bases of which are perforated by numerous secretory pores similar to those of erigonine spiders. Furthermore, the eye region is modified by variously sized raised mounds with distinctive bunches of long hairs. In addition, male pahorines have a stridulatory organ in which a small pick on the posterior margin of the carapace rubs against a series of ridges on the abdomen.

Subfamily Physogleninae

This subfamily is distinguished from the Pahorinae by the greatly elongated abdomen of the male and a more prominent stridulatory organ. Although this is positioned between the abdomen and carapace (as in the pahorines) the difference is that the pick consists of an erect lobe on the pedicel rather than a projection from the posterior margin of the carapace. In fact, this structure is very similar to the one found in *Cambridgea* (Stiphidiidae) which, coincidently, also builds a somewhat similar web. The four genera exemplify a typical Gondwana distribution pattern with *Physoglenes* in Chile, *Tupua* and *Paratupua*

Fig. 12.32a (below left): Pahora *species (Synotaxidae) construct distinctive snares of dense dome-shaped webs supported above by numerous threads which serve to 'knock down' flying insects onto the dome. Spiders wait below. These snares are often built within the upper knock down threads of* Cambridgea *webs.*

Fig. 12.32b (below right): This Pahora murihiku *male waits for an opportunity to approach a female. Notice the long legs and patterned abdomen.*

in Tasmania and Victoria, and *Meringa* (Fig. 12.33) in New Zealand.

However, the most widely distributed, native synotaxid *Mangua* (Fig. 12.34) is found not only on the three New Zealand islands but in Subantarctic, Auckland and Campbell Islands as well. Unlike other members of this family, male *Mangua* species not only lack a stridulatory organ but also have an abdomen similar in shape to the female's. Almost invariably this is black with a number of pale patches. These spiders construct an irregular snare under logs in the forest or amongst grasses and moss on the forest floor. In the Subantarctic Islands the web is found at the base of grasses and low shrubs. Females of many of the fourteen known species have been observed carrying their eggsac suspended beneath their palps while moving about the web. This behaviour is not shared with any other synotaxids. Moreover, as the eggsacs are carried even after the spiderlings hatch and can be seen moving under the silk, these spiders show a remarkable similarity to the eggsac-carrying, long-legged pholcid spiders (see Figs 12.25a, c) although they are not related.

Fig. 12.33 (above left): Meringa otagoa *(Synotaxidae) makes a small web amongst moss and litter close to the forest floor or under logs. This male was found in a forest remnant near Taieri river mouth, Otago, South Island.*

Fig. 12.34 (above right): This small synotaxid spider is Mangua medialis *and the female shown here is carrying her large eggsac with her fangs.*

Cyatholipidae

A recent family

For a long time members of this family were confused with theridiids because of some similar characteristics. For example, it was once thought that *Tekella* might be related to the theridiid *Rhomphaea* because both spiders had a triangular-shaped abdomen. Subsequently, when we examined *Tekella* more closely, we realised that it was not a theridiid but belonged to a new family group only distantly related to the theridiids. This new family became known as Cyatholipidae.

How the Cyatholipidae was named

In the 1930s Bryant said she thought *Tekella* was related to the South African *Cyatholipus*, but this notion was never followed up. Some sixty years later these spiders came under scrutiny again in three different parts of the world. Charles Griswold was collecting and studying them in South Africa; Valerie Davies from Queensland and Jörg Wunderlich of Germany were working on Australian material, while in New Zealand we were examining the large collection which had accumulated over the last two decades, including some specimens from Australia. After sixty years of obscurity these spiders were featured in a number of scientific papers, resulting in the establishment of the Cyatholipidae. Based on the group name Cyatholipeae designated for South African representatives by Simon in 1894, the descriptions included many previously unknown species of which six were new to New Zealand.

Tekella examined

One of the things soon observed about *Tekella* is that the posterior respiratory system possesses two spiracles – presumably retained from the original pair of lungbooks – instead of the single one which usually eventuates after booklung spiracles migrate medially. Furthermore, each spiracle leads into a thick bundle of tracheae which extends into the abdomen and through the pedicel into the cephalothorax. This observation supports the view that, not only did such tracheae develop directly from the leaves of the booklungs (as in the Orsolobidae, Oonopidae and Segestriidae for example), but also that this state of affairs has arisen independently in these groups, for none of them is related.

Separation of southern cyatholipid populations

Despite the wide distribution and the undoubted separation of these southern populations for many millions of years, their structure and habits are remarkably uniform. In fact, with one exception, the comments made when describing our own cyatholipid fauna apply almost verbatim to those spiders living elsewhere. The exception is a strange group of species found on the forest floor in various parts of Australia. Known as *Matilda*, this group is quite different from other cyatholipids in their external form, although not in their internal anatomy. In appearance they look very much like lungless anapids with their abdomens protected by hard plates, and they are also much smaller than their soft-bodied relatives. Three genera, *Tekella*, *Tekelloides* and *Hanea*, are known from New Zealand although only the first two are commonly found. The original species described by Urquhart in 1889 was placed in *Linyphia*. But five years later he decided that these spiders were new and named a second species *Tekella absidata* and so we use his name *Tekella* today for five of the eight species we now know from New Zealand.

Characteristics

These spiders, as well as the rare *Hanea paturau* from Nelson, possess normal ovoid abdomens but the North and South Island species of *Tekelloides* have quite distinctive, triangular-shaped abdomens (Fig. 12.35a). All species are restricted to a humid forest environment and construct their snares in shaded situations. Webs are generally found on the trunks of trees although they are also built on steeply sloping banks and in low shrubs. Each consists of a horizontal sheet of closely meshed fine silk strung out from the substrate by lateral and ventral threads. Below the sheet is a network of threads to which the spider clings upside-down (Fig. 12.35b). The spherical eggsacs laid by *Tekelloides* are ornamented with small tubercles and each contains very few eggs, usually from four to six (Fig. 12.35c). As many as four or five sacs may be laid in quick succession and these are attached to ventral silk threads near the substrate.

Fig. 12.35a *The triangular shape of the abdomen of* Tekelloides *is a feature of this genus of Cyatholipidae. The male* Tekelloides australis *depicted here has a creamy white abdomen with a strong black pattern, although sometimes these spiders are pale orange with less distinct black markings.*

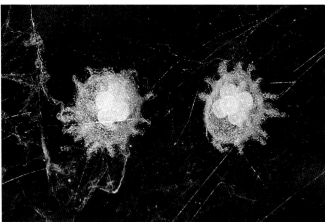

Mimicry

When the habits of *Tekella* were originally described we suggested that they might be social spiders because our field notes referred to numerous adult spiders that had on occasions been collected from a single web. However, a closer examination of the preserved specimens on which these notes were based showed that each time several specimens were collected together from a single web only one of these spiders was in fact a *Tekella*. The others turned out to be theridiids, although as yet undescribed, which apparently live as symbionts or kleptoparasites in the *Tekella* web. These spiders are remarkably similar in size, shape and colour pattern to the species of *Tekella* with which they are associated. A number of different species of these theridiid mimics are now recognised and each is associated with a different species of *Tekella*. Occasionally such mimics are found in *Tekelloides* webs, but in these instances the mimicry fails because their ovoid abdomens do not duplicate the triangular abdomen of *Tekelloides*.

Neolanidae

History of a new family name

In 1959 Richard Marples was doing fieldwork in central Canterbury when he noticed some unusual webs at Hanging Rock. Found mainly on damp limestone overhangs, these small webs were about 60 mm wide and were anchored away from the rock face by a series of threads attached to the margins. The owner was found resting on a smaller sheet below the main sheet and it was here that the biconvex egg cocoons were found. However, after examining this spider closely, Marples decided that it was identical to the species earlier described by Bryant as *Ixeuticus janus* from inland Canterbury.

The next step came in 1967 when Pekka Lehtinen decided that the 'Marples' spider could not be a species of *Ixeuticus* and established a new genus called *Marplesia* so it became known as *Marplesia janus* (Bryant). This conclusion was based on the assumption that *Ixeuticus janus* and *Marplesia janus* were one and the same species. However, when we examined the original specimens studied by Bryant we realised the 'Marples' spider was not Bryant's *Ixeuticus janus* and that another new name was needed, at the same time noting that *Ixeuticus janus* also needed revision. Thus the 'Marples' spider from Hanging Rock became *Marplesia dugdalei* (Forster) after John Dugdale, an entomologist from Landcare, who had originally found the spider at Cass in Canterbury and, as the original *janus* could not be retained in *Ixeuticus*, it became *Neoramia janus* (Bryant).

At the same time as Marples named his *Ixeuticus janus*, he had described a similar species from Rotorua–Taupo (North Island) as *Ixeuticus dalmasi*. Unfortunately, the habits of this spider were not known until much later but in 1973 this species and two others were described from the North Island and placed in a new genus *Neolana*, which then became the basis for a new family name,

Fig. 12.35b (above left): A Tekelloides *spider is seen here hanging from threads below a dense sheetweb. Large numbers of these webs are often found in close proximity although there appears to be no link between them.*

Fig. 12.35c (above right): Two eggsacs, each with four to five eggs, photographed in a Tekelloides australis *web. The silk covering is thin but the sacs themselves are edged more thickly and ringed with short silk spokes.*

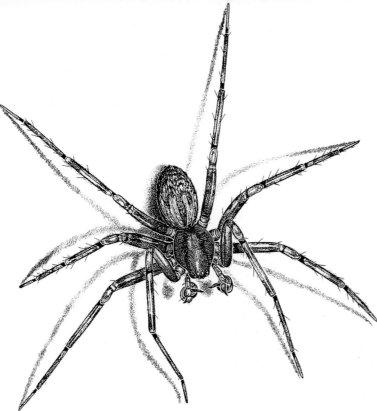

Neolanidae. These spiders also construct a similar type of sheetweb but mostly on tree trunks in the forest. As far as we know neolanids are found only in New Zealand with *Marplesia* (Fig. 12.36) in the South Island and *Neolana* (Figs 12.37a, b) in the North although it is likely that a third genus, *Waterea*, based on a single female found in Te Araroa on the East Cape of the North Island, also belongs in this family. Nevertheless, it would not be surprising if, sometime in the future, further examples of this family were discovered in Australia or even in those other southern continents, South Africa and South America.

Fig. 12.36 (above): This Marplesia pohara *male belongs to a recently named family, Neolanidae. A grey spider with black and white markings,* Marplesia *builds a small and unusual sheetweb that is anchored to the substrate by a series of threads. Below the main sheet is an even smaller sheet where the spider rests.*

Fig. 12.37a (right): A medium-sized spider, Neolana dalmasi *is found in the North Island. The legs and carapace are amber-coloured and the abdomen is grey. The one shown here is female.*

Fig. 12.37b (below right): The conspicuous white sheet webs of Neolana dalmasi *are usually constructed on mossy tree trunks. They are about 10 cm across and are guyed on to irregularities in the bark so that they are raised from the surface. Spiders sit underneath waiting for prey to be ensnared in the web.*

CHAPTER THIRTEEN
MIDGET SPIDERS

Here and there, while discussing other families, we mention some particularly small spiders such as the tiny male of the moth-catching *Celaenia* (Araneidae). There are, however, entire families of spiders, as well as groups within families, in which all species and both males and females are minute. Seven families represented in New Zealand come into this latter category: the Anapidae, Hahniidae, Holarchaeidae, Mecysmaucheniidae, Micropholcommatidae, Mysmenidae and Pararchaeidae. Moreover, in other families where most spiders are of moderate to large size, there are also groups – e.g., Erigoninae (Linyphiidae) and *Phoroncidia* (Theridiidae) – with small and often grotesque members. Indeed, the most commonly seen miniature spiders are the introduced Erigoninae or money spiders which balloon across country during the summer and autumn months. These little spiders have, in the absence of a vigorous, open-country native fauna, taken over much of the grasslands and more particularly the modified pasturelands. Here they thrive and, in fact, often become more abundant than in their original homes in the Northern Hemisphere.

Apart from the Erigoninae, the abundance of minute spiders in this country is a comparatively recent discovery. It was made in the last three or four decades during detailed studies of the moss and litter fauna of the forest floor, as well as that of the tussock and subalpine regions. The discovery of these fascinating and largely unknown groups of spiders has done much to initiate and encourage investigations into the cryptic fauna of all the southern continents. This is because such faunal interrelationships are presumed to reflect the way these now widely separated lands, including New Zealand, were linked together as the ancient Gondwana in the long distant past.

Anapidae

Only a few decades ago anapids and the related symphytognathids, with which they were originally placed, were thought to be among the rarest of all spiders. Only a handful of species or even specimens were known and these came from widely separated parts of the world. However, as interest has grown in forest floor ecology – that world of moss, fallen leaves and debris that is home for a vast array of cryptic animals – it has become obvious that the rain and temperate forests in southern countries are important habitats for anapids and that here they are quite abundant.

Structure and behaviour

Anapid spiders are characterised by an upgrowth in the carapace in both male and female, more particularly in the head region (Fig. 13.1). There are usually eight eyes on the elevated frontal region although in some species the front pair (AM) may be missing. Usually these spiders are a deep reddish-brown or black with a thick cuticle which may be slightly punctate or rough. Hard plates often adorn the abdomen, particularly in the male. Body length, which differs little between the sexes, rarely exceeds 1.5 mm and is sometimes barely 0.5 mm. In New Zealand, anapids are an important element of the forest floor, preying on small insects and other tiny arthropods which abound there. The limited number

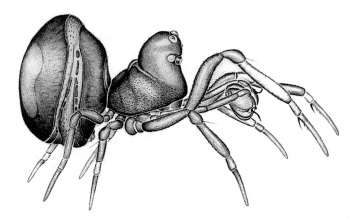

Fig. 13.1 Novanapis spinipes *is just one of thirteen species of Anapidae known in New Zealand at the present time. This tiny male spider, only about 1.3 mm when adult, builds a miniature orbweb amongst the leaf litter and moss on damp forest floors throughout the country. Despite its small size it is one of the largest anapids known.*

of species is compensated for by the enormous number of individuals often present.

Despite their lack of affinity either in size or appearance with the sizeable garden spiders which build those familiar orbwebs, anapids are also orbweb spiders. Their minute orbs are carefully constructed in amongst the litter and debris on the forest floor and some are even found clinging delicately to moss on the trunks of trees. Often only 5 mm in diameter, orbs are usually strung up horizontally with the spider clinging at rest beneath the hub. Because the silk is extremely fine and the webs so delicate, finding spiders in their webs is made easier by using a fine water spray or dusting a likely area with talc or starch. This enables webs to be seen against the darker background. Those few eggsacs found so far have held only a small number of eggs, often just three or four. The eggsacs themselves are lightly spun with fine white silk and shaped like tiny 'double saucers' bound together. They are either attached to the margin of the web or to a nearby surface.

Another peculiarity of anapids is the tendency for the female pedipalps to be reduced in size and in the number of segments. So in many species only the coxal segments remain as seemingly useless knobs, while in others even these have gone. Interestingly, female spiderlings of the latter species have no sign of these appendages when they first hatch. That females can manage without this pair of appendages is not particularly surprising because most of the normal pedipalp functions can be taken over by the front pair of legs. But what does give food for thought is why there are so few other species where the female's pedipalps have been modified in this way. Of course, the male's pedipalps are normal in every way because, as in all mature males, they are used to transfer sperm to the female during mating (see chapter 1).

Disadvantages of being small

Smallness brings with it certain problems of which the most serious is the loss of body moisture through the skin. As an animal becomes smaller, the area of the outside skin surface becomes greater in proportion to the bulk of the organs within, so that a correspondingly larger proportion of water is lost through evaporation. Unless some specific steps are taken to reduce the loss of body fluids, dehydration leads to the loss of internal body functions and the spider dies. Many minute animals found in forest litter avoid this fate by living where the humidity is so high that there is little water loss through the skin. But their fate hangs in the balance as they are liable to perish if by some mischance their home dries out.

Many midget spiders, including most anapids, utilise a morphological control of water loss by having their abdomens covered with hard plates or tough and thickened skin. Another way that spiders may lose moisture is through the open spiracles which lead to booklungs and tracheae, but more particularly through the booklungs. These organs, by virtue of their lamellate structure, expose a large internal surface to the incoming air. This has led to a trend in True Spiders for the posterior pair of booklungs to be replaced by tracheae. However, in anapids as well as a few other (mainly small) spiders, a further solution involves reducing the number of leaves in the anterior pair of booklungs. Occasionally, an elongation of the remaining leaves and a reduction in their width results in typical tracheal tubes. Indeed, because of this tendency in minute spiders, they were originally linked together by taxonomists as Apneumonomorphae, based on the assumption that this trend reflected a relationship rather than a reaction to physiological pressures. So we see that one of the very interesting characteristics of anapids is the complete loss of both pairs of booklungs in most of them and their replacement by tracheae. These respiratory changes can be explained by pointing out that they relate to water loss prevention but it must be admitted that experimental proof to support this logical explanation is so far lacking.

Distribution

New Zealand and Australia appear to be the main headquarters for this family. One probable reason is that, because of their southern isolation, anapids were never subjected to competition during their early evolutionary history from more vigorous spiders such as the money spiders, some of which occupy similar habitats in the Northern Hemisphere. Whatever the reason, anapids deserve mention as being one of the most interesting sections of our forest fauna and as long as one can 'think small enough' they are easy to find. You might suppose that such spiders, little more than half a millimetre in length, would be difficult to see amongst the leaf litter and of course this is true. However, the best way to find them is to sieve litter and moss into a white dish (see chapter 18) where they can be picked up with a small damp paint brush as they move about.

Micropholcommatidae

Minute spiders

In contrast to the length of their family name all spiders in the Micropholcommatidae are minute. This family was established for a tiny Australian spider named *Micropholcomma caeligenus*. Another similar and congeneric spider from Victoria, *Micropholcomma longissima* (Fig. 13.2) is often found in moss or leaf litter in beech forests.

In New Zealand the most widespread and abundant species belong to the genus *Textricella* (Fig. 13.3). These extremely small spiders, all less than 1 mm long, stretch their small sheetwebs amongst the moss and liverworts on the forest floor or on the trunks of trees. It is thought that they feed mainly on springtails and other minute insects. Most *Textricella* species are golden-brown but a few are black. The entire carapace is elevated so that the lateral profile is almost a square, contrasting with

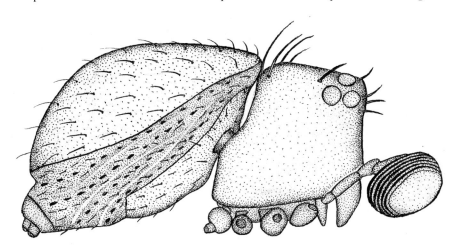

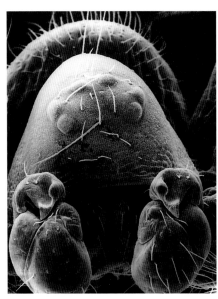

Fig. 13.2 *(below left): A lateral view of* Micropholcomma longissima *(Micropholcommatidae) from Victoria, Australia, shows the squarish profile of the cephalothorax and the elevated abdomen above the pedicel. This diminutive spider has all the features and internal organs that we showed in Fig 1.6 functioning inside it.*

Fig. 13.3 *(below):* Textricella salmoni *is a spider so small (less than 1 mm long) that we do not have a regular photograph of it. This greatly magnified, Scanning Electron Micrograph is a frontal view of a male showing the palps in front, the six eyes situated on the lofty carapace and the rounded abdomen behind.*

the sloping profile and lofty frontal region of anapids. The AM eyes are always small but in some species they have been lost so that only six eyes remain.

Textricella were first recorded from Tasmania by Hickman. A keen observer, he soon gathered knowledge about their habits but was very puzzled by their internal anatomy, about which little was known in the 1940s. Eventually, however, textricellid spiders ended up in the Micropholcommatidae. We know now that these little *Textricella* spiders are found not only in Australia and New Zealand but also in Papua New Guinea. Indeed, many more species of these minute spiders are known from all over New Zealand and, surprisingly, they are also quite common amongst the tussock on the windswept Subantarctic Islands, Auckland and Campbell, to the south of New Zealand.

In addition to *Textricella* there are two other genera of micropholcommatids in New Zealand. One of these, *Parapua*, is easily recognised by the prominent 'punctures' in the carapace (Fig. 13.4) and its bright reddish-brown coloration, while the second, *Pua*, looks rather like *Textricella* except that the carapace is somewhat lower. Moreover, a species of *Parapua* is reported to have been collected in New Caledonia.

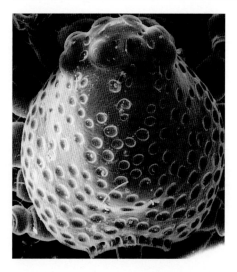

Fig. 13.4 *Another micropholcommatid spider found in New Zealand is* Parapua punctata, *a name reflecting the circular indentations or 'punctures' in the carapace. This Scanning Electron Micrograph is a dorsal view of the carapace showing these punctures as well as the eight eyes found in this species.*

Mysmenidae

Australasian group

Only a single Australasian species, *Mysmena tasmaniae*, of this family of minute soft-bodied spiders has ever been officially recorded (Hickman, 1979) and that one was from Tasmania. But despite the fact that many species are now known, no distinctive southern scientific names are available for any of them because no such nomenclature has been published. At present, these species are referred to by Northern Hemisphere generic names, *Mysmena* and *Trogloneta*, the two groups with which the New Zealand and Australian species seem most closely related.

Even though this family has no formal status in New Zealand, we know that members are found here, commonly living in our forests, in moss and litter, and occasionally in litter at the base of tussock in open country. These small spiders, usually much less than 1 mm long, can be recognised by the oval abdomen rising at right angles to the carapace while the hind region is usually blackish and marked with pale spots or patches. The sexes differ little except that the eyes of the males are positioned slightly higher and the stout bristles on the metatarsi of the first pair of legs are probably used to clasp the female during mating. It is virtually impossible to find these little spiders just by looking, and most of them have been discovered by sieving leaf litter and moss or using a Berlese funnel (see chapter 18). Little has yet been recorded about any of our native species.

Habits of Tasmanian species

By keeping *Mysmena tasmaniae* alive in damp conditions in his laboratory, Hickman spent some years studying their behaviour, and his observations may well apply to New Zealand species. Watching them under a microscope, he noticed that their small webs which they spun amongst grass stalks, consisted of a few irregular threads strung in a horizontal plane. None of the threads were sticky and so did not entangle prey but whenever a suitable prey item, such as a springtail, touched a thread the spider would dash across to it. Then, without further ado, it would bite the springtail. It did not wrap up its victim in silk, as is the habit of many other spiders such as araneids or theridiids, but left it hanging on the web by a few threads until required.

According to Hickman these spiders mate in spring or early summer. Upon taking up a position immediately in front of the female, the male stretches out his palps beneath her to insert first the left one and then the right one, to opposite genital openings. Palps are applied alternately about eight times to each side for up to ten minutes. Three days after mating the first egg clutches are laid, mostly in small batches of eight to ten eggs at a time. Each clump of eggs is loosely tied together with a few threads of crinkled silk and hung beneath the web. Egg-laying continues from October through to January. Observations over two-and-a-half months revealed that one female laid six eggsacs from a single fertilisation. Incubation lasts from ten to fifteen days after laying and spiderlings undergo at least four instars before reaching maturity some thirty to fifty days later.

New Zealand representatives

Several species related to the Tasmanian *Mysmena* have been found throughout New Zealand, mainly living on the forest floor but also amongst the litter around the bases of tussocks. These species are very small, most having a body length of less than 0.5 mm. This makes them almost the smallest spiders in the world, a record held at present by the Samoan symphytognathid *Patu marplesi*. The abdomen rears up at right angles to the carapace, as in the anapids, so that the pedicel appears to be inserted halfway along the abdomen. The posterior surface of the abdomen is usually dark with several conspicuous pale spots while the front face is paler. In males, the ocular region is only slightly elevated in marked contrast to the second group found in New Zealand.

Primarily because of the extraordinary arrangement of the male spider's eyes, this second mysmenid group has been linked with the northern genus *Trogloneta*. In these males, the frontal carapace is projected forward as a long, curved and slender rod with six eyes positioned at about two-thirds of its length and the remaining two eyes at the very tip (Fig. 13.5). Females have no such 'eye stalks' and their ocular area is little different from those of the related *Mysmena*.

Common spiders

Trogloneta spiders are quite common. Although they occupy a variety of humid habitats such as low ferns and shrubs in rain forests, they are more commonly found in clumps of sedges and grasses growing in marshy ground. Observations show that several adult spiders may live together in a loose web, little more than a tangle of threads with no hint of an orbweb structure. Nevertheless these spiders have not been closely studied and no doubt, like much of our fauna, a plethora of interesting information will come to light with careful study. So little is known of these spiders that we are yet to discover at what stage during the male spider's life history the eyes and carapace are modified. Immature males from collections show no signs of partly developed eye stalks, which suggests that a sudden change takes place during the last moult. It is difficult to envisage the function of this strange modification but no doubt further studies will reveal that, as with the equally unusual modifications in micryphantids, it facilitates some essential role during mating.

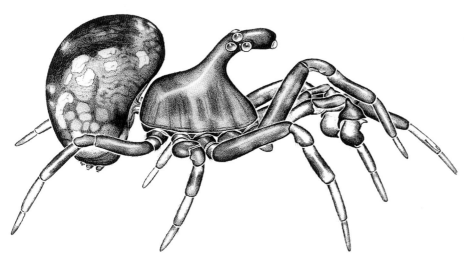

Fig. 13.5 *New Zealand representatives of the family Mysmenidae can only be referred to at present by Northern Hemisphere genera, Mysmena and Trogloneta (details in text). The male 'Trogloneta' depicted possesses features that typify these midget families except that here the carapace has been projected forward as a stalk bearing the eight eyes.*

Some overseas workers say that mysmenids construct an orbweb while others, notably Hickman, have described the web as an irregular network of threads somewhat like the well-known theridiid web. In fact many of the spiders now attributed to Mysmenidae were once placed in Theridiidae because of this belief. This is just one of the many mysteries about spiders waiting to be solved.

The Archaeid group

Many spiders now assigned to the three families Holarchaeidae, Pararchaeidae, and Mecysmaucheniidae were originally placed in the Archaeidae, a family group which has had a long and interesting history. It was first established in 1854 when six fossil species were found embedded in pieces of ancient Baltic amber from Europe. So good was their state of preservation that they could have been entombed for a few decades instead of thirty-million-odd years. These small archaeid spiders had greatly elevated carapaces particularly in the head region but, unlike anapids and *Trogloneta*, the chelicerae had moved upwards with the carapace to remain near the eyes. As a result the chelicerae had become greatly elongated so that the spiders' fangs could reach the mouth. Moreover, instead of the double row of cheliceral teeth found in most spiders today there were numerous peglike bristles.

At first these archaeids were thought to exhibit yet another evolutionary experiment in specialisation and that they had then become extinct. To most arachnologists' surprise, however, in 1881 the Rev. F.O. Pickard-Cambridge, a well known arachnologist, described a living species from Madagascar (Fig. 13.6). In the next 80 years twenty or more species were discovered from a wide range of Southern Hemisphere countries, namely South America, South Africa, Madagascar, Australia and New Zealand. All were considered to belong to the Archaeidae on the basis of having high carapaces and peg teeth on their elongated chelicerae. Later, this view was rejected because marked differences in the genitalia and various sensory structures such as the tarsal organ and trichobothria were revealed. As a result three new families were established in addition to the Archaeidae.

True archaeids

The presence of 'true archaeids' in Australia was subsequently validated and we now recognise three species from the mainland. But many of the earlier recorded spiders with similar modifications of the carapace and chelicerae, while

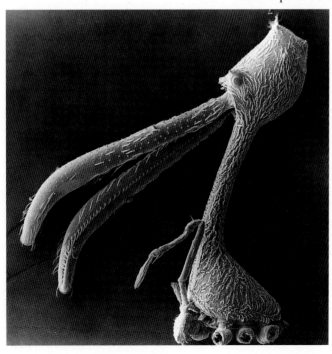

Fig. 13.6 *Living today in the forests of Madagascar, this strange-looking spider,* Archaea gracilicollis, *is an extreme example of the evolution of the Archaeidae (see text). The carapace has been greatly extended to form a 'neck' that bears the head region. One pair of eyes is near the base of the chelicerae, which are also elongated. Three other pairs of eyes (not visible) are found on the upper part of the head. One fang can just be seen at the end of the right chelicerae; it is closed inwards and lies against the row of bristles which runs all the way up the inner part of the basal segment of the chelicerae.*

distantly related, do not belong with archaeids. We now also know that 'true archaeids' do not occur in New Zealand or South America and so those previously misplaced spiders have been separated into three families, Holarchaeidae, Mecysmaucheniidae and Pararchaeidae, all with representatives in New Zealand. Behavioural studies of true archaeid spiders such as the Madagascar species, suggest that, like mimetids, they are all araneophages – that is, they prey on other spiders. One explanation is that the peculiar modification of the carapace and the long mobile chelicerae enable these spiders to capture their prey from a safe distance. Nevertheless, similar although less pronounced modifications found in the three New Zealand families do not necessarily imply similar habits.

Holarchaeidae

A single species in New Zealand

In New Zealand at present the Holarchaeidae is represented by only a single species, *Holarchaea novaeseelandiae* (Fig. 13.7). This particular species lives in moss in Fiordland but others live in unmodified forests from the Three Kings Islands in the north to outlying islands off Stewart Island in the south. These species will persist as long as their habitats are not destroyed. For many years Holarchaeidae was known only from New Zealand, but in 1981 Hickman described a new species from a single 1.5 mm long female collected from moss in the dense rain forest of south-west Tasmania. This he named *Zearchaea globosa*. Subsequently it was realised that the Tasmanian spider was really another species belonging to *Holarchaea* and it is now named *Holarchaea globosa*. No other species have yet been found elsewhere in Australia so that New Zealand and Tasmania are the only known localities of this unusual spider.

Mecysmaucheniidae

Despite the fact that this family was only established in 1967 by Pekka Lehtinen, its first representative from Chile, *Mecysmauchenius segmentatus*, was described in 1884 by Eugène Simon. Because of the high carapace and relatively long chelicerae studded with peg teeth, this spider was originally placed in the Archaeidae. *Mecysmauchenius segmentatus* remained a lone oddity for some sixty years until C.L. Wilton found a similar but much smaller spider in Wairarapa bush in the North Island of New Zealand. This he described as *Zearchaea clypeata* (Fig. 13.8), noting that, although markedly different from the Chilean spider, it was undoubtedly closely related. Subsequently these spiders have been sought in Australia, South Africa, southern South America as well as New Zealand. Surprisingly, there has been no sign of them in South Africa or Australia and so their distribution complements but does not overlap that of the Archaeidae in southern lands.

In New Zealand a relatively large species, almost 3 mm in length was found living deep in the Fiordland forest. It was recognised as being quite different

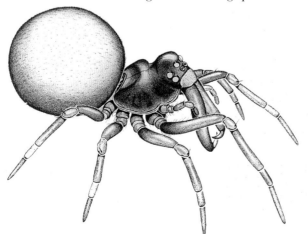

Fig. 13.7 (below left): Only one species of Holarchaeidae is known in New Zealand. This is Holarchaea novaeseelandiae, *which lives in moss in unmodified forests. In the female, seen here, the rounded abdomen, the elevated anterior of the carapace and elongated chelicerae are distinguishing features.*

Fig 13.8 (below right): Zearchaea clypeata *was the first New Zealand spider in this group to be placed in the family Mecysmaucheniidae. Except for its small size, it is not modified in ways that we see in other midget spiders.* Zearchaea *is restricted to moss and leaf litter in native forests.*

Fig 13.9a (far left): *This tiny male* Aotearoa magna *(Mecysmaucheniidae) crouching amongst liverwort is about 3 mm long. Its features include long legs, large palps, greatly enlarged black chelicerae (only just visible in front) as well as distinctive body coloration and patterning.*

Fig. 13.9b (left): *As the* Aotearoa magna *male approaches his mate, she grasps one of his fangs with hers. This ensures her safety and keeps them both in the right position so that he can apply his palps. (Photographs Frances Murphy)*

Fig. 13.9c (above): *In* Aotearoa magna *peg teeth are associated with the fangs, as seen in this greatly magnified micrograph. These peg teeth are characteristic of the Mecysmaucheniidae as well as other families such as Archaeidae, Pararchaeidae and Mimetidae.*

from *Zearchaea* but closely related to the original *Mecysmauchenius* from Chile. This colourful orange-reddish patterned spider was first called *Zearchaea magna* but is now a true New Zealander bearing the name of *Aotearoa magna* (Fig. 13.9 a). When these spiders mate, they adopt the mating stance used by *Scytodes*, mygalomorphs and many six-eyed families (Fig. 13.9b). While further species related to *Zearchaea clypeata* have been found in most parts of New Zealand, mainly in forest but occasionally in grassland and swamp, no more species of *Aotearoa* are known so it appears that the only species is restricted to the Fiordland National Park. Recent field surveys in Chile and nearby islands by Platnick have resulted in fourteen new species one of which is related to our *Zearchaea* while the other thirteen are related to the original Chilean *Mecysmauchenius* and the New Zealand *Aotearoa*.

Although sometimes found in silken retreats while moulting and laying eggsacs, these delicate little spiders do not construct prey-catching webs. Indeed, the only spinnerets remaining are the anterior pair, the other two pairs being represented by just one or two spigots. It is thought that mecysmaucheniids may be araneophages partly because they are related to archaeids and partly on the basis of a similar morphology. For example, the chelicerae of *Aotearoa magna* have peg teeth (Fig. 13.9c) in the vicinity of the closed fangs and this suggests that they function in the secure and rapid clamping of prey. However, in captivity they are known to catch and devour a wide range of insects as well as spiders – as do most other spiders. The eggsac is usually a flattened cocoon shaped like a miniature poached egg, which is spun onto the surface of leaves or moss.

Pararchaeidae

Pararchaeidae consists of seven similar species, five of which are recorded from Australia and two from New Zealand. All are placed in the genus *Pararchaea*. In appearance they are like *Aotearoa*, having similar high carapaces and elongated chelicerae but, in addition to several less prominent differences, they possess six fully developed spinnerets. Despite this, they make little use of silk since they do not construct snares and only occasionally build thin silken retreats for protection during moults. No detailed studies of *Pararchaea* have been undertaken and the eggsac is unknown. In New Zealand these spiders are found mainly beneath fallen logs on the forest floor, in leaf litter and moss, while in Australia they are found in similar habitats but also under the loose bark of trees. The two New Zealand species are strikingly different in appearance. One of them, the South Island *Pararchaea alba*, has a pale cream abdomen while the second, the North Island *Pararchaea rubra* (Fig. 13.10), has conspicuous orange-red and yellow abdominal patterning.

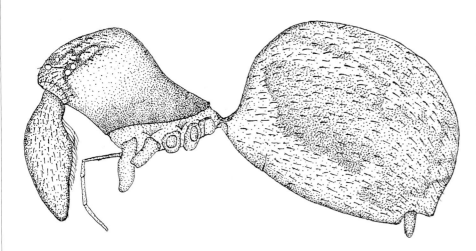

Fig. 13.10 Features that characterise the Pararchaeidae include the elevated carapace and the associated upward movement of the chelicerae. Because the fangs have to be able to reach the mouth, the chelicerae had to become longer. This lateral drawing of a female Pararchaea rubra *shows the much enlarged basal segment of one chelicera (fang obscured). Cheliceral teeth are replaced by a row of bristles.*

Erigoninae (Linyphiidae)

Money spiders

All families so far discussed in this chapter are native to New Zealand but the Erigoninae or money spiders, while abundant, have all been introduced both here and in Australia. As often happens when vertebrates or invertebrates are introduced into a new environment, these spiders have prospered. With the right environmental conditions and an absence of predators or parasites to control their numbers, they have rapidly become even more abundant than in their original home. Erigonines are little shiny black spiders, which as children we called money spiders because it was thought that to be kind to one of these active little creatures would bring good luck, hopefully in the form of money. Sometimes they are called rain spiders in the belief that if one is accidentally squashed then rain follows. In his *World of Spiders* W.S. Bristowe recounts many of these superstitions long held in rural England so it is clear that such myths, as well as the spiders, were brought out by early English immigrants.

Although some species are more abundant in New Zealand than in their homeland, the variety of species is not as great, and habitats are less diverse than those occupied in the Northern Hemisphere. In their European homelands the number of different kinds of erigonines is bewildering. Indeed, anyone who hopes to become an expert in this group is likely to have a full-time occupation. Most of these spiders live on the ground where they spin little sheetwebs to catch small soil insects such as springtails. Still more are found in specialised habitats, such as caves, and their pigment and even eyes are lacking, losses which may have occurred as an adaptation to cave life. All are small, ranging in body length from 1 to 2 mm. As we know, many young spiders move about by ballooning, but erigonines are small enough to balloon as adults. Upon landing, large numbers of them often festoon a grassy field with sheets of glistening silk (see chapter 2).

The commonest species

Of the seven species currently recorded in New Zealand, the most abundant is *Diplocephalus cristatus* which has not only taken over pastureland here but also occurs well above the bushline in mountainous terrain and even invades disturbed patches of native forest. Those minute shiny black spiders seen running over lawns and paths or landing on your arm while gardening usually belong to this species (Fig. 13.11). The small sheetwebs these spiders make to trap their equally small prey of springtails and other tiny insects are built in soil or amongst grass close to the ground. These webs are all but invisible – except in heavy mist or fog when hundreds of them suddenly appear as small white patches all over the ground. Native hahniids are another group of spiders which construct similar webs often in the same habitats and these are mentioned below. *Diplocephalus*

Fig. 13.11 *The minute* Diplocephalus cristatus *(Erigoninae, Linyphiidae) is one of the most widespread species in New Zealand. Best known of the seven species of money spiders, it was introduced into this country during the early days of colonisation. (Body length 2 mm)*

eggsacs are tiny and show up as circular patches of white silk domed in the centre and attached to stones or pieces of debris. In *Diplocephalus* males, the frontal head region is elevated and divided into two distinct lobes (Fig. 13.12a) but this species is unremarkable when compared to some related overseas species in which the ocular region is adorned with extraordinary spines, mounds and bumps. Not all species have such distinctively modified carapaces but since most genera are recognisably different we have illustrated four of the seven species found in New Zealand (Figs 13.12 a–d).

The history of *Erigone wiltoni* is interesting because it reveals that not only is much of the spider fauna of New Zealand still unknown but that this is also true of Northern Hemisphere groups such as the money spiders. *Erigone wiltoni*, a common spider in this country and found as far south as the Auckland Islands, was originally collected by Wilton some years ago. When he first tried to identify it, he could not trace it in any world literature and concluded that, although obviously an erigonine and hence not native to New Zealand, it had not even

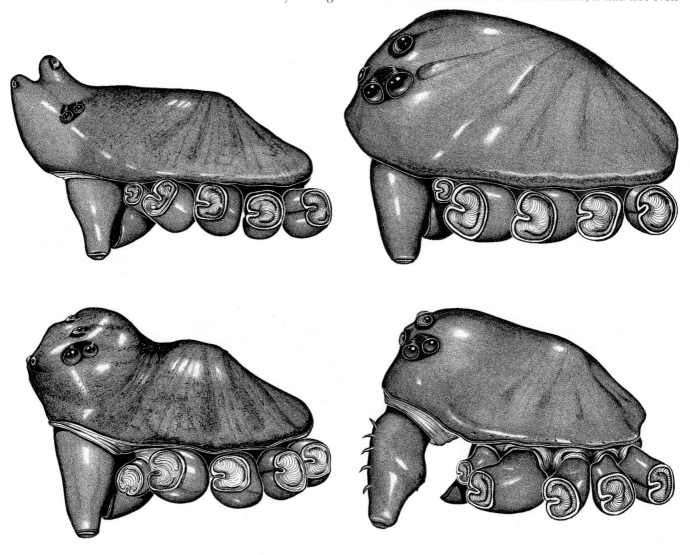

been recorded in its home country, wherever that might be. Eventually described by Lockett, it was named after Wilton who first drew attention to its presence in this country. So widespread is this species that no one is really sure of its original homeland.

Studies of the mating habits of species in New Zealand show that in those species where males have modified heads, this structure is held by the chelicerae of the female as mating is accomplished. Moreover, there has been a suggestion that during this time the female feeds on a fluid exuded from glands associated with these lobes and knobs.

Hahniidae

Look-alike sheetwebs

We mention these spiders here because they may be confused with money spiders, not because of any anatomical likeness but because they spin look-alike sheetwebs in much the same places. All 27 species in New Zealand are native and tend to live in open country areas not converted to pasture, although occasionally they are found in habitats occupied by *Diplocephalus*. But hahniids are larger than money spiders, being between 2 and 3 mm long, are usually pale greyish in colour and often have chevron markings along the top of the abdomen. However the most striking hahniid character concerns the spinnerets. Six of these are present but they lie in a straight row across the back of the abdomen (Fig. 13.13) instead of in a group of three pairs.

Small sheetwebs, slightly larger than those of the money spiders, are spun across small depressions in the ground or at the base of clumps of grass or tussock, while yet more may be found in rock screes on mountain slopes. Their fine silken webs are inconspicuous so, like erigonine snares, they become highlighted after heavy dew or fog when drops of moisture condense on the silk. Small insects such as springtails, of which there are plenty, make up their prey. The eggsac is a white lenticular structure laid down on the surface of the ground or a stone near the margin of the web. It contains few eggs, rarely more than ten to twelve.

Fig. 13.12 (opposite): Side views of carapaces of males of four of the seven introduced money spider species (Erigoninae, Linyphiidae). Differences in the position of the eyes help to distinguish these spiders, all of which are found in pastures and gardens.

***Fig. 13.12a** (top left):* Diplocephalus cristatus. *In male spiders, the front of the carapace is divided into two lobes. The eight eyes are positioned as follows: two situated on either side of the two frontal lobes and the remaining two pairs on either side of the head region.*

***Fig. 13.12b** (top right):* Microctenonyx subitaneus. *This spider can be recognised by the eight eyes all on the frontal part of the carapace.*

***Fig. 13.12c** (bottom left):* Araeoncus humilis. *Six eyes in this spider are positioned on a slightly elevated and enlarged anterior of the carapace. Two smaller eyes below are facing forwards.*

***Fig. 13.12d** (bottom right):* Erigone wiltoni. *In this spider, the eight eyes are clustered together on the front of the carapace.*

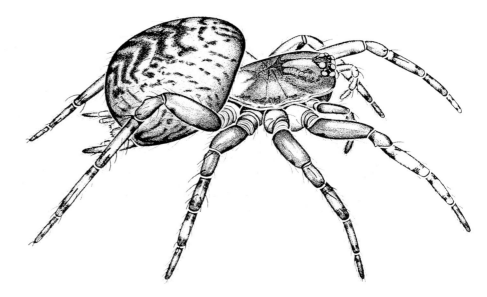

Fig. 13.13 (below): Hahniidae is a family of native spiders often found in similar habitats to the Erigoninae where they spin almost identical small sheetwebs. However, the spiders themselves are not only a little larger but they also have a major distinguishing feature in that the six spinnerets are positioned in a row at the back of the abdomen as in this species, Tuata insulata, *shown here. In most other spiders, spinnerets are grouped in pairs.*

Phoroncidiinae (Theridiidae)

Phoroncidiinae is a subfamily of theridiid spiders but as they are always small, (about 2 mm long), these spiders can be mistaken at first glance for other small spiders we have described. Like them, the skin of the abdomen is thickened and invariably adorned with strange lobes and knobs. Moreover the male's eye region is often protruding in much the same way as *Trogloneta*. Although many species from other countries as well as New Zealand have been known as *Ulesanis*, this name has now been replaced by *Phoroncidia* (Figs 13.14a, b). Detailed knowledge of their habits was unknown until Brian Marples observed some of these spiders in Samoa, and then began a study of the local species in Otago. Several years earlier Hickman described two species from Tasmania and noted that some spiders were found on the underside of stones in association with small irregular webs. Here too, were the spherical eggsacs enclosed in a tough brown parchment-like silk. They had short basal stalks and were suspended from long threads in these webs.

Single thread snares

Although Marples invariably found the New Zealand species on clumps of grass and twigs in association with single silken threads, their small size prevented him from seeing exactly what was happening. So he set them up in his laboratory under a microscope. He saw that spiders first spun a single thread about 10 cm long, which might be inclined at any angle. Part of this thread was covered with globules of sticky droplets much larger than those normally present in orbwebs and easily seen with the naked eye. While most webs consisted of single cross-threads, some also supported one or two additional vertical threads, likewise coated with sticky droplets. Then, resting on a twig, the spider would grasp the free end of one thread with an upraised front leg, while the other end remained fixed. This kept the web under tension and when prey hit it and was trapped by the sticky section, the spider, after first attaching a thread behind itself, released its hold on the twig. The spider then rolled up the original thread while paying out new thread behind until it reached the prey. Here, the spider turned its back and threw silk over the prey with the aid of its hind legs. Once wrapped, the spider bit the immobilised prey and carried it back to the twig where it was eventually consumed. This example of a spring-type web was of particular interest to Marples who had earlier, in England, studied an orbweb spider, *Hyptiotes* (Uloboridae), which employed a similar technique but used two or three sectors of an orbweb instead of a single thread.

Fig 13.14 Phoroncidiinae is another group of small odd-looking spiders. As far as is known all of them catch prey in the same way. A single thread coated with sticky droplets is spun and held under tension until prey strikes it, then it is released so that prey becomes entangled.
Fig. 13.14a (below left): Phoroncidia puketoru.
Fig. 13.14b (below right): Phoroncidia quadrata.

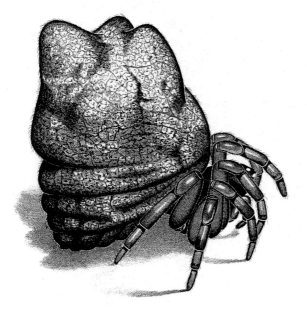
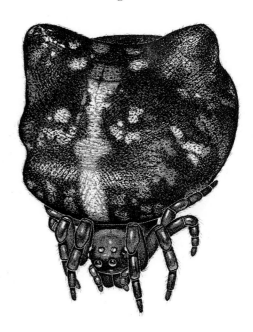

CHAPTER FOURTEEN
SEASHORE SPIDERS

In keeping with our extensive coastline there is a wide range of spiders found only on our beaches and, as a high proportion of them belong to the family Desidae, it is appropriate to discuss the entire group, including those terrestrial members not found on the seashore, in this chapter. The intertidal spider (*Desis marina*) from which the family name is derived belongs to this remarkable assemblage of spiders which has become adapted to life on our coastal beaches and it has become restricted to the region between the high and low tide marks. One seashore spider belonging to the Salticidae, *Marpissa marina* (Fig. 14.1), is discussed in chapter 9 with other jumping spiders.

History of name changes
Several family revisions have resulted in additions to the Desidae. For example, first a little Tasmanian spider (*Toxops montanus*) and later 26 New Zealand species all with unusual eye arrangements were assigned to a new family, Toxopidae, but a later revision led to some of the toxopids being included in Desidae. Moreover, another genus, *Toxopsiella*, originally also included in Toxopidae, has been transferred to the Cycloctenidae (see chapter 8) and the family name Toxopidae has been abandoned. In recent years, 68 new species from New Zealand have been added to the Desidae, now a prominent and varied group of spiders found in practically all habitats.

Cribellate and ecribellate spiders
Spiders now grouped together in the family Desidae include both cribellate (having hackled silk) and ecribellate (without hackled silk) members. The difference between these silk producers is described in chapter 1.

Fig. 14.1 This is Marpissa marina, *a seashore jumping spider found living on rocky shores along Otago coasts. Similar species are found on rocky beaches all round New Zealand.*

Desidae

Characteristics of the Desidae

These spiders are linked together by a few uniform characters although there is considerable variation in both structure and behaviour. All of them, both terrestrial and seashore species, have branched tracheal systems and share simple types of palpal organs with hooked median apophyses (see chapter 1). The outer surface of the tibia of the male palp possesses prominent lobes or processes. In the female, the epigynum is weakly developed and the internal genitalia associated with it are generally tubular. Both cribellate and ecribellate genera are known and, as a result, the presence or absence of a cribellum and calamistrum is associated with a behavioural change, namely from prey capture using a hackled web snare (see chapter 15) to the loss of this web and the adoption of a free-living lifestyle.

An intertidal marine spider

The first intimation that a truly marine spider occurred in this country was in 1877 when C.H. Robson, then resident Cape Campbell lighthouse keeper, reported his observations to the Wellington Philosophical Society.[1] Robson, like so many of the people who take up these lonely jobs, was an ardent and observant naturalist, and we can do little better than repeat the notes he recorded at that time:

I found a veritable spider at home under the water having a nest in an old Lithodomus hole, of which the rocks here are full. All of the spiders of this kind which we have found have had nests in these holes and always under water at all times of the tide. Over the mouth of the hole the spider spins a close web, which when finished looks like a thin film of isinglass, and is waterproof; and behind this film is the nest and eggsac, which last is of various shapes and contains a large number of eggs!

When the spider is disturbed it goes to the bottom of the pool and if a small stick or straw is extended to it, it at once gets ready for a fight, advancing its long and powerful fangs for that purpose. When a small fish is placed in a bottle of water with one of the spiders, the latter will attack at once, driving its long, sharp falces into the fish near the head and killing it instantly!

Naming this spider

When this information was published, James Hector noted that the habits of this spider were similar to the water spider, *Argyroneta aquatica*, found in streams and ponds in England. He concluded that they were related and proposed that the name *Argyroneta marina* be adopted for the New Zealand species. The next year, Llewellyn Powell pointed out that the spider was not related to *Argyroneta* but had features in common with spiders from New Guinea and Singapore that were known under the generic name *Desis*. He suggested *Desis robsoni* as its name, so that as well as the spider being correctly placed, the name of its discoverer would be recognised. Despite these praiseworthy intentions, the rules governing the allocation of scientific names insist on the acceptance of the first species name bestowed and so the spider became officially known as *Desis marina* (Fig. 14.2).

Fig. 14.2 Desis marina *(Desidae) is a medium-sized spider with a dark grey abdomen and reddish-brown carapace and legs. The large projecting chelicerae are its most noticeable feature.*

An unusual habitat

The intertidal spider, *Desis marina*, one of the most unusual seashore spiders, is found between high and low tide levels around all our coasts and along the shores in the Chatham Islands. Their silk retreats can be found in tubeworm burrows and empty seashells but, on the more exposed seashores, they are found in hollows in the holdfasts of bull kelp *Durvillea antarctica*. This genus, *Desis*, is also found in Australia, Asia, South Africa and Japan and all its species behave in much the same way. *Desis* is a striking spider, some 8 to 10 mm long with a bright

---

[1] Later, the Wellington Branch of the Royal Society of New Zealand.

reddish-brown carapace and a creamish-grey abdomen, but the really eye-catching features are the large chelicerae, almost as long as the carapace, which project out in front (see Fig. 14.2). Until recently, little attention was paid to these interesting spiders.

As we have already said, a characteristic of the Desidae is that they all have complex branched tracheal systems (see chapter 1), and, in most of them, these tracheae extend into the prosoma. *Desis marina* is in this latter category, as is *Matachia*, a related terrestrial species, so it can be seen that *Desis* has no physical respiratory adaptations to a semi-aquatic life. Moreover, in exposed coastal situations, *Desis marina* builds its silken nests in hollows under the holdfasts of bull kelp which are regularly submerged. Two Canadian biologists, McQueen and McLay, curious as to how this intertidal spider was able to remain under water for long periods – during high tides, for example – conducted a series of experiments to find out what made this possible. Their field studies showed that spiders were located as far down as 77 cm below mean sea level and, periodically at this depth, they had to survive up to nineteen days of tide-induced submergence. By measuring the amount of air needed for spiders to survive under these conditions (taking their weight into account) and comparing this to the air space available in and around nests during submergence periods, McQueen and McLay concluded that the air available was sufficient for their respiratory requirements. They also found that spiders emerge and feed when the tide is out and their main diet consists of marine isopods and amphipods and other small invertebrates which live between the tides rather than fish, as originally suggested by Robson.[2] Other experiments showed that the respiratory rate was slower in *Desis marina* than in other similar terrestrial spiders, thus suggesting that they are physiologically adapted for spending long periods under water.

Seashore relatives

A second desid species associated with the seashore was first noticed by Goyen in 1889. This spider, found living under stones between the high and low tide watermarks in Otago Harbour, was first given the name *Habronestes marinus*. Now known as *Myro marinus* (Fig. 14.3), it is related to a number of species recorded from the French Subantarctic Islands, Kerguelin, Heard and Crozet, where they are not restricted to the seashore. When we first searched for this spider in the Otago Harbour we did not find any of them living between the tide marks but discovered that they were common under stones and driftwood well above the high tide level. Subsequently we located a few below the high water mark, just as Goyen described, showing that the spider is equally at home both above and below the high tide levels. Its distribution is apparently limited to the eastern coast of the southern part of the South Island, having been recorded only from Otago and Stewart Island.

Once we knew where to look, we soon located *Myro marinus* on the undersurface of stones, either above or between the high and the low tide watermark, at full tide being immersed in from 10 to 30 cm or more of water. But it has never been found any great distance from the sea. It is very like the bits of wood and debris adhering to these stones so that it is difficult to distinguish it from them. If detached from a stone while still under water *Myro marinus* rises to the surface like a cork and floats but if prodded will run across the surface of the water. However, it cannot dive under the water, only succeeding in going below the surface if it can climb down a rock or stone. In this respect, it is very like *Dolomedes aquaticus* (see chapter 6). Moreover, this spider does not seem to construct any cocoon for itself under water but during immersion it makes use of the air held by the hydrofuge hairs which cover its body, a respiratory system known as a physical gill.

The eggsac, flat on the bottom and rounded above, is firmly attached by the

Fig. 14.3 Another desid species which lives under stones in the intertidal zone as well as above the high tide level is Myro marinus.

[2] It is possible that small fish in tidal pools are prey for these spiders in some localities.

Fig. 14.4 (above left): This coastal spider, Otagoa nova, *is mostly found in the shingle banks above the high tide level.*

Fig. 14.5 (above right): A rather squat and solid spider, Amaurobioides picunus *(Anyphaenidae) is distinguished by the characteristic chevron pattern on its abdomen.*

flat side to the underside of stones, and is also under water for long periods as the tide comes and goes, but when they hatch the spiderlings are as much at home in that element as the mother spider. The covering of the eggsac is thin but tough and is apparently impervious to water. About twenty large pale-yellow eggs are attached to the top of the cocoon, which itself is inflated with air. The young spiderlings remain in the cocoon for some time after hatching, and this imprisoned air is no doubt intended for their use.

Myro marinus is replaced on other New Zealand and Chatham Island beaches by species of a related genus, *Otagoa* (Fig. 14.4), similar in appearance to *Myro* but with a different eye pattern and somewhat different habits. However, unlike *Myro marinus*, these spiders are largely restricted to the banked-up shingle above high tide level or otherwise in crevices in the cliff faces within the spray zone. Although these spiders do not construct a snare to capture prey, they are often found in tubular retreats that are open at each end. The eggsac is a flattened disc similar to that of *Myro* and usually attached to the undersurface of a stone.

Anyphaenidae

The large and distinctively patterned spiders of the genus *Amaurobioides* (Fig. 14.5) have, in the past, been placed in a separate family – the Amaurobioididae – one of the few families then believed to consist of a single genus, but nevertheless distinctive by its restriction to a maritime habitat. Today these spiders have been placed in the Anyphaenidae, a widespread family which otherwise does not occur in New Zealand. The first species in this genus, *Amaurobioides maritimus*, was described by Pickard-Cambridge in 1883 from specimens sent to him from Allday Bay in Otago. Subsequently other species were described from South Africa, Campbell Island and Australia and more recently from Navarino Island in South America. When the New Zealand fauna was studied in detail it was realised that a number of new species were involved, each separated one from the other along the coastline. These findings indicate that all these spiders are ancient relicts of the original Gondwana fauna which have speciated over the millennia but which have nevertheless retained many of their basic characteristics.

Coastal habitat

In New Zealand, *Amaurobioides* species occur on any part of the coast where conditions are suitable. These spiders, noted for their dark-brown abdominal patterns, live in permanent tubes of tough grey silk usually constructed in cracks or crevices on rock faces in the splash zone, the area just above the high tide mark. Sometimes, however, they can be found on small ledges where sand has accumulated but always near high tide level. Moreover, in the subantarctic Campbell and Auckland Islands, the retreat is often found under stones well below the high tide level. Just why these spiders should be restricted to this particular habitat is puzzling because they still thrive in captivity well away from

the sea and are not particularly selective regarding their food. In their normal habitat, however, they feed mainly on isopods which abound in cracks and crevices. Perhaps it is their preference for this prey, which is undoubtedly restricted, that limits the distribution of these spiders. The white silken entrances to their retreats are easily seen against the darker rock faces (Fig. 14.6), but in early summer when the eggs are laid, the opening is sealed, and so it remains until the spiderlings hatch. The eggsac is lenticular, although not as clearly defined as in the Desidae, and attached to the wall of the tube retreat.

Fig. 14.6 (above left): Amaurobioides maritimus *makes use of cracks in the rock face to build its retreat. Here, the silk which lines the retreat is clearly visible against the darker rock surface.*

Fig. 14.7 (above right): The Agelenidae is another family represented on the seashore. Oramia littoralis *is a large spider some 12–14 mm in body length, and generally found among piles of stones above the high tide mark.*

Agelenidae

A number of other families of spiders are represented by species restricted to the seashore. *Oramia* (Fig. 14.7) is a genus of large agelenids mostly found among the stones piled up above the high tide mark on shingle beaches, although some species have also been recorded away from the seashore. Moreover, several *Oramia* species are known from the South Island and the offshore islands including the Chathams. They are cribellate spiders which construct a sheetweb amongst the stones, feeding mainly on amphipods and kelp flies. The eggsac, a large spherical white sac, is placed within the retreat associated with the web.

Terrestrial relatives of Desidae

Several terrestrial desid spiders are very similar in structure and appearance to the intertidal spider *Desis marina*, and, as they are also members of the Desidae, we have included them in this chapter. They are usually found a long way from the sea, mainly in forest and scrubland.

Burrow dwellers

Matachia (Fig. 14.8a) and its close relative *Notomatachia* are both cribellate spiders which have adapted to life in small and narrow burrows. These spiders do

Fig. 14.8a (below left): Many spiders belonging to the Desidae are also found some distance away from the sea. Shown here is Matachia australis *which has become adapted to a burrow-dwelling lifestyle.*

Fig. 14.8b (below right): Early stages of a web being built by Matachia australis. *The rectangular nature of the cribellate silk spanning two supporting threads can be seen here.*

not make their own burrows but utilise old insect larval holes in twigs and branches vacated by various insects after they reach adulthood. In forests they particularly favour burrows leading into galls produced by larvae of the moth *Morovia subfasciata* which feed inside the stems of *Muehlenbeckia*, a woody creeper. After adult moths emerge they leave behind conveniently sized burrows which several different kinds of spiders including *Matachia* take over as ready-made homes. In more open scrub country *Matachia* favours a number of shrubs which harbour boring insects. Burrows are lightly lined with silk by the spider and this forms a safe retreat which the spider occupies.

In keeping with the physical limitations of this habitat, spiders are markedly slender and they rest with the first three pairs of legs directed to the front. Using this burrow as its base, the spider attaches a prey-capture web to the branch and twigs immediately outside its hole. The basic pattern of these snares (Fig. 14.8b) is typical for most cribellate members of the Desidae, as well as the related Dictynidae, and may be compared with the lower snare portion of the web built by the cribellate gradungulid *Progradungula carraiensis* (see Fig. 5.6).

A simple snare

The snare, in its simplest form, consists of two threads from 10 to 15 cm long joined with cross-threads of calamistrated silk produced from the cribellum. These two threads initially radiate out from the burrow entrance but become almost parallel in places and are ultimately attached to neighbouring twigs, often diverging at this stage. The technique used to lay down the calamistrated cross-threads is quite straightforward. Starting at a distant point, the spider moves from one line across to the opposite line, letting out a calamistrated thread as it does so. This thread is attached to the second line. After walking a short distance along the second line, the spider then moves back to the first line where the thread is attached again, then walks along that line and crosses to the second line, and so on until it is back to its retreat. The result is a long narrow, somewhat ladderlike, sticky sector, the two supporting lines merging at the entrance to the burrow.

On the first night, the spider usually constructs two or three of these sectors, sometimes adjoined, sometimes apart. At first, these sectors radiating out from the burrow give the impression of being cut-away segments of an orbweb. Further sectors are added each night until it becomes difficult to see the primary calamistrated sectors and the web appears as a convoluted network of silk. Moreover, sectors are not always built in the same plane nor are they necessarily additions to previous webs. Prey capture further confounds the original design. Moths and beetles and a variety of flies are caught in the web but spiders only emerge at night to secure them. Generally the smaller prey is dragged within the burrow and eaten there. Larger insects are consumed in the web where they are trapped but, when the meal is over, the remains are dropped to the ground.

Web additions benefit the spider

A closer look at these webs reveals the new sectors, which now begin to be placed rather haphazardly in relation to the original ones. The fresh silk is clean and glistening and can be distinguished from older parts so that the basic design can be clearly seen. Often these additional sectors are built at different angles, or at the edge of previous ones mainly to take advantage of available twigs, so that the result may be a three-dimensional structure. Only a day or two is required to reduce the silk to dusty, less-retentive threads so that regularly adding fresh sectors would enhance the chances of trapping prey. Deep inside the burrow *Matachia* rests with the first three pairs of legs towards the entrance but at night the spider moves forward so that the anterior pairs rest on threads attached to the opening of the burrow. From this position, it can quickly detect vibrations emanating from struggling insects. Once prey is intercepted by the

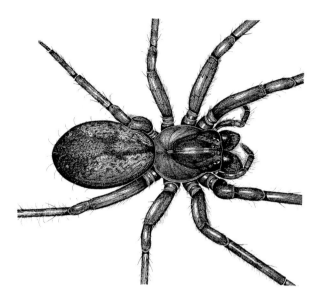

Fig. 14.9 Related to Matachia *is a spider called* Nuisiana arboris, *which is not a burrow-dweller but nevertheless constructs a similar web. In this spider only the front two pairs of legs are directed forwards.*

web, its efforts to escape further entangle it in both retentive and not-so-retentive silk.

Mature males begin to appear in November and they mate soon after. As seen in those cribellate spiders which depend on snares, males lose the ability to produce calamistrated silk at maturity. This means they cannot make snares and hence catch prey, and so they do not live very long after mating. Females start laying eggs during December. The eggsac, which is attached to an inside wall of the burrow, contains only a few eggs (on average about ten or twelve) held together with a thin cover of white silk. Like many tunnel-dwelling spiders, the newly emerged spiderlings continue to live in the burrow with the mother until they are about half grown – often up to the fourth instar. Presumably during this time they share with her the food trapped in the web. When spiderlings decide to take up an independent life they settle nearby, so that it is usual to find these spiders in quite dense aggregations. This means that although such colonies are scattered through the forest they may be quite numerous in one area but absent from another, even though potential homes are abundant.

There are many desids related to *Matachia* and its fellow burrow dwellers but the most closely related spider, *Nuisiana arboris* (Fig. 14.9), which constructs a similar snare with calamistrated silk, is not a burrow dweller. This species is usually found crouching beneath loose bark on the trunks of forest trees and stretching out from here is the typical ladderlike sector web structure. Although this spider possesses the fundamental anatomical features which link it to its *Matachia* relatives, only the two front pairs of legs are directed forwards, as in all other desid spiders.

Ancient and modern lineages

Spiders similar to our two genera, but placed in a separate genus *Paramatachia*, are found in the eastern states of Australia and in Tasmania where they live in hollow twigs and empty insect burrows in branches, in much the same way as their New Zealand relatives. Moreover, a fossil spider found in the Oligocene Baltic amber of Europe dating back by some 30 million years and named *Eomatachia* by Alexander Petrunkevitch is believed to be a direct ancestor of *Matachia*. If this is so, then it is likely all these Australasian spiders are relicts of a once much more widely distributed group at one time living all over the world. Undoubtedly they are an ancient part of our New Zealand fauna for, in addition to the cribellate *Matachia* and *Notomatachia*, we have at least seven other related genera of which six have lost the cribellar silk glands and spigots. In their place is a flattened plate or colulus and, correlated with this loss, the row of metatarsal

Fig. 14.10 (above left): *This brightly coloured desid is a female* Rapua australis, *typically found in shrubs and low vegetation. Its eight eyes are in two rows at the front of the carapace.*

Fig. 14.11 (above right): *Many of the tree-loving desids are quite colourful as is this male* Goyenia fresa. *The ability to move freely on leafy surfaces is because these spiders have developed large flattened tenent hairs beneath the claws of each leg.*

bristles or calamistrum used to manipulate the calamistrated silk has also disappeared. Although some of these spiders, such as *Gasparia*, hunt their prey on the forest floor, most are now hunters on the foliage of low forest shrubs.

Hunting desids

Many of our desids have lost the ability to produce calamistrated silk and are primarily hunters, foraging mainly on the leaves of shrubs although a few seem to manage on the forest floor. Most of those foliage hunters now possess specialised expanded hairs beneath their claws, which assist them in gaining purchase on the smooth leaf surfaces. Unlike spiders in the few other families which have developed these specialised hairs, usually in the form of thick tufts (scopulae), they have retained the single unpaired claw as well as having the other two claws. In *Rapua australis* (Fig. 14.10), typical of many shrub inhabiting desids, the eight eyes are set out in two rows of four, and may at first be mistaken for one of the arboreal clubionids, although they are not given to hopping about in the manner of those spiders.

The most common arboreal desids belong to *Goyenia* (Fig. 14.11) which, like many other arboreal desids, have these large flattened tenent or gripping hairs beneath the claws of each leg. *Goyenia* is of particular interest because nine of the ten species we know from throughout the country have lost the ability to produce calamistrated silk. This means they have abandoned the production of a snare in favour of direct hunting, a behavioural change clearly related to the development of tenent hairs. However, the tenth species, found just once in the mangrove swamps at Kohukohu in Northland, is cribellate and produces the retentive entangling silk, which it presumably still uses to construct a snare. No one has yet revisited these swamps to find out if this is true. Within this genus, therefore, as in some other New Zealand genera, we have a range of species which possess both primitive and derived characters.

Shrub dwellers

Perhaps the most delicate and indeed the most attractive desid spiders are the shrub-inhabiting species of *Lamina* (Fig. 14.12a). These small (3 mm long) spiders are characterised not only by their distinctive green colour, but also by the separation of the lateral eyes and the strong procurve of the posterior row. But by far the most striking feature is the translucent skin, which allows the spider's beating heart to be seen. Clearly visible in the middle of the upper surface of the abdomen, this transparency enables the actions of the heart and movement of body fluids in other organs to be tracked quite readily. These spiders favour the native Horopito or pepper tree (*Pseudowintera colorata*) and are usually found resting on the leaves, although they blend in so closely with the background they are very difficult to see (e.g., *Lamina minor*, Fig. 14.12b). *Laestrygones* (Fig. 14.13), a group known for a long time, has been placed in a wide range of families but is now firmly established in the Desidae. It is not a forest dweller but

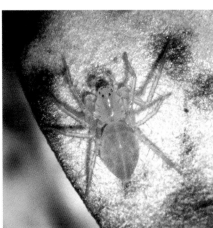

Fig. 14.12a (above): Life among the shrubs is favoured by this little pale green spider, Lamina montana, *also a member of the Desidae.*
Fig. 14.12b (left): *The native Horopito or pepper tree is the ideal habitat for* Lamina minor, *a spider whose green coloration is lightly spattered with purplish-pink, thus merging with the leaves of this tree.*
Fig. 14.13 (below): *The legs of this small ecribellate spider,* Laestrygones otagoensis, *have long and conspicuous spines. A pastoral dweller, it is occasionally found in shrubbery.*

Fig 14.14 Hapona otagoa *is a hunting spider favouring damp undergrowth and ferns in its forest habitat. The pale patterned abdomen and dark legs banded with pale yellow are distinguishing features.*

favours grasslands and is occasionally found in shrubs. However, another genus, *Hapona*, is a forest group living in ferns and shrubs. Although *Hapona* (Fig. 14.14) is known only from a number of species from New Zealand, it has been found in bush near Sydney, and is probably widespread along the east coast of Australia.

Grey house spider

The best known member of Desidae in New Zealand is the Australian grey house spider, *Badumna longinqua* (Fig. 14. 15a), once known as *Ixeuticus martius*. This spider was first recorded last century by Eugène Simon from specimens collected in the South Island by an early French Expedition to New Zealand. Now accepted as an Australian spider, it was introduced either during the very early days of colonisation or arrived here by ballooning. One reason for this conclusion is that this spider (and others) are found almost entirely in modified habitats and not in native habitats such as forest or tussock. *Badumna* is blamed, with some justification, for those unsightly webs collecting dust on

Fig. 14.15a This Australian desid spider, Badumna longinqua, *is now one of the commonest spiders in New Zealand. Its untidy webs festoon our houses, fences, vehicles and vegetation.*

the outside walls of houses but despite their obvious abundance the spider itself is rarely seen. This is because it is a night feeder, and does not generally leave its web and wander into the house as do other spiders common about our houses and gardens.

Those ubiquitous webs

Although *Badumna* makes use of cracks and crevices as a base from which to construct a web it is not an insect-burrow dweller like its relative *Matachia* and so can manage with a chunky body and short legs. Like other cribellate Desidae, the web consists largely of calamistrated silk (Fig. 14.15b). The spider first lines a suitable crevice with silk, at the same time weaving a tunnel-like entrance with additional threads attached to nearby structures. From this base, the spider lays down several long, partly parallel lines of silk, which spread out from the entrance. These lines are then linked by a series of ladder-like cross-threads which are made from the special adhesive silk of the cribellum. The next and following nights, the spider spins more such ladder-like webs, often in different directions. These extensions may result in the web becoming a three-dimensional structure, much depending on whether there are suitable attachment spots. However, on the walls of houses, for example, webs simply expand along the flat surfaces in all directions. As they readily attract dirt and dust as well as insects, they soon look untidy.

Each evening, before making any additions, the spider roams around existing web structures, its dragline often serving to fill gaps left by prey capture. Because the web is occupied for long periods and has regular additions, sectors tend to merge so that the structure eventually appears to be a fairly large untidy sheetweb arising from a conspicuous opening. If an insect becomes trapped in the sticky web, the spider rushes out, bites and entangles it further but it is not usually eaten until nightfall. Prey consists of a wide range of insects, such as houseflies, blowflies, craneflies and moths as well as items such as wasps and even bumble bees often much larger than the spider itself and which the spider is usually able to overcome. Its eggsacs are somewhat flattened and attached within the retreat, and they contain on average about one hundred eggs.

Another related species, *Badumna robusta*, is found mostly in the North Island, although it is not as common. Recent studies in Australia confirm that these two related species, *Badumna longinqua* and *Badumna robusta*, are both present in Australia and are part of a larger, typically Australia–Papua New Guinean group, to which the generic name *Badumna* is applied. This name was

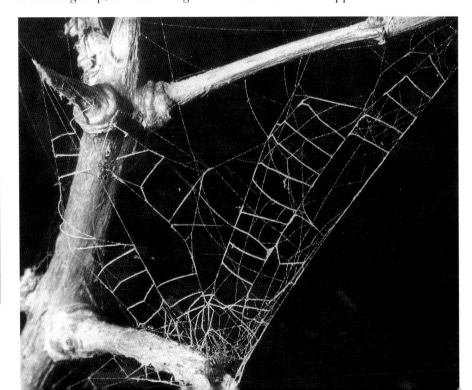

Fig. 14.15b The Badumna *web is very like that of* Matachia *in the early stages of its construction. However, the spider adds more sectors nightly and the web soom assumes its more easily recognised tattered state.*

established before *Ixeuticus*; furthermore, because the grey house spider is identical with an Australian species described even earlier under another species name, it now bears the name *Badumna longinqua*. *Badumna robusta* is known as the black house spider in contrast to the better known grey one and has similar habits. Probably, it requires a higher temperature and more arid conditions than are found in most of New Zealand to breed, and so it has not spread very widely. However, the grey house spider is firmly established not only in New Zealand but also in California, where it was mistakenly described as a native of the United States with the new name of *Hesperauximus sternitzkii*. More recently it was reported from Hawaii and also from Yokohama in Japan. The suspicion is that it may have been imported to Japan with pine logs from New Zealand – a strong possibility as these spiders are often found in crevices on the bark of trees in pine plantations.

CHAPTER FIFTEEN
HACKLED-SILK SPIDERS

A number of spiders possess two closely linked structures, the cribellum and calamistrum. Both of these are necessary to produce the crinkled opaque thread variously known as cribellate, hackled silk or calamistrated silk. It is believed that the cribellum was derived as follows: ancestral spiders originally had four pairs of spinnerets, and, in some of these spiders, the anterior-median pair became modified to form the cribellum, a flattened plate studded with spigots. Spiders which have a functional cribellum also possess one row or more of evenly spaced bristles or hairs, collectively known as the calamistrum, on the metatarsal segment of the fourth pair of legs (see Figs 1.8d and 1.23). To begin with, the spigots on the cribellum exude a swathe of viscous fluid from their tips and this is then 'combed out' by the hairs of the calamistrum as the spider alternately brushes its hind legs across this area, thus producing the tension which reforms the fluid into a bundle of crinkly threads. At the same time, the spider uses these legs to manipulate this calamistrated bundle so that it coats a double axial thread being produced by the posterior-lateral spinnerets in the usual way (Fig. 15.1a, b). Clearly, as these structures evolved, the spider's behaviour too was developed so that the fourth legs were used in the appropriate manner.

In the course of time, however, the cribellum became redundant in some spiders and so lost its silk-producing function. As a result, the flattened plate was modified as a plate or nodule and is known today as a colulus. This means that spiders with a colulus have been derived from ancestors which once produced hackled silk. In most of these instances, the calamistrum too has been modified or lost entirely. This category of spiders is known as ecribellate.

Although spider families sharing these easily distinguished features are often grouped together under the convenient title of 'cribellate' or 'hackled-silk' spiders, they are not necessarily closely related to each other. Six families of these spiders are found in New Zealand but most cribellate species fall into two families – the Dictynidae and the Amaurobiidae. The four others discussed in this chapter are the Agelenidae, Nicodamidae, Oecobiidae and the Amphinectidae. Another family, Uloboridae, is also cribellate but these spiders are unusual in that they are also orbweb spiders and so are discussed in chapter 11, while Desidae are discussed in chapter 14. It is important to note, though, that most of these family groups have both cribellate and ecribellate spiders as members.

Fig. 15.1 Shown here are two Scanning Electron Micrographs of cribellar silk (greatly magnified).
Fig. 15.1a (below left): In Badumna longinqua (Desidae) *the hackled silk consists of bunches of irregularly scalloped cribellum fibres. The axial cores are not visible.*
Fig. 15.1b (below right): *The hackled-silk capture threads produced by* Waitkera waitakeriensis *are formed from thousands of fine looped fibrils which appear as 'puffs' along the axial fibres. (Micrographs Brent Opell)*

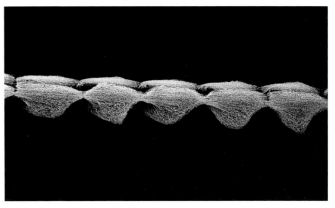

Dictynidae

Background

In the Northern Hemisphere these spiders are abundant and their cribellate webs are easy to find, but in New Zealand webs are either less prominent or absent. The family name comes from *Dictyna*, the northern genus within which the few close relatives from New Zealand were earlier placed. Hence, the first of our species, named by Compte de Dalmas in 1917, was *Dictyna cornigera*, but more recently this has been changed to *Arangina cornigera*. Today, true dictynids are represented in this country by nine species belonging to three genera, *Arangina* as already noted, and the two more colourful genera, *Viridictyna* and *Paradictyna*. A common characteristic of these spiders is that all males share the peculiar bow-legged structure of the chelicerae (Fig. 15.2). We suspect these unusual chelicerae have the same function as their overseas relatives whereby the male uses them to clasp and disable the female while he inserts his palps during mating.

Riverbed Arangina

Although most of the northern hemisphere species construct ladderlike snares which are often conspicuously placed on tufts of grass or the tips of low shrubs, the only New Zealand dictynids which spin webs are the riverbed *Arangina* species (Fig. 15.3) and these webs are sparse by comparison. All *Arangina* species are small spiders (about 4 mm long) and generally greyish in colour. This base colour is speckled with white hairs on the carapace and abdomen with the result that their overall appearance closely matches that of the riverbed stones. Various riverbeds in both North and South Islands host several species which, although quite common, are difficult to find until you become familiar with their webs. The first sign is a number of radiating silk lines bridged irregularly with zigzag threads of typical cribellate silk extending from a retreat usually found in a crevice between two closely spaced stones. Round eggsacs, loosely bound with white silk, are found at the end of a retreat where they are guarded by the mother.

Colourful genera

The two genera *Viridictyna* and *Paradictyna*, both represented by a number of species throughout the country, have similar habits. They are hunting spiders found on the leaves of low-growing shrubs within the forest. Although they are occasionally associated with sheetwebs strung across the surface of leaves it is not certain whether these are moulting retreats or snares. Spiders are active during daylight hours and may capture their prey while they roam. Eggs are contained in a flat silk disc plastered onto the surface of a leaf (Fig. 15.4a) and often protected by a layer of silk above under which the spider rests. While these two groups of spiders are morphologically similar they differ markedly in their pigmentation. *Viridictyna*, as the name suggests, is predominently green

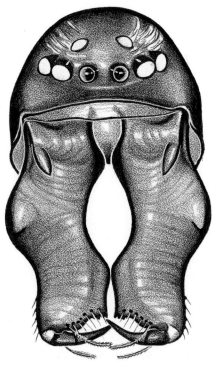

***Fig. 15.2** (top):* A frontal view of the chelicerae of a male Arangina cornigera *(Dictynidae). The hollowed-out central portions of these large chelicerae are used to encircle the body of the female during mating.*

***Fig. 15.3** (second from top):* Arangina pluva *is a riverbed species which builds its meagre web amongst the stones. This female has a mottled grey body sprinkled with white hairs which help to disguise it in this habitat.*

***Fig. 15.4a** (right):* Viridictyna kikkawai *(Dictynidae) is a hunting spider found on low vegetation in the forest. The disc-like eggsac is laid directly onto a leaf and is covered by a thin sheet of silk under which the spider rests.*

(Fig. 15.4b), although there are yellowish patches on some specimens. *Paradictyna* (Fig. 15.5), undoubtedly one of our most handsome spiders, is extremely variable with red, green, yellow and brown pigmentation set against a black background.

Fig. 15.4b *(above left): As its name suggests,* Viridictyna kikkawai *is a predominantly green spider relieved only by pale yellow chevrons down the abdomen. A dictynid, this spider builds a flimsy sheet web across the surface of a leaf and rests beneath it.*

Fig. 15.5 *(above right): Although* Paradictyna rufoflava *is basically a green dictynid spider, it is easily distinguished by the patches of red, yellow and black on its abdomen.*

Agelenidae and Amaurobiidae

Two old spider families

The Agelenidae and Amaurobiidae are two of the oldest named spider families in the world, originally having been based on European spiders described at the beginning of the nineteenth century. Since then numerous species from all parts of the world, presumed to have some affinity with either of the founding genera *Agelena* or *Amaurobius*, have been added to these families. But neither of these genera has any striking characteristics which sets it aside from all other spiders, so these two groups have become convenient 'holding' places. As a result many of our own web-dwelling ground spiders, particularly those found in the forest, have been placed temporarily at least – in either the Agelenidae or Amaurobiidae.

We mention this because, in time, many New Zealand spiders now placed in these families will probably be located within 'future' families which are likely to have their main representation in southern continents. In fact, a number of spiders, including all the amaurobiids that we discussed in our 1973 volume, are now found under new family names such as the Stiphidiidae, Amphinectidae, and Desidae. Until we have a more complete knowledge of the diversity of these southern spiders and the relative importance of their structure and behaviour, these two family names provide a utilitarian home for information as it is slowly accumulated.

Features

In general all these spiders are a dull greyish or brown colour, are of moderate to medium size, have eight eyes in two rows, and possess three claws at the tip of their legs. None of them seems to have developed highly structured webs but when snares are present they consist either of a small sheet or a number of irregular threads spreading out from a retreat. Here the spider remains until an insect comes close enough to be captured. The eggsacs are generally spherical and attached near the retreat or sometimes inside it.

Until recently it was thought that the presence of a cribellum-plus-calamistrum was evidence of a unique evolutionary development which separated these spiders from those which lacked this structural combination. Now we realise that most True Spiders without these structures almost certainly had a cribellate ancestry but lost the specialised silk glands and spigots, as well as the row of bristles, during their evolution. Clearly this process of loss is still part of the modifications slowly being undergone by our spiders because we find within our fauna examples of closely related species-pairs of which one is cribellate and the other is ecribellate. Because *Amaurobius* is cribellate and *Agelena* is not, it was once customary to place only conforming species within each of the two respective families. But with our new understanding of spider evolutionary processes, this distinction is no longer followed and either family may now include both cribellate and ecribellate species.

Agelenidae

House spider

Just over one hundred species of agelenids have been recorded from New Zealand but only one is introduced. This is the ubiquitous house spider *Tegenaria domestica* (Fig. 15.6a) which was originally native to Europe but is now worldwide in its distribution. As its scientific name suggests, it has become so closely associated with human habitation that it is seldom found far from buildings. *Tegenaria domestica* has a light brown carapace and a somewhat paler abdomen with five distinctive brown chevrons on the posterior half of the dorsal surface. The body of an adult spider is only about 12–13 mm in length but the legs are relatively long and noticeably hairy. A further characteristic of *Tegenaria* is that the hind pair of spinnerets are conspicuously longer than the other two pairs and, with the exception of *Aorangia* (see below), this serves to distinguish the house spider from all other agelenids in New Zealand.

***Fig. 15.6a** (left): The only introduced spider belonging to the Agelenidae in New Zealand is* Tegenaria domestica *which is now widely distributed throughout the world. Although this pale brown spider is seldom seen, it can be recognised by the dark and distinctive chevron pattern on the abdomen.*

***Fig. 15.6b** (below): The web of* Tegenaria domestica *is often built across a corner with the retreat tucked into a crevice above it. Its dense silk web attracts dust and grime so that it generally looks dirty. When an insect lands on the web, the spider runs across the top to seize it.*

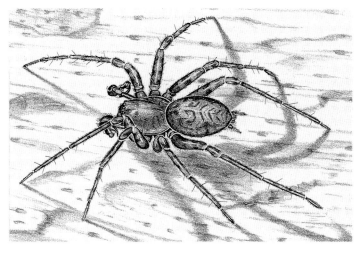

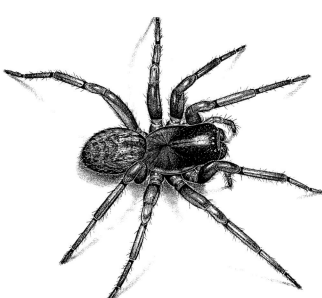

The *Tegenaria domestica* web is a thick horizontal sheet leading to a short tunnel retreat wedged into a corner or crevice but always on the upper surface of the web (Fig. 15.6b). By contrast, similar sheetwebs built by the indigenous *Cambridgea* species (see chapter 12), always have the retreat located on the underside. When an insect alights on its sheetweb, *Tegenaria* dashes out from its retreat, bites it and then drags it back to the retreat where it is eaten. Unlike *Cambridgea*, which walks on the underside of its sheetweb, *Tegenaria* always runs or walks on top of its web. Within or near its retreat, the female spider guards its dingy white eggsac, which is laid in middle or late summer.

The exception we referred to above is *Aorangia* (Fig. 15.7), the only one of our native genera in this family to have its posterior spinnerets so elongated that they extend behind the abdomen as in *Tegenaria*. *Aorangia* is a widespread forest floor genus whose sixteen species are found in both North and South Islands. These active little spiders, about 5–6 mm in length, mottled with black and perhaps a reddish flush on top of the abdomen, lead secluded lives on the forest floor where they build small multiple sheetwebs beneath stones and logs.

Differing lifestyles

The widespread *Orepukia* genus, which contains some 25 ecribellate species, is more typical of most of the agelenids at present recognised in New Zealand. So far, no web of any substance has been associated with any of these spiders and it appears that they are primarily nocturnal, forest-floor, hunting spiders. Their eggsacs are usually attached to the underside of rocks or rotting logs on the forest floor and are guarded for a short period by females. Moreover, the cribellate *Neoramia* (Fig. 15.8), with over twenty known species from bush, subalpine screes and meadows, is found not only over all New Zealand islands but also in the far south on the Subantarctic Campbell and Auckland Islands.

Fig 15.7 (above left): Aorangia ansa *is a native spider noted for its long posterior spinnerets, which extend well beyond the abdomen. It is a forest floor species, building small sheetwebs under logs and stones. Its present placement in Agelenidae is uncertain.*

Fig 15.8 (above): More than twenty Neoramia *species are known from forests, subalpine screes and pastures. These cribellate agelenids construct shapeless webs which extend outwards from retreats under logs or stones. The one shown here is* Neoramia alta.

Unlike *Orepukia*, *Neoramia* constructs a distinctive if somewhat shapeless web, usually extending from a retreat built in a hole or under a log or perhaps beneath a loose stone. The eggs are laid within the retreat and are lightly covered with silk. The closely related *Oramia*, which has similar habits, is restricted to the seashore (see chapter 14).

By contrast, the five species of the smaller and related, but more brightly coloured, genus *Tararua* live on low shrubs or in moss on tree trunks and the forest floor. These spiders, found throughout our forests, represent interesting examples of the change from prey catching lifestyles centred on the use of entangling calamistrated silk to ones based on cursorial hunting. Although undoubtedly closely related to *Neoramia*, only a few species retain functional cribella. Those that have done so do not construct a snare but hunt their prey in much the same way as their ecribellate sister species. This change in behaviour is demonstrated by the fact that, when present, the cribellum is only weakly developed in some and functionless in others, thus paralleling trends in the unrelated agelenid genus *Mahura*.

Amaurobiidae

Family changes

On checking our earlier account of New Zealand amaurobiids we found to our surprise that not one of the eight spiders we discussed then are now considered to belong in this family. Some of these species are currently placed in Desidae and others in Stiphidiidae and Amphinectidae. In the meantime this gap in the amaurobiids has been filled by the discovery of another 34 previously unknown native species. These belong to five genera but we need mention only three here.

Three new genera

The commonest of these new genera is *Pakeha* (Fig. 15.9), which is similar in general appearance and distribution to the agelenid *Orepukia* but from which it can be readily distinguished by the compactness of its eight eyes as compared to the more widely separated eyes in *Orepukia*. Furthermore, the habits of these two groups of spiders are similar and, indeed, they are often found sharing similar habitats on the forest floor. The second genus, *Paravoca* (Fig. 15.10), is a more robust spider with the carapace raised as in Zodariidae, and is found only in Otago and Fiordland in the South Island. These spiders also live on forest floors and are particularly common inside rotting logs where they feed mainly on the larvae of insects.

The third genus, *Otira*, a small amaurobiid which looks like a young *Pakeha*, is also a denizen of forest floors and at first glance has very few eye-catching features. Nevertheless, this spider has an unusual tarsal organ or taste receptor on each leg. This can only be clearly seen when magnified more than 1,000 times by means of SEM (Scanning Electron Microscope) techniques. Usually,

Fig. 15.9 (below left): This amaurobiid spider comes from the Fiordland area and is called Pakeha subtecta. *It is a medium-sized spider with the eight eyes forming a compact group on the front of its head.*

Fig. 15.10 (below right): Paravoca otagoensis *is another species belonging to the Amaurobiidae. It is a robust spider found only on the forest floors in Otago and Southland.*

Fig. 15.11 *(above): This dark-coloured spider,* Megadictyna thileniusi, *relieved only by reddish hues on the legs and carapace, belongs to the Nicodamidae. Although cribellate, it makes very little use of hackled silk; its small sheetwebs are found near the forest floor.*

the tarsal organ consists of a small recessed chamber only visible at the surface of the tarsus. In *Otira*, however, this organ is extended into a long erect rod, quite unlike that seen in spiders anywhere else in the world except for another *Otira* species subsequently found in Tasmania. This finding substantiates yet another link in the relationship of our Australasian fauna.

Nicodamidae

Name changes

In 1906, a German arachnologist, F. Dahl, described a cribellate spider from New Zealand which he believed to be a large dictynid. He named this rather dark-looking spider with its mottled brown abdomen and pale brown carapace streaked with black down the middle, *Megadictyna thileniusi* (Fig. 15.11). Nothing more was heard about this spider for fifty years until Richard Marples found some specimens of it in a collection. He overlooked the obscure, and by then practically forgotten, description by Dahl and renamed the spider *Ihurakius forsteri*. However, its rebirth into scientific literature drew the attention of Pekka Lehtinen, then studying cribellate spiders of the world. After investigating further specimens, Lehtinen concluded that this rather insignificant-looking animal was actually one of the most primitive True Spiders which he felt should be placed in a special family of its own, Megadictynidae. This name has now been replaced by Nicodamidae, but because Dahl first described this spider it retains the name of *Megadictyna thileniusi*.

Megadictyna thileniusi

As often happens with the sudden prominence of a spider, we realised that very little was known about its habits although it is widespread in North and South

Island forests. In spite of the well-developed cribellum, it appears to make little use of cribellate silk – although it does construct small sheetwebs on or near the forest floor. The large eggsac may contain up to a hundred eggs held together with a thin covering of silk suspended near the web, usually under a log (see Fig. 15.11).

Although we came to the conclusion that there were no other cribellate relatives of *Megadictyna* anywhere else, it was noted that there was a very interesting group of Australian ecribellate spiders called *Nicodamus* (Fig. 15.12) which, despite having lost the ability to produce cribellate silk, actually shared many other characters with *Megadictyna*. Eventually it was decided that these two groups of spiders should be put in the same family, but as Simon had already provided the name Nicodamidae for *Nicodamus*, this meant that *Megadictyna* underwent a third change and it too became a member of that family.

Oecobiidae

An introduced spider

The tiny cribellate oecobiids, never more than a few millimetres long, have adapted to living on the walls of houses and some species have now achieved a worldwide distribution. They weave small sheetwebs, only 2–3 cm wide, across cracks and corners and rest flat against the surface beneath them with their legs spread out to one side (Fig. 15.13a). It seems certain that there are no native species and that the one found in the North Island is the typical Northern Hemisphere species, *Oecobius navus* (Fig. 15.13b). This spider is a pale yellowish-brown with several black patches which show up clearly against the paler colouring. However, the feature which makes members of this family so distinctive is the relatively large anal tubercle above the spinnerets. The spider itself is so small you need to look at it under a microscope and when magnified the anal tubercle is seen to be fringed with two conspicuous rows of long hairs.

Fig. 15.12 Nicodamus bicolor *is an Australian relative of* Megadictyna thileniusi *but is ecribellate. Its red cephalothorax and black abdomen make this a striking spider.*

Fig. 15.13a *(above left):* This tiny spider, Oecobius navus, *is just 3–4 mm long. It belongs to the Oecobiidae and is now found worldwide. This female has built its small sheetwebs between the ceiling and a wall, and is resting between them with its legs stretched out.*

Fig. 15.13b *(above):* Oecobius navus *has a very pale, translucent abdomen marked with black patches.*

Amphinectidae

Many species

The Amphinectidae was a family established in 1973 to accommodate about a hundred species of New Zealand spiders, some of which had been previously placed, albeit temporarily, in the Amaurobiidae. There are also some recently described Australian amphinectids in this family. Moreover, it is likely that more will be recorded from distant southern continents as the spider fauna of those regions becomes better known.

We know that there are two genera of large ecribellate amphinectids which wander on the forest floor at night because they are often found in pitfall traps. One of these is *Amphinecta* (Fig. 15.14), which has a body length of 8 to 12 mm while the other is *Mamoea* (Fig. 15.15a), which is only about 5 to 9 mm in size. Both groups are generally dark brown in overall coloration but *Mamoea* tends to have a chevron pattern on the abdomen while *Amphinecta* bears a row of pale patches along the dorsal surface. Like those of many other members of this family, the eggsacs are spherical or purse-shaped and are attached to the undersurface of logs or stones, with the females sheltering nearby.

Fig. 15.14 *(left):* This large ecribellate spider, Amphinecta pika, *has a dark amber carapace and legs with a lighter abdomen marked by pale chevrons. It is a nocturnal hunting spider of the forest floor.* (Amphinectidae)

***Fig. 15.15a** (above):* Mamoea rufa *is a large ecribellate spider with a very distinctive appearance. This female was found hiding under bark (temporarily removed) with its eggsac.*

***Fig. 15.15b** (above right): Pale and semitranslucent after its moult, this male* Mamoea rufa *waits until its new exoskeleton hardens and darkens before moving away. Its cast-off skin can be seen to the left above.*

Mamoea *moulting*

On one occasion we were able to see *Mamoea* moulting. First a few threads of silk were spun underneath a rock outcrop, the spider hanging upside down as it held the web with its claws. It appeared to be resting partly on its side. During the moulting sequence the carapace split around the side, the spider pulled its cephalothorax clear, then the abdomen was freed, and finally the legs were withdrawn from their outer skins (Fig. 15.15b). Whereas *Amphinecta* is distributed over most of the country, *Mamoea* is more common in the South Island, being replaced by the smaller *Paramamoea* (Fig. 15.16a, b) in the north and on Three Kings Islands.

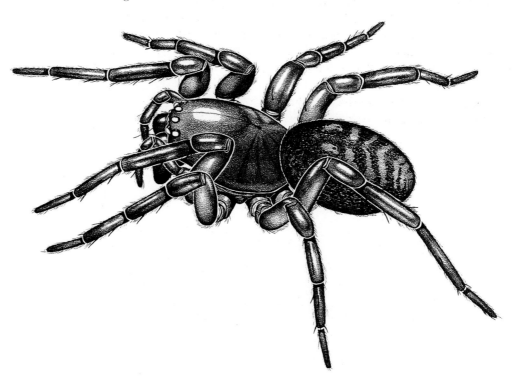

Fig. 15.16 Paramamoea *is a genus of Amphinectidae found only in the North Island and on Three Kings Island. These spiders are smaller than* Mamoea *and live amongst the leaf litter.*

***Fig. 15.16a** (above): A female* Paramamoea parva.

***Fig. 15.16b** (facing page, top): A male* Paramamoea waipoua.

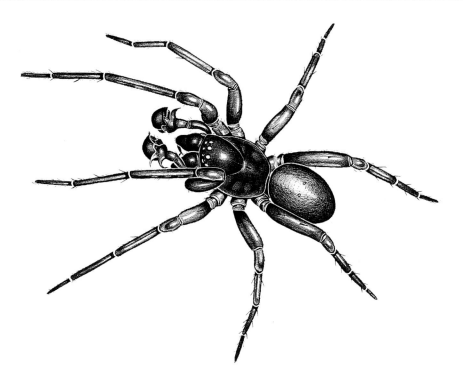

A cribellate genus

The commonest amphinectid spiders in the forest belong to the cribellate genus *Maniho* (Fig. 15.17). These agile spiders have a brown carapace and legs, and dull-greyish abdomens are sometimes relieved by a double row of pale spots. They construct a small sheetweb under loose bark, logs or hollow trees. Nothing is known of their habits other than the kind of web they make; even the shape and structure of the eggsac have not been recorded. Moreover, cribellate silk plays little part in their lives and the cribellum is always small. It is, therefore, not surprising that the genus *Amphinecta* seems similar to its closest relative, *Maniho*, the difference being mainly that the former has gone a stage further in having no trace of the cribellum or the associated calamistrum.

Fig. 15.17 (left): The cribellate Maniho gracilis *spiders are of medium size, up to 10 mm in length, and are commonly found beneath fallen logs or under loose bark. Here they build their small sheetwebs.*

A subterranean spider

Whereas *Maniho* spiders apparently lead simple lives and do not draw much attention to themselves, a group of closely related – but at one time rare – species placed in *Akatorea* puzzled us for many years. This generic name comes from Akatore in Otago, where the first specimen was found in a fissure in a cliff face near the beach. This spider, which had many of the characteristics of *Maniho* and seemed to construct a similar web, lacked the dark pigment of *Maniho*, being a pale straw colour, while the eyes were small and the legs relatively long and slender. Later, a second species was discovered in Fiordland, which lived deep in rotting logs and was only rarely found out in the open. It was evident from the loss of pigment, reduction of eyes and longer-than-normal legs that these spiders normally live somewhere in the dark. Although the Otago species, *Akatorea otagoensis* (Fig. 15.18), was occasionally found in rotting logs like its Fiordland relative, it was surprisingly rare until a sudden emergency with drains on our Dunedin property required an excavation. A metre or so below the surface of our lawn the likely answer to the true home of these spiders was revealed. There, lining the cracks and crevices in the clay subsoil, were webs (and eggsacs) all inhabited by these pale straw-coloured spiders, which proved to be the previously rare *Akatorea otagoensis*. Creatures which live most of the time in such subterranean habitats, but are not confined to them, are said to be trogloxenes.

Fig. 15.18 *The pale straw colour, small eyes and long legs of* Akatorea otagoensis *suggest it is a spider that lives in the dark for part of its life, at least. This was shown to be so when it was found living in a subterranean habitat, a metre below the surface, where its webs and eggsacs lined the cracks and crevices in the clay.*

CHAPTER SIXTEEN
FOUR FAMILIES

Four families – the Zodariidae, Ctenidae, Pschridae and Huttoniidae – do not readily fit within any of our previous groups. Each of these families is poorly represented in New Zealand and of limited distribution. Zodariidae, for example, is represented by a single species whereas Ctenidae has three genera living here, two of which are native but the third may well be introduced although there is little to indicate from which country it came. The single species belonging to one endemic genus of the Pschridae is found only on low-growing vegetation near one of the South Island lakes, while a second genus has been seen only in open country. The family Huttoniidae, on the other hand, has only ever been found in New Zealand, no relatives having yet been found anywhere else in the world. To date, nothing links these four families together or with any other indigenous families.

Zodariidae

The zodariid family group has been used from time to time as a holding place for a wide range of spiders whose placement is uncertain. For example, *Zodarion*, the species on which the family name is based, encompasses a group of Northern Hemisphere spiders which specialise in the capture of ants. These spiders construct a dome retreat on the ground near pathways used by the ants on which they prey.

A single species

In New Zealand, however, the lone species, *Forsterella faceta* (Fig. 16.1), is very different from the typical northern *Zodarion* spiders both in appearance and probably also in habit. It is recorded only from the North Island and would perhaps be more appropriately associated with another family based on the Australian genus *Storena*, which are relatively large spiders, often conspicuously patterned and are generally found on the ground under stones or logs while a few actually live in burrows. Although *Forsterella faceta* has been known for a long time, it has only recently been named by Joqué who undertook a world survey of this group. Those few notes made when the specimens were collected suggest that *Forsterella* spiders lead a vagrant life and do not construct a snare, but they are found so rarely that observations in the wild have not yet been possible.

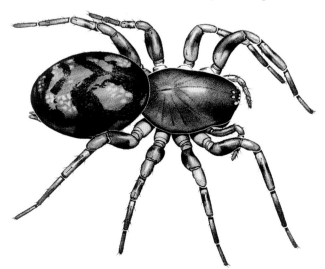

Fig. 16.1 Forsterella faceta, *shown here, is the only species belonging to the Zodariidae found in New Zealand so far. A wandering vagrant spider, it usually lives in the forest under stones or logs.*

***Fig. 16.2** (above): This native ecribellate spider,* Nemoctenus aurens *(Ctenidae), lives amongst rocks in subalpine zones in the South Island.*

***Fig 16.3** (above right): It is very likely that this ctenid spider (*Horioctenoides *sp.) has been introduced into New Zealand from Australia. Mostly confined to the North island, it is usually found in fields and gardens.*

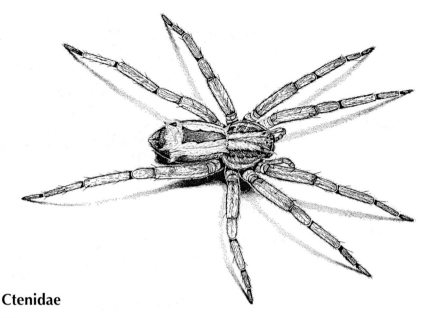

Ctenidae

Three genera

In New Zealand, three genera are placed in the family Ctenidae. All are wandering or free-living spiders which have two claws on the tarsi, are ecribellate and bear no relationship to any other native spider families. *Nemoctenus aurens* (Fig. 16.2) and *Zealoctenus cardronaensis* are both found high on the mountain side. *Nemoctenus* has been found living amongst the rocks in the subalpine zone of the South Island as high as 1,500 metres while *Zealoctenus* lives in the high-country tussock. *Nemoctenus* constructs its flattened eggsac beneath a stone where it lives. While it is certain that these two genera are native, the third genus in this family, *Horioctenoides* (Fig. 16.3), is strikingly different and, in contrast to the high country species, has been found only in suburban habitats around houses and paddocks.

Introduced species

Most sightings of *Horioctenoides* sp. have been in the North Island but some spiders have been seen recently in Marlborough and Nelson. This matches the distribution pattern generally exhibited by introduced spiders and may mean that *Horioctenoides* is slowly spreading to the south. Furthermore, a specimen which seemed identical to those found in this country was intercepted with plants imported from Australia, a circumstance which led to the notion that the species is really an Australian importation. Tentatively, this spider has been put with the Australian genus *Horioctenoides* but because it is slightly different from any of the Australian species it has not yet been given a species name.

Psechridae

Worldwide family

Two New Zealand genera, each with a single species, are tentatively assigned to the worldwide family Psechridae. The two species in question, *Poaka graminicola* (Fig. 16.4a, b) and *Haurokoa filicicola* (Fig. 16.5) are hunters which do not construct snares. One of these, *Haurokoa filicicola*, is cribellate and the posterior eyes are strongly curved. It lives in wet Fiordland forest areas and so far has only been recorded from Lake Hauroko. Little is known of the habits of this spider except that it hunts on the ferns and low-growing shrubs where it is found. It has occasionally been seen sheltering in a thin silken retreat. The second species, *Poaka graminicola*, is very different. It is not a cribellate spider and the eyes are set in two straight rows and not curved as in *Haurokoa*. These spiders are not found in the forest but live in the open on grass and tussock and also on low scrub. Unlike *Haurokoa*, *Poaka* is widely distributed, being found in both North and South Islands.

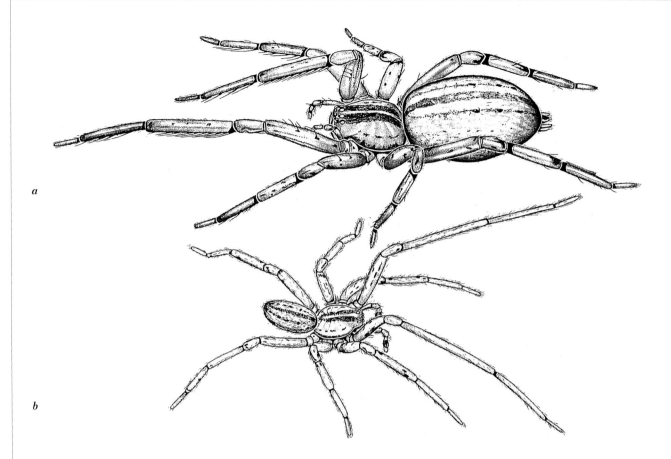

Fig. 16.4 (above): This South Island hunting spider (Poaka graminicola) *is placed in the worldwide family Psechridae and lives in open countryside habitats such as tussock and grasslands.* **a** *Female,* **b** *Male.*

Fig. 16.5 (left): Usually found hunting in amongst ferns and shrubs beside Lakes Hauroko and Te Anau, this little amber-coloured spider is called **Haurokoa** filicicola *after one of its favourite haunts.*

Huttoniidae

Hutton's spider

At present, huttoniid spiders are known only from a single named species, *Huttonia palpimanoides* (Fig. 16.6) which was collected from Dunedin last century by Captain F.W. Hutton who sent it to Pickard-Cambridge for identification. This spider puzzled the well-known English authority who decided it was not previously known. In 1879 he established a new genus named after Hutton and then decided the specific name would be *palpimanoides* which means that he thought the spider was related to the widespread family Palpimanidae. The identity of Hutton's spider remained in limbo for over a hundred years until in 1984 it was decided that a separate family was necessary to classify it adequately. However, it was also concluded that, as Pickard-Cambridge believed, this new family did indeed belong within the palpimanoid group of families.

Fig. 16.6 *The Huttoniidae is named for a single species,* Huttonia palpimanoides, *the only named member of this family to date. Although quite small (4 mm), its reddish colour and darker markings make it quite distinctive.*

Favourite habitat

Huttonia palpimanoides is actually quite a common spider in the bush around Dunedin, if you know where to look. The spider is fairly small, about 4 mm in length, and the carapace and legs are a handsome reddish colour while the lighter-coloured abdomen is also figured with reddish chevrons. We found out by chance that the favourite habitat for these spiders in the Dunedin area is amongst the dead fronds of low-growing ferns where they may be found in loosely spun silken retreats. The spiders are quite active and capture small invertebrates on the move, so have no need for a snare. As yet neither the mating habits of these spiders nor the structure of the eggsac has been observed.

A New Zealand family?

We know that, despite only a single species being named, there are perhaps twenty or so awaiting description from other parts of both the North and South Islands, all of which look superficially like *Huttonia palpimanoides* and are easily recognised as huttoniids. This means that the family is far more widely distributed than at first supposed. However, it seems that Huttoniidae spiders do have a claim to fame because – despite intensive collecting in southern regions such as Australia, South Africa and South America – they have been found only in New Zealand.

CHAPTER SEVENTEEN
HARMFUL SPIDERS

To many people the mere mention of a spider conjures up images of spider bites and their dire effects, thus leading to a conviction that they are animals to fear and shun. No doubt the fact that spiders live by predation and use poison builds up a sense of repugnance but once this is accepted as their way of life, an appreciation of their fascinating behaviours and lifestyles develops.

Almost all spiders use venom to subdue their prey but in only a few is this substance actually harmful to humans. To begin with, many spiders have fangs that are too small or too feeble to penetrate human skin. In other spiders, the poison may be too dilute to be effective on anything other than the spider's prey, or alternatively, its components are not toxic to humans. Others which are physically capable of biting do not do so because the conditions under which they bite – that is, while catching prey – are not present. Of course, some people may be more allergic to a spider's bite than others and may experience a reaction. By chance, some spiders have venom that can be harmful to humans but fortunately relatively few spiders fit into this category. We will look first at one of the spiders whose impact on humans is legendary.

A spider legend

The Tarentella dance

Tales of the effects of spider poison on humans are legion but not all are based on fact. Worldwide, there are probably no more than twenty or so species whose bites cause more than short-term discomfort. The most famous story of so-called spider poisoning is associated with the lively dance of Italy, called the tarentella. In the vicinity of Tarento in Italy during the seventeenth century, a craze which had been known spasmodically for some centuries reached its peak. Anyone believed to have been bitten by *Lycosa tarantula*, a common wolf spider of that area, danced wildly to the accompaniment of music hour after hour until exhausted and in this way the poison was supposed to be worked out from the system. This phenomenon became so widespread at that time, that special music was prepared and during the harvest roving fiddlers would move from field to field waiting to be hired when someone was bitten. Some readers might conclude when viewing present-day entertainment that perhaps a spider bite was not really necessary to produce this craze. Recently, numerous experiments have been carried out to test the toxicity of *Lycosa tarantula*, even to the extent of having them bite volunteers. Their responses ranged from 'less inconvenience than a gnat bite' to 'insignificant' and there is no reason to believe that the toxicity of this spider is any different today than it was some centuries ago.

In general, most wolf spiders are harmless but the bite of a few of them including the notorious *Lycosa erythrognatha* of South America may cause extensive local lesions. None, however, seem to produce the overall effects which might account for the dancing mania. Many theories attempt to explain this extraordinary epidemic – some suggesting it was part of a tourist promotion of those times, another supposing that all the dancers, adherents of the ancient cult of Bacchus, used the spider bite explanation to conceal the real reason for

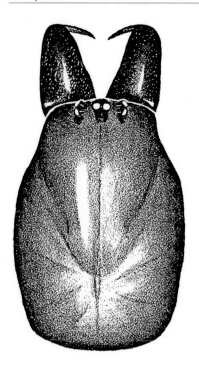

Fig. 17.1 This spider bites when its two diaxial fangs close towards each other.

their behaviour. Probably some cases at that time resulted from spider bites but the symptoms suggest that it was more than likely bites were the work of the local species of *Latrodectus,* spiders related to the New Zealand katipo.

While the wolf spider blamed for the dancing epidemics of Europe is frequently called a tarantula, this name really belongs to the large mygalomorph spiders, the so-called bird-eating spiders, which abound in the tropics. This term is often extended to include the trapdoor spiders and their kin. Despite their large size and somewhat fearsome appearance, only a few of these mygalomorph spiders have been branded as dangerous to humans.

Spider venom

The bite of a spider is inflicted by its two fangs (Fig 17.1) which either pierce downwards (e.g., as in funnelweb spiders), or close towards each other (e.g., as in katipos) on the body of the victim. Venom is squeezed through slits on the side of each fang and enters the site of these wounds. Spider venom is composed of different substances, the greatest proportion of which consists of neurotoxic polypeptides and proteins. In addition to these neurotoxins, a number of enzymes are also present. Although many of these toxins have the same action, each may show a different selectivity. For example, Stefan Schultz points out that the extract a-latrotoxin obtained from *Latrodectus* spiders, has been shown to act exclusively on vertebrates whereas a-latroinsectotoxin affects only insects. What has been found is that a-latrotoxin is a large protein of high molecular weight, whereas the components of most other spider venoms are of small molecular size. For this reason spiders, in general, are harmless to humans, as their toxins have evolved mostly to capture small invertebrates. However, the composition of their venoms is very variable depending largely on the type of prey hunted by particular spider groups, the method of capture and how silk is used for restraint.

Because of their small molecular size and considerable potency, some spider venoms are now seen as having a great deal of potential in the development of insecticides and pharmaceuticals, according to Donald Quicke. Attention, however, is being paid to araneid or orb-weaving spiders whose venom is found to block the transmission of nerve impulses to muscles in insects and crustaceans. One such venom component is named argiotoxin after the genus *Argiope* and other venoms of araneid spiders are named accordingly. These toxins represent the simplest acylpolyamines and arouse interest because of their biological activity as in, for instance, the selective blocking of glutamate receptors. However, although all acylpolyamines possess insecticidal properties because their main role involves the subduing of prey, not all show the same effects.

Funnelweb spiders

Funnelweb spiders belong to the worldwide group known as Mygalomorphae. There are a number of species of funnelweb spiders in eastern Australia, the most well known and dangerous ones being *Atrax* spp. and *Hadronyche* spp. (Fig 17.2), tree-dwelling funnelweb spiders. In the area around Sydney, for instance, *Atrax* may be found in eucalyptus forests or, in urbanised situations, in association with logs or shrubs where they build their nests. Further north, according to Robert Raven, *Hadronyche* is found in moist cool rain forests as well as eucalyptus forests. Both are large, black spiders although the female is paler with a brownish tinge. The head region is shiny and without hairs while the long legs possess a number of prominent spines. In both spiders, the major active component of their venom has been shown by Dieckmann and his colleagues to be atraxotoxin and, although *Atrax robustus* has been responsible for several deaths, there are no records of *Hydronyche* causing injury to humans.

Both males and female funnelwebs are aggressive but males are more likely to be encountered as they wander in search of females. When disturbed, spiders rear up with the front legs and palps lifted high in the air before striking down-

Fig 17.2 The funnelweb spider shown here, Hadronyche infensa, *looks very similar to* Atrax robustus, *which is well known in the Sydney area. (Photograph Robert Raven)*

wards with their fangs so producing two puncture wounds about 3 mm apart. The bite of either of these two species is extremely painful partly because of the large and powerful fangs and the pain may persist for several days. According to Struan Sutherland, the venom of the male is far more toxic than that of the female. The main fraction of the venom is a protein known as atraxotoxin, one of several components which act directly on the nerve fibres causing the widespread release of acetylcholine, adrenaline and noradrenaline. A number of deaths, often occurring within a few hours, have been attributed to these aggressive spiders. Symptoms may include extreme pain, vomiting, sweating, muscular spasms, respiratory difficulties, dryness of the mouth and joint pains. In some cases, however, no symptoms appear and no treatment is required. An antivenom for the funnelweb bite is available from the Commonwealth Serum Laboratories in Melbourne, Australia.

Fortunately, neither *Atrax robustus* nor *Hadronyche infensa* have ever been found in New Zealand although our common tunnelweb spider, *Porrhothele antipodiana*, often alarms Australian visitors because its appearance and habits are so similar to their own species.

Tunnelweb spiders

In New Zealand, several people have received bites from the tunnelweb spiders *Porrhothele antipodiana* and *Hexathele hochstetteri* although the effects have always been slight. Sometimes a bite is just two punctures of the skin, at other times there is a slight swelling or a raised blister or lump. Only one case lists any serious symptoms, which were due perhaps to a secondary infection of the bite.

Spiders that sometimes bite

Orbweb spiders

Most spider bites, other than by the katipo, recorded in New Zealand have occurred when the spiders have become entangled in clothing. In such cases, spiders are usually provoked by being squeezed or are frightened because they cannot escape. Of those cases known, five have been by orbweb spiders and, by coincidence, the identified species were found to have been introduced from Australia where they are widespread. Four bites were by *Eriophora pustulosa* and these resulted in local swelling and the development of a small blister about an hour later. The slight pain in the region of the bite disappeared within a few hours. The culprit in the fifth case was the large *Eriophora transmarina* formerly known as *Aranea transmarina*. The victim was an elderly lady who was bitten on the finger by a female of this species. She suffered considerable pain which spread over the hand and arm and pain remained in the joints for several days.

Among the most potent raw venoms studied in Australia include those of *Nephila edulis* and *Isopeda immanis* – the former a spider which has been seen occasionally in northern areas of New Zealand. Tests by Atkinson and Wright showed that these venoms were capable of causing skin loss and tissue damage in the cultured skin of humans and mice. *Nephila edulis* is a large and spectacular spider which makes huge orbwebs. They are not aggressive, live quietly in their webs, and bites have not been recorded.

Web building spiders

Cambridgea foliata is a large sheetweb spider found in the North Island. Chamberlain (1947) reports on the symptoms of a bite by this spider, as follows:

The punctures were in the left thigh about midway between the hip and knee joints. Only slight pain was felt at the bite and no symptoms appeared for about six hours. At this time an intense itching developed, and a swelling commenced at the site of the punctures. By the end of the third day the swelling covered an area roughly 18 cm in diameter, and was causing considerable pain. Vesication was observed at the site of the punctures consisting of a lymph-filled blister about 12 mm in diameter, and surrounded by a narrow erythematous area. On the fourth day the swelling began to subside. A scar tissue formed under the blister and the upper layer of skin was sloughed. At the end of the week the condition of the patient had returned to normal.

There are numerous species of *Cambridgea*, many with huge fangs, which build their webs in hedges and outbuildings and these are often found in the house. If these spiders are capable of inflicting such drastic bites then it is surprising that so little is heard of them. Our observations suggest that they strike only at insects caught in their webs and none of those we have had in captivity has shown any inclination to bite.

Badumna longinqua is another Australian immigrant which has become widespread throughout the country. It is responsible for most of those webs festooning the outside of houses, sheds, cars and fences, etc. Several bites have been reported, most effects being limited to local pain on being bitten and some swelling and redness at the site. In Australia, symptoms have included sweating, nausea, vomiting, headache and giddiness. No treatment is available and most effects clear up within a few days. Fortunately, despite the prevalence of this spider, bites are unlikely as it stays within its retreat during the day and rarely ventures away from its web. However, *Badumna insignis* (previously known as *Ixeuticus robustus* and commonly called the black window spider) is not so benign. Although bites are rare, reports indicate that the bite may be very painful and causes swelling, discoloration, numbness, nausea and muscular pain. *Badumna insignis* (Fig 17.3) has been recorded in the North Island of New Zealand.

Fig. 17.3 Badumna insignis *is often known as the Black window spider in Australia. Now occasionally found in the North Island of New Zealand, this black hairy spider may be recognised by a faint marking on the abdomen. (Photograph Robert Raven)*

Hunting spiders

Four records implicate *Dysdera crocota* (the slater-eating spider) in bites which resulted in swelling and severe pain. Several bites from this spider have also been recorded from Australia. Although an analysis of the venom revealed two histochemically different substances, these spiders are quite aggressive and have large fangs so perhaps the injuries received are mostly caused by them.

Steatoda is represented in New Zealand by at least four species, only two of which, *Steatoda grossa* and *Steatoda capensis*, are commonly found. Both are introduced and live mainly in and around houses, scrubland and the seashore. Bites from these spiders are not unknown although in most cases they have been mistakenly attributed to the katipo. Katipos are never found inland but people who have received a spider bite often claim that a katipo is responsible when it is, in fact, the 'look-alike' *Steatoda* which does belong to the same family, Theridiidae, as the katipo.

Overseas studies by Maretic *et al* have shown that the bite of the European *Steatoda paykulliana* is toxic to mammals such as guinea pigs. These animals exhibited signs of redness, swelling, and some necroses followed by extreme restlessness and muscle cramps after being bitten. These signs disappeared in about four to six hours. Moreover, it has been shown by Korszniak *et al* that venom gland extracts from *Steatoda capensis* cause effects that suggest the presence of at least two active components. The effects shown included the release of substances from the nerves and a subsequent increase in heart rate and blood pressure.

When venom extracts from two huntsman spiders, *Delena cancerides* (Fig 17.4) and *Isopeda montana* were tested on rats, Korszniak *et al* found they produced pain, an increase in heart rate and blood pressure as well as local inflammatory reactions. Although *Delena cancerides* has been in New Zealand for a number of years, no bites have been recorded. Fortunately, it is not an aggressive spider but the results above suggest that perhaps the elderly and the very young might be affected by a bite.

In the miturgid family, *Cheiracanthium* species (Fig 17.5) have been reported as giving very painful bites but the arrival of one of these in New Zealand has, to our knowledge, resulted in only one person being bitten. A boy aged two-and-a-half was admitted to Christchurch hospital overnight following a bite on the hand by a spider said to be *Cheiracanthium stratioticum*. The boy was said to be playing with spider webbing when the spider rushed out and bit him. Australian records suggest that *Cheiracanthium* is capable of administering a painful bite which increases the heart rate and blood pressure, as well as producing inflammatory reactions at the bite site. Other reports indicate that symptoms include headache, nausea and sometimes even coma.

Fig. 17.4 *(left): The Avondale spider,* Delena cancerides, *is an Australian immigrant which has made its home in the Auckland area.*

Fig. 17.5 *(above): A miturgid spider suspected of giving a painful bite is this* Cheiracanthium *sp. from Australia. (Photograph Robert Raven)*

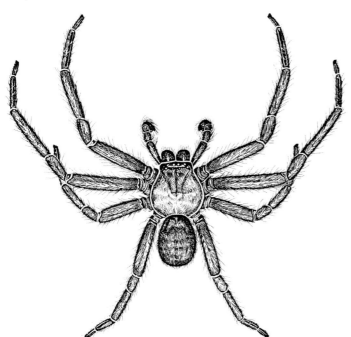

Miturga is a large ground-dwelling spider that was once considered to belong in the Clubionidae but today is in the family Miturgidae. An entomologist collecting arthropods at about 1,500 m up on the Grampian mountains, South Canterbury, was bitten by one when he turned over a stone and it ran onto his hand. While he was trying to capture it, the spider bit him on the top of the second finger. The bite resulted in immediate pain described as being similar to a wasp sting. Two punctures about 2.5 mm apart were clearly visible. The pain persisted and the area around the bite became swollen and hard. Some twenty minutes later, the victim's knee joints began to stiffen and walking was painful. Two aspirins were taken at bedtime and free perspiration occurred during the night.

The brown recluse spider, *Loxosceles reclusa*, is believed to have found its way into Australia. Because of the distinctive pattern on the carapace, this spider is often known in North America as the fiddleback spider. The bite itself is usually painless and generally not noticed, but severe pain, swelling, ulceration and even necrosis occur some hours after envenomation. This spider has not been reported in New Zealand but as it has a liking for warmth and darkness, people's houses provide suitable hiding places.

The Latrodectus group

Latrodectus is a world-wide genus of over thirty species which are implicated in serious cases of poisoning, some even leading to death. Consequently, most of them have their own common, if not entirely popular, names wherever they are found. The best known name is undoubtedly the 'black widow' in North America for the common species of that region *Latrodectus mactans*. The name black widow refers to the generally held, but erroneous belief, that the female invariably feasts upon the male on the completion of his nuptial duties. Another Northern Hemisphere species found in Italy and France, *Latrodectus tredecimguttatus*, is called *malmignatte* (bad gnat) while in Russia the species found there is known as the *karakurt* or 'black wolf'. In South Africa *Latrodectus indistinctus* is known as the 'shoebutton spider' or *knoppiespinnekop* and yet another species from Madagascar is called the *menavodi*. In Australia, *Latrodectus hasselti* is known as the 'redback' while across the Tasman, New Zealand boasts two species, *Latrodectus katipo* (commonly known as the katipo) and *Latrodectus atritus* usually known as the black katipo because of the absence of the red abdominal stripe. So it is obvious that the poisonous propensities of these spiders rarely go unnoticed.

Historical records

The presence of the katipo in this country was recorded as long ago as 1855 by T.S. Ralph. In fact, the spider and its poisonous capabilities were well known to Maori who gave it the name 'katipo' which means 'night stinger', distinguishing it from all other spiders which were grouped together and called 'Punga-were-were'. It is interesting to read in the account by Rev. Chapman, an early missionary, that Maori generally avoided sleeping on the sea beach, but have no fear of the katipo half a stone's throw inland from the beach line, suggesting that Maori also used the name katipo for the black species which lives a bit further away from the beach and that it was not poisonous.

In a land remarkably free from harmful animals, it is not surprising that so much attention has been given to its only native poisonous creature, and we find that some of the earliest scientific papers in New Zealand discussed this spider. For example, one reported on four cases involving bites in the Auckland area in the earliest days of colonisation. Since then, numerous other accounts have appeared in journals both in New Zealand and overseas covering some forty cases. In spite of the often outlandish treatments prescribed in the past, the great majority of these unfortunate victims recovered after a few days although the effects were often severe.

How to recognise the katipo

Although not trying to minimise the possible effects of other bites by spiders in New Zealand, we believe that *Latrodectus katipo* (Fig 17.6) is the only native species to be treated with caution. Although this spider is well known by repute, it is not so well known by sight. Most people are surprised to find that it is quite small. The abdomen of a fully grown female is about the size of a garden pea, round and shiny and usually black with a bright red stripe down the back of the abdomen and two narrow reddish stripes marking the boundaries of the 'hourglass' underneath. This spider is very distinctive and easily recognised by the bright red stripe on top.

Distribution of the katipo

The fact that the katipo is restricted to warm sandy beaches explains to a certain extent why there are so few recorded bites although its retiring nature is another contributing factor. This means, however, that its preference coincides with that of the majority of surfers and bathers, all being present in large numbers during the summer. Once near Christchurch, when collecting these spiders from the base of nearly every tuft of marram grass, a concentration of three to four webs for every square metre, we had to wend our way through sunbathers stretched out on the sand. If this proximity of spiders and people is multiplied along the warmer coastlines and compared with the number of katipo bites, it is clear that the chances of being bitten are very remote.

Only the female can bite

Only when the female becomes adult is it capable of inflicting a bite. It is at this stage when it is most conspicuous with its black body and legs and bright red stripe. Juveniles are more noticeably patterned with black and white patches on the body and the red stripe is less distinctive. The male is much smaller than his mate and has the appearance of a small juvenile.

The 'black katipo'

Most people are surprised to learn that there are actually two species of 'katipo' in New Zealand. In the northern half of the North Island, in sandy dunes further back from the beach, there lives a second species now known as *Latrodectus atritus*. For a long time, it was thought this was a black variant of the katipo because its appearance was identical except for the fact that it did not have a red stripe. Later it was thought to be a species of *Steatoda* (see chapter 12) but more recent studies showed that it was a species of *Latrodectus*. Examination of the reproductive system revealed close similarities to the katipo and other *Latrodectus* species and, moreover, that it was able to mate with the katipo, although the eggsacs produced were infertile. *Latrodectus atritus*, as it is now accepted, is restricted to warmer climates than the katipo, and its habitat, although identical, does not overlap. This means it can be described as an allopatric species to the katipo. No poisonous bites have been recorded from this species.

Fig 17.6 A small, mostly black spider, *Latrodectus katipo*, *is easily recognised by the red, diamond-shaped stripe down its back.*

Latrodectus geometricus (Fig 17.7)

This is one of the most widespread spiders belonging to the *Latrodectus* group, being found in many countries including, more recently, Australia. Far more colourful than most other species with its patterns of orange, pale red, and brown, it is also less likely to be recognised as *Latrodectus*. However, the underside of the abdomen still has a pale reddish hourglass mark and this is the most readily identifiable feature. Although studies by McCrone showed that its venom is one of the most lethal of the *Latrodectus* group, the risk of being bitten is minimal as this spider is of a retiring nature and not in any way aggressive.

The Australian redback spider

Few spiders have been as newsworthy as the redback when it was first discovered in New Zealand. At the same time as a mature female was located near Wanaka in the South Island, others were being intercepted by quarantine officers at various ports around the country. The redback spider, *Latrodectus hasselti*, is closely related to the katipo but since it is larger and not confined to the coastline, potentially there is a much greater chance of it coming in contact with humans. It has probably become established in a few places in this country, but because of our somewhat cooler and damper climate, it is unlikely ever to be present in the same numbers as in Australia.

In general appearance the redback is very similar to the katipo. Differences are that its coat is more velvety black in contrast to the shiny black covering of the katipo while the dorsal abdominal stripe is wider and of a much deeper red.

Fig. 17.7 Latrodectus geometricus, *better known as the brown widow, is paler with more red coloration than other 'widow' spiders. The knobbly eggsacs are characteristic of this species. (Photograph Robert Raven)*

If you care to look on the underside of the abdomen you will find that the redback also possesses a very distinctive red hourglass shape, much more conspicuous than the thin reddish stripes of the katipo. This hourglass is a characteristic feature of all *Latrodectus* spiders although its size, shape, hue and prominence varies greatly from species to species. Because the redback favours dry, sheltered places to live and breed, it has readily adapted to human outbuildings, its first preference in early pioneering days in Australia being the outdoor privy. It is not hard to appreciate, either, that such a habitat provided an abundance of suitable prey.

The effects of katipo and redback poisoning

Because the katipo and redback belong to the genus *Latrodectus*, the effects of being bitten by either are much the same. However, because the redback is larger, it is likely that the quantity of venom injected is greater. Antivenom was developed in Australia in the mid-fifties and has proved effective against bites by both species. This antivenom is available at most major hospitals and as it works for 4–8 hours after a bite is received, there is no need to fear if one is bitten.

Bites from either species may occur when a spider gets caught up in clothing or if one is handling an object under which the spider is hiding. Occasionally, the spider bites spontaneously as in the case of one woman who reported being bitten by a redback as she watched it walk over her foot. Another person was bitten by a katipo as she lay on the sand reading. Only later, when symptoms arose, did she find two small punctures on her leg. Originally, in the days of outside privies in Australia, men were most at risk of being bitten. The poison is a neurotoxin and particularly affects the nervous system and, because of its subsequent effects throughout the body, remedies of the past have rarely proved of much use.

Latrodectism

In humans this is a syndrome related to bites from *Latrodectus* spiders and there are a number of characteristic symptoms. The bite itself is usually very slight and may not even be felt, and it may take from ten minutes to an hour for symptoms to arise. First, a reddish area or a lump at the site may be observed and red streaks may spread out from there. Pain may occur in the abdomen, waist, thighs and spread to other areas, becoming more intense. Other symptoms may include: cold perspiration, an increase in blood pressure, a mild rise in body temperature, nausea and anorexia, vomiting, slight twitching or spasms of the muscles of the extremities, malaise, priapism, urinary retention, constipation, speech defect, insomnia, local oedema, annesia, difficult breathing, restlessness, cyanosis, depression, vertigo, chills, prostration or shock, paralysis, a macular skin eruption, jaundice, convulsions, tremors, muscle twitching, anxiety, and an increase in the polymorphonuclear leucocytes and delirium.

Treatment

In the event of a bite, do not apply a tourniquet. The patient should rest and a cold pack could be placed over the bite – this may delay the onset of other symptoms. Many patients recover without further treatment. If, however, more severe symptoms develop, the patient should seek medical help. An antivenom is available in Australia and New Zealand and instructions for its use are included. Once administered, symptoms usually disappear within 24 hours. Young children and the elderly are more at risk and if bitten should be seen by a doctor as soon as possible. No deaths have been recorded since antivenom became available.

The white-tailed spider

This spider, *Lampona cylindrata*, known since its arrival in New Zealand more than a hundred years ago has been the subject of much media speculation and bad publicity in recent years. A survey of this group by Platnick (in press) showed,

Fig. 17.8 A juvenile Lampona cylindrata *is distinguished from the adult by the more conspicuous white patches on its back.*

however, that there is also a second species, *Lampona murina*, in this country. Both have come from Australia and it is likely that this second one has also been here for a long time since differences between them are not readily detected.

The adult *Lampona*, some 12–15 mm in length, is easily recognised by its dark metallic grey or black colour marked with a conspicuous white patch at the rear of the abdomen. In juvenile spiders (Fig 17.8), there are usually two other rows of white spots on the abdomen and sometimes these have not completely disappeared in the adult. Better known for its nocturnal habits, this spider is sometimes seen inside houses in the daytime although it hunts mostly at night. One of its favourite meals is *Badumna longinqua*, a web-building spider found in silk-lined cracks and crevices on the outside of houses and elsewhere. Although it has long been known that *Lampona* feeds on *Badumna*, it has been reported that this spider will actually enter the *Badumna* web in order to obtain its prey.

Lampona has been blamed for large necrotic lesions supposedly following its bite, although the evidence for this is largely circumstantial. However, in cases where the spider was 'caught in the act of biting' there is little doubt about the symptoms that arise. Pain at the site occurs almost at once, followed by pain and soreness in areas beyond the bite. Swelling and tenderness may occur and the patient feels distinctly unwell. A day or so later a blister may form at the bite site with a small area of necrosis developing in the immediate vicinity. In no instance where the spider has been seen to bite has a large necrotic lesion occurred. Generally, any minor lesion heals by itself a week or so later.

Following sensational stories about lesions in Australia with victims blaming *Lampona*, studies have been undertaken to test the lethality of the venom. For example, Atkinson and Wright showed that *Lampona* had a venom of low toxicity and that it could not cause the massive lesions described. Two alternative suggestions arose. One, that the lesions were the result of *Mycobacterium ulcerans*, a soil-dwelling microbe that was picked up by the fangs of *Lampona* in the course of its wanderings. When the spider bit someone, that person became infected with this microbe, one which has already been shown to cause necrotic lesions. Sutherland, who took a sample from a bite victim was able to isolate *Mycobacterium ulcerans* from the wound although he could not prove it came from *Lampona*. Clearly, any insect or spider could be the vector or a victim might pick up an infection from another source. The second suggestion was that when the spider injects venom, whether into prey or a human victim, it is likely to exude digestive juices from its mouth at the same time. These digestive juices, designed to liquefy the tissues of the spider's prey, could be responsible for destroying human tissues also. These theories are being tested in Australia.

Treatment

No antidote is yet available for a *Lampona* bite and if necrosis does not occur, then keeping the wound clean is the most important requirement. If necrosis occurs, the Commonwealth Serum Laboratories in Melbourne recommend testing for acid-fast baccilli which would show whether *Mycobactrium ulcerans* was likely to be present. Medical treatment is required.

Most spiders are not harmful

There are about two and a half thousand different kinds of spiders in New Zealand. The message is that most spiders are harmless to humans. Those that bite and cause harm only do so because they are frightened, perhaps because you are, too. They are small fragile animals and may feel trapped if handled by large 'creatures'. Walking on an alien skin surface, surrounded by strange hairs and strange smells may seem a very hostile environment to them. Treat them with care. If you are studying them, move them gently from one container to another with a small paint brush (see chapter 18), not your fingers, and you will come to appreciate their remarkable skills and diverse lifestyles.

CHAPTER EIGHTEEN

HOW TO FIND AND STUDY SPIDERS

Spiders are found wherever there is food for them to eat, but in New Zealand many of them are active at night so are not often seen. In this chapter we will tell you where they live and how you can find and study them. Of course, you will soon want to know precisely what animals you have found, and a good way of doing this is to make a collection of preserved spiders which you can look at under a hand lens or a microscope. Remember though that you will need only one adult male and female so try not to preserve more than this. Chapter 1 explains how to tell males and females apart. As you get to know them, you will find that most species can be recognised in life if you look carefully – by the way they move, the webs they make, their colour patterns or reactions when disturbed.

Watching spiders

While it is most rewarding to watch these animals in their natural habitats, this cannot always be done. Ways must be devised to keep them alive and happy in captivity so that you can see and note the important features of their life histories and behaviour. You might find an unknown eggsac and wish to rear the spiderlings, or perhaps you want to observe the mating behaviour and eggsac construction of a particular species. It is often easier to carry out observations at home particularly as much of the activity may be at night. However, if these observations are being made at school, students can take it in turns to take the cage or containers home and make notes on any activity occurring in the evening or early morning. Release your spiders in a suitable habitat once you have made your discoveries. Then, armed with your new knowledge, you will be able to pick the right time to check your findings in a natural environment.

Building up and maintaining a collection

Preserving spiders

Unlike insects, spiders cannot be pinned out but should be bottled in 75 per cent or 80 per cent ethyl alcohol to preserve the soft tissues of the body. 'Rubbing' alcohol (isopropyl) can be used as a substitute and is available from a chemist. Other substances, such as antifreeze and methylated spirits, can also be used. A few drops of acetic acid may be added to help preserve the colours. Never use formalin because this makes the specimens brittle and they may lose some of their more delicate structures. Only specimens of a single species which have been collected at the same time should be kept in the same vial.

Recording

Label and store each container carefully with appropriate information as shown (Fig. 18.1a, b). Basic information about the specimen, plus scientific name if known, should be written in pencil, or waterproof ink, on a label included in the vial. Field notes (corresponding to a number on the label) on observations about the specimen in the wild can be written in a notebook. Take good care of your notes otherwise your specimen has little value.

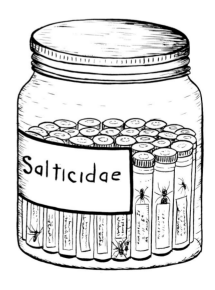

Fig. 18.1a *(below): Labelling your specimens is one of the most important tasks. Use firm paper and waterproof ink or pencil.*
Fig. 18.1b *(bottom): One way of storing your specimens is shown here.*

TRITE AURICOMA (URQUHART).
FEMALE WITH EGGSAC.
N.Z. DUNEDIN, LEITH SADDLE,
UNDER LOG IN FOREST.
20TH DECEMBER, 1999.
COLL. R. SCOTT.
DET. R.R. FORSTER.
FIELD NOTES NO. 648.

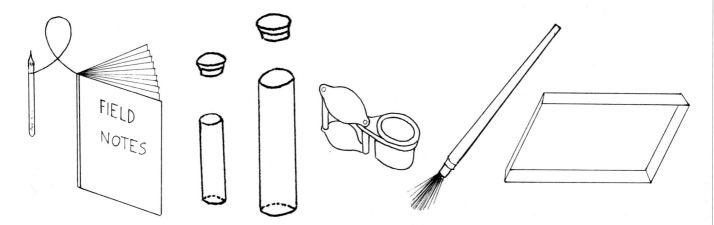

Fig. 18.2 *Before you set out on your field trip, you will need some equipment. The most important items are a notebook and pencil, a selection of vials, a hand lens, a small paint brush and a tray.*

Under a microscope

To examine preserved spiders under a microscope put them in a small glass dish in which they can be submerged in alcohol. It is necessary to get the spider in the right position to see what you want and one way of doing this is to smear the base of your glass dish with a blob of Vaseline before pouring in the alcohol. Press the specimen into this blob lightly but be careful not to smear it with Vaseline. Another way is to hold the specimen in place with blue tack or tiny glass beads or clean sand.

Collecting methods

To find out about the fauna of a particular area you can use several collecting methods which result in a varied collection of creatures to be sorted, identified to a particular group, and watched or preserved.

Collecting requires several skills, such as a sharp eye, patience, and quick reactions. To be successful you will need some basic items of equipment (Fig. 18.2). Look around your home to find used plastic containers and small glass jars with lids. Don't forget that empty food pottles are often just what you need. If there is no lid, you can use plastic film and a rubber band to hold it in place. A small paint brush is useful to coax a spider into your container. To find out what sort of spider you have, it is often necessary to use a hand lens and this can be purchased at stationers and nature shops. However, the most useful hand lens should have a magnification of ten times (x 10).

As you become a keen collector you will begin to add all sorts of things to your collecting gear. Other items you might find useful are a butterfly net, narrow elastic (better than rubber bands), string or twine, a small cutting tool, coloured strips of plastic (to mark spots you intend to return to), a variety of plastic bags and containers, and even a small trowel if you intend to look for trapdoor spiders or set pitfall traps. Spider enthusiasts the world over carry around a wide assortment of homemade devices to catch the spiders they are especially interested in.

If you live near native bush you can collect moss and leaf litter in plastic bags (recycled supermarket bags are great for this). Then when you are back at school or home, you can put a little at a time in a tray and sort through it for spiders and other invertebrates. A white tray or dish with a lamp directed onto it is best, for then you can see things moving more easily. A more elaborate method involves putting the leaf litter through a Berlese funnel, an apparatus that makes use of the fact that spiders dislike light and that they also move from dry to moist areas.

Berlese funnel

This can be made from a new or used plastic bucket (about 20 cm diameter) open at the top and having a plastic funnel glued into the base into which a circular hole has been cut to fit it (Fig. 18.3). Make sure that the hole is slightly smaller than the rim of the funnel to provide firm support for it. A circle of nylon or metal mesh (5–6 mm) is cut to a larger size than the funnel opening and glued across the top. Hang up the bucket with your leaf litter – a little at a time – and place an anglepoise lamp above it. The light, plus the gradual drying out of the litter cause the animals to move down to the mesh barrier. They fall through and then drop down the funnel where they are caught in a container placed below.

Most invertebrates that you find will be quite small and you should first divide them into groups such as spiders, insects, snails, larvae, beetles, millipedes, etc., before attempting to identify them further. It is best to keep them separate to prevent them eating each other.

If you intend to make a long-term study of a particular behaviour, then you might find it necessary to set up a special container. You can buy a variety of plastic wares from specialist shops at quite reasonable prices. Clear plastic is usually best but make sure the plastic is flexible enough for you to cut entry holes for food and water. These can be plugged with small corks or cottonwool.

Shaking

For this method hold your tray under a branch or shrub or clump of grass and shake the foliage vigorously. Spiders (and other small creatures) resting on stems or leaves will drop onto the tray and can then be captured. Use a damp paint brush to transfer small spiders to a container. If time is short, the whole collection, debris and all, can be shaken into a plastic bag and brought home to check at leisure. Spiders rarely attack each other amongst the debris but be sure the bag is not left in the sunlight or everything will be cooked.

Sweeping

This is a similar method often used to catch grass and tussock dwellers. Here you will need your butterfly net which is swept through grass and low foliage, catching animals as they get knocked from their perch. You can either empty the contents into a tray before scooping up your catch into vials or wait until spiders crawl up the sides when the net is held with the bag downwards. Your small paint brush is helpful in encouraging spiders into vials.

Searching

Shaking and sweeping will catch most spiders living above the ground, but some live in empty insect burrows in the twigs and branches of trees and shrubs. Fortunately, all spiders use silk in some way, even if only as a dragline. So you only need to find a hole where there is silk and then open it carefully. In addition, just turning over stones and logs and looking under leaves carefully often leads to success.

A large number of forest species live on the forest floor among the leaf litter and moss. To capture these, two main methods are used.

Sieving

This simply requires shaking the animals contained in a sample of leaf litter through a sieve (5–6 mm mesh) onto a white tray. A lot of fine debris will come through the sieve as well as spiders and other small creatures. At first, all tend to lie quite motionless, making them practically impossible to see. It is best to bring the litter sample home to sieve and put the sievings under a strong light. Soon everything 'comes to life' and will be scurrying about to find a dark spot to hide.

Pitfall trapping

In open country, particularly grassland, it is difficult to find spiders which do not make webs and it is then that pitfall traps are most effective. Several plastic pottles (with holes punched in the bottom to drain away water) can be buried in the ground with the opening flush with the surface and left for a few days. As

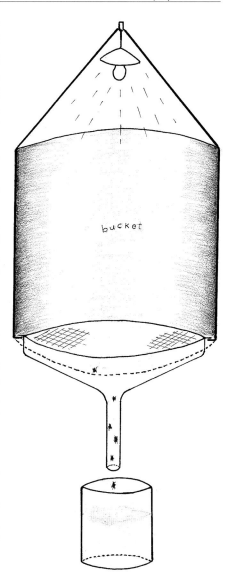

Fig. 18.3 Diagram of a Berlese funnel (not to scale). A bucket (cut-away in this illustration) is hung from a hook, the funnel is fitted into hole in the base, and light shines onto moss and leaf litter causing small creatures to fall into container below. Remove live material from container and house it separately. (Details in text.)

ground spiders wander at night, they drop into the pottle and are not able to crawl out again. If you want to preserve the spiders, omit the holes in the bottom and pour in a small quantity of preservative. Plastic plates may be placed on sticks above the pottles to deflect the rain. Traps should be checked every day for live spiders although they may be left in place longer if they contain preservative.

Daylight collecting

If you hunt for individual spiders during daylight, remember that nocturnal species will be sheltering somewhere in the dark. Look under bark, or logs and stones, or at the base of shrubs. Many of the web builders hide near their webs, either in a special retreat or at the end of a silk thread. One way to bring out web spiders is to touch the web with a vibrating tuning fork. Once the spider is out, block its retreat with a twig, leaf or a paint brush. But remember that the spider may drop to the ground so place your net, or your tray, or a large white plastic sheet below it, to catch the spider.

Night collecting

Perhaps the most exciting and rewarding collecting technique involves the use of a light at night. Then you are most likely to see spiders behaving in a natural manner. The best battery torch is one which fits on your forehead. This has the great advantage of leaving your hands free and not dazzling the wearer. It also throws a beam of light in line with your vision so you can catch the glow back from the reflecting layer in the eyes of many spiders. The spider's eyes show up as minute green lights which can be seen many metres away, long before you can see the spider itself. In addition, webs show up white and are more easily seen than in daylight.

Collecting techniques for particular situations

Forest

Most forest dwellers are nocturnal and the use of a light an hour or two at night will soon give you an idea of the wealth of species which are present, even in the smallest patch of bush. Most spiders become active some time after nightfall – in the summer not until about 9 or 10 pm – and peak activity is reached by midnight. It is during this time that the web builders begin to repair their webs or start new ones, and the hunting spiders actively seek food. The first thing you will notice is the mass of silk gleaming from every shrub and fern, but as many of these threads are draglines left behind during hunting forays or while ballooning, a search for an associated spider will often be in vain. However, a close look at some of the single threads may reveal minute spiders using them as a kind of aerial tripline. The spider may be found clinging to the thread or at the base.

The owners of the larger snares will often be sitting in their webs waiting for victims to become entangled. If you want to catch these spiders, you will need to use a small twig to block their retreat before you try to catch the spider itself. As some spiders drop to the ground when disturbed, place your tray (or sheet) on the ground underneath the spider. Hunting spiders can be seen running about or resting on the trunks of trees, the leaves of shrubs, or the forest floor.

To find all these animals during daylight, shake low shrubs and ferns, turn over logs and stones, sieve the litter and moss from the forest floor or carefully trace the signal thread from the web to the spider's retreat. A pitfall trap sunk in the forest floor provides much information on the movements of ground dwellers, but will also capture the males of many web builders wandering in search of a mate.

Grassland

Sweeping with a net will bring to light many of the spiders which construct their snares among clumps of grass but the only way to be sure of locating some of the vagrant spiders is to set out pitfall traps. As with many hunting spiders, the males of various trapdoor spiders will also fall into these traps. But females remain in their burrows. To catch the females you first need a sharp eye to distinguish the outline of a burrow lid or a small pile of excavated soil. Then it is necessary, carefully, to dig out the spider from its hole.

Riverbeds

You might think that riverbeds, continually subjected to sudden flooding, would be barren. But shingle riverbeds in New Zealand have a varied fauna, apparently derived from the subalpine region but which is peculiar to this habitat alone. Most riverbed spiders hunt at night and may be seen with a light, but the most effective, if strenuous, way of finding them is to turn over the loose stones under which they shelter during the day. Some species will only be found close to the water, but generally they are concentrated on higher levels less prone to flooding. Pitfall traps are effective, too, as long as the river does not flood.

Seashore

A surprising number of species are found only near the sea, and many of these occur in very restricted places. The sandy area with marram grass above high tide level is the home of the katipo and pale wolf spiders. The same region beyond a pebbly beach supports the black jumping spiders and a number of dark-coloured wolf spiders. It is best to look for wolf spiders at night with a light, but a pitfall trap is also effective. Some spiders live only among the piles of stones which accumulate above the high tide level. To find these, clear a 'face' and work slowly into the pile by removing each stone carefully. If there are bare rock faces within the splash zone, various species of the trapdoor spider *Migas*, the tube dwellers *Amaurobioides*, or the six-eyed spider *Ariadna* will be found in their retreats attached to the rock or perhaps built into cracks.

There is only one species which can be truly called a marine spider. This is *Desis marina*, a most conspicuous reddish-coloured spider which lives in the area between high and low tide levels. To find it, look in cracks and crevices, tubeworm cases, in the holdfast bases of bull kelp, or turn over loose stones.

Keeping spiders alive

Housing your pet spiders

Plastic containers make suitable homes for many spiders, particularly web-building ones. A little plasticine or blue tack can be used to hold sturdy twigs in place so that your spider can build a snare. Take careful note of the way in which webs are shaped and attached in the wild and prepare your twigs accordingly. Very large cages are not always satisfactory if you feed spiders regularly with house flies because flies often remain on the sides or top of the container and seldom get caught in webs. Holes can be made in the plastic to insert food and damp cotton wool. If you need to use a cage large enough for web-building, you will need to catch moths, or small wetas which can be placed in the web itself.

Observing spiders

An observation cage (Fig. 18.4) can be useful for watching behaviour such as web-building, courtship or mating. The one shown here has a sliding glass front which can be removed when required. A hole that can be plugged with a cork is useful for introducing food or a mate. But always try to observe spiders in the wild first and then make your cage to suit that situation. When you are supplying the food, spiders do not require as much cage space as needed outside. If your spider is spinning snares and catching food, then you may be sure that the conditions are satisfactory.

It is not a good idea to keep your spider in a cardboard box. Such containers absorb moisture very readily and your pet spider will become dehydrated and

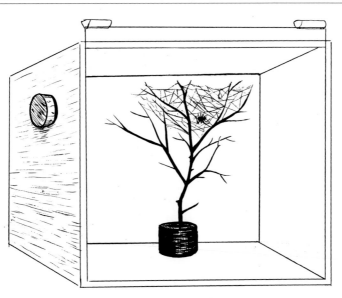

Fig 18.4 *An observation cage. Note that the branches should not touch the sides or the top of the cage, or the spider will be able to climb into a corner and build its web there.*

soon die. Glass jars with plastic lids are best for hunting spiders which find it hard to walk on plastic. Their 'feet' readily grip wood and textured materials but many plastics lack a gripping surface.

With care and attention to food and water, it is quite easy to keep your spider alive for a long time. You will find that a young spider moults every seven to ten days or so during the lighter, warmer months, but that it will be sluggish and eat less during cold weather.

Rearing spiderlings
Spiderlings are harder to rear, the most critical time being the first few days and early weeks of their lives. They should be placed in individual tubes after hatching, with a wisp of damp (not wet) cottonwool, and then fed with a wingless fruitfly the next day. Once they have moulted they are usually able to tackle something larger. If you put a few houseflies or other insects in a plastic bag or pottle in a refrigerator for ten to fifteen minutes, you will find that you can easily grasp one (gently) with padded tweezers and put it in with the spider. Remember, though, that spiders thrive best on a varied diet so you will need to provide supplementary food on a regular basis.

While it is necessary to maintain humidity in containers holding spiderlings or larger spiders, care must be taken to see that moisture does not gather on the inside or spiders will be trapped in the surface film and die. Make a few small holes in the lid or sides so that moisture can escape and fresh air can enter.

If you have completed your observations or for some reason do not wish to keep your spiders any more, release them into the wild in places similar to the ones you captured them in.

How to get an orbweb spider to build an orbweb

If you want to watch an orbweb spider building a web (Fig. 18.5), first you need to find (or make) a Y-shaped branch. The forked sections should be at least 20 cm long, separated at the top by about 25 cm. The handle can be from 10 to 20 cm long. Next find an orbweb spider, preferably a subadult or adult female not only because they are larger but also because they are more likely to build a web. A good choice is *Eriophora pustulosa*, an orbweb spider that is common around houses and gardens and is found all over the country. (For other possibilities, look in chapter 11.)

Turn your Y-shaped frame on its side and place the spider on the upper fork. Wait until it settles down a bit, then tap the fork firmly behind the spider once or twice. The spider will drop on a line so make sure the second fork is below it so that it will land there. Then turn your frame upwards again. Do all this slowly

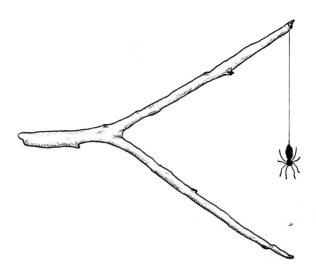

Fig. 18.5 *A Y-shaped branch showing how the spider is gently shaken from the top to the bottom arm of the Y-frame. When this bridge line is complete, turn the Y-frame up again and wedge the base firmly, perhaps with plasticine. Place this in a shallow tray of water which should be large enough so that if the spider drops down from the branch, it touches the water and immediately climbs back up its silk line again.*

and quietly. The spider is most likely to run back across this now horizontal line again.

Place your frame upright in a firm position, perhaps in a weighted pot that has a lid with a hole in the middle in which to wedge the frame handle. Then put this in a tub or a large shallow tray in which water can be held. If the spider drops from the Y-frame and touches the water, it will quickly run back up its dragline again. Make sure that the spider cannot reach a firm surface if it should drop from the Y-frame. With luck, the spider will build an orbweb within the frame but you may have to leave it all in place for several days before it does so. Cover this arrangement with a large fish tank, or equivalent, if it has to be left unsupervised. You may think of some other method but it is wise to prepare all this before releasing your spider onto the frame.

Photography

Photographing spiders can be fun and it is also a useful way to build up permanent records to supplement your notes. The photographs illustrating this book were all taken with a 35 mm single lens reflex camera while most of the black-and-white drawings were based on colour slides projected on to drawing paper to provide the original outline. Some photographers suggest a sojourn in the refrigerator or a mild dose of anaesthetic keeps the animal quiet but a little patience, combined with a few less drastic techniques, will produce satisfactory results.

Sharp focus

To obtain good photographs there must be sufficient depth of sharp focus to show the features you wish to highlight. For this reason, it is best to use electronic flash equipment rather than depend on available light. Be sure the extension cord from the camera to the flash is long enough to use the flash while not attached to the camera. All photographs should be taken with the aperture of the lens fully closed, preferably to f.22, so that the correct exposure is entirely dependent on the distance the flash is held from the subject. However, a ring-flash is a good option too, for it provides all round direct illumination and leaves few shadows. It is good for colour photography because it provides depth.

Close-up shot

This is achieved by moving the lens further away from the film in the camera, either by means of a series of simple tubes or with a bellows attachment. But as a result, less light will reach the film. To compensate for this loss, you must increase the light on the subject and this can be done by moving the flash closer to it or using a second flash.

Flash distance
This is largely a matter of trial and error, so until you have experience it is best to take a series of shots at different distances. The background will make a difference so if possible choose one that does not contrast too greatly with the animal you are photographing.

Autofocus
Most present-day cameras possess this facility but if you wish to record a moving subject, the focus changes are often not fast enough to be useful. It is better to persist with manual focus and adjust other variables accordingly.

Recording
If you wish to record particular features of the spider, it is best to photograph it while sedentary. In this case, you can place your camera on a tripod and put your spider in a situation from which it cannot readily escape.

Behavioural records
These are best achieved by photographing the spider in its natural habitat. First you should watch it carefully so that you know exactly what you want to achieve. You may need extra lights, a pair of scissors to snip off bits of foliage that interfere with the picture, some suitably coloured cloth or cardboard to shield distracting background, some aluminium foil to highlight the spider, an atomiser to spray the web, a small paintbrush to tickle a recalcitrant spider, a couple of extra hands and lots of patience.

Videotaping
A camcorder and VCR add immeasurably to the ease with which you can record behaviour. To begin with, trial and error is far less expensive although your time budget may be more costly. But you should find out all you can about a spider's particular behaviour before commencing to videotape. With a camcorder and close-up lenses you can tape outdoor scenes and supplement them with carefully prepared indoor set-ups. All the best film units do this and, with careful editing, indoor and outdoor taping can be successfully merged. This book has not the space to explore these possibilities. All we can say here is that if your interest is great and your wallet big enough, then this is the direction in which you should go.

APPENDIX I
WORLD LIST OF SPIDER FAMILIES

Actinopodidae
*Agelenidae
*Amaurobiidae
Ammoxenidae
*Amphinectidae
*Anapidae
Antrodiaetidae
*Anyphaenidae
*Araneidae
Archaeidae
Atypidae
Austrochilidae

Barychelidae

Caponiidae
Cithaeronidae
*Clubionidae
*Corinnidae
Cryptothelidae
*Ctenidae
Ctenizidae
*Cyatholipidae
Cybaeidae
*Cycloctenidae
Cyrtaucheniidae

*Deinopidae
*Desidae
*Dictynidae
Diguetidae
*Dipluridae
Drymusidae
*Dysderidae

Eresidae

Filistatidae

Gallieniellidae
*Gnaphosidae
*Gradungulidae

*Hahniidae
Halidae

Hersiliidae
*Hexathelidae
*Holarchaeidae
Homalonychidae
*Huttoniidae
Hypochilidae

*Idiopidae

*Lamponidae
Leptonetidae
*Linyphiidae
Liocranidae
Liphistiidae
*Lycosidae

*Malkaridae
Mecicobothriidae
*Mecysmaucheniidae
*Micropholcommatidae
Microstigmatidae
*Migidae
*Mimetidae
*Miturgidae
*Mysmenidae

*Nanometidae
*Nemesiidae
*Neolanidae
Nesticidae
*Nicodamidae

Ochyroceratidae
*Oecobiidae
*Oonopidae
*Orsolobidae
*Oxyopidae

Palpimanidae
*Pararchaeidae
Paratropididae
*Periegopidae
Philodromidae
*Pholcidae
Pimoidae

*Pisauridae
Plectreuridae
Prodidomidae
*Psechridae

*Salticidae
*Scytodidae
*Segestriidae
Selenopidae
Senoculidae
Sicariidae
*Sparassidae
Stenochilidae
*Stiphidiidae
Symphytognathidae
*Synotaxidae

Telemidae
Tengellidae
Tetrablemmidae
*Tetragnathidae
Theraphosidae
*Theridiidae
*Theridiosomatidae
*Thomisidae
Titanoecidae
Trechaleidae
Trochanteriidae

*Uloboridae

*Zodariidae
Zoridae
Zoropsidae

* A family found in New Zealand

APPENDIX II

HISTORICAL NOTES ON EARLY ARACHNOLOGISTS

Berland, Lucien. Deputy-Director of the Paris Museum. Produced many papers on spiders including some from New Zealand between 1923 and 1947.

Bonnet, Pierre. Pioneered work on the life cycle of spiders. Published the first volume of *Bibliographia Araneorum* in 1945, followed by another six volumes in later years.

Bristowe, W.S. A chemist by training and a former administrator in the Industrial Chemical Industries, Bristowe's contribution to the study and understanding of spiders is widely recognised. He began publishing in 1921 and his interest in spiders was lifelong. *The Comity of Spiders, Parts 1 & 2* were published in 1939 and 1941, while his book *The World of Spiders* was published in 1958.

Bryant, Elizabeth. Born in USA in 1876 and based at MCZ, Harvard. Worked mainly on the spiders of the Americas. Examined Urquhart's collection from Canterbury Museum. Published three papers in their *Records* between 1933 and 1935. Added new records from this museum's collections.

de Dalmas, Comte Raymond. Born in Paris 1862. Wealthy. Sailed his yacht to New Zealand for salmon fishing. Made a large collection of spiders for Eugène Simon. Was persuaded to work them up himself and published paper of 100 pages titled *Araignées de Nouvelle-Zélande*.

Gillies, R. Farmed in the Oamaru area in the 1870s and later became a prominent businessman in Dunedin. Became interested in the habits of trapdoor spiders. His collection is housed in the Hope Museum, Oxford.

Goyen, P. School inspector in Otago. Keen interest in spiders led to six papers in *Transactions of the New Zealand Institute* between 1887 and 1892.

Hickman, V.V. Former Professor of Zoology, at University of Tasmania. Published a large number of papers on Australian spiders from 1926 to 1979.

Hogg, H.R. An Englishman who became interested in spiders while living in Australia and worked on the spiders of both New Zealand and Australia after returning to England in 1892.

Hutton, Captain F.W. In 1873, Hutton was appointed Professor of Natural Science and Curator of the Otago Museum. With wide interests in natural history, he collected spiders to send to Pickard-Cambridge.

Marples, Brian J. Born 1907. Professor of Zoology at Otago University 1937–1967. Published a number of papers on New Zealand's spider fauna. Encouraged students in advanced study. Important paper on cribellate spiders by his son, R.R. Marples.

Millot, J. Best known for developing a microtechnique for the section-cutting of spiders (1926). This led to a great improvement in our knowledge of the internal structure of spiders.

Parrott, A.W. Biologist at Canterbury Museum, then Curator of Insects at Cawthron Institute. Built up a collection of spiders and in 1946 published *A Systematic Catalogue of New Zealand Spiders*.

Pickard-Cambridge, Rev. F.O. Born in England in 1860, resigned his curatorship of St Cuthbert's Church to specialise in scientific drawings. Published widely on spiders, including some from New Zealand, from 1889 to 1914.

Powell, Llewellyn. Medical practitioner in Christchurch. Interest in spiders led to paper on the poisonous katipo in 1871. Produced paper on taxonomy of some jumping spiders and short note on intertidal spider *Desis*.

Simon, Eugène. Born in Paris 1848. Greatest arachnologist of his time. Published the initial part of his *Histoire Naturelle des Araignées* at the age of sixteen. Produced 319 books and papers which established basis for present-day classification. His work covered a wide field (birds, insects, fungi, etc.) throughout the world. He died in 1924. His spider collection is in the Museum of Natural History, Paris.

Urquhart, Arthur Torrane. Born Switzerland, 1839. Migrated with his family to New Zealand in 1856. Settled on a farm in Karaka, south of Auckland. His interest in natural history resulted in eighteen papers on the taxonomy of New Zealand spiders between 1882 and 1897. Aroused the interest of authorities in England and Europe. Part of his collections are placed in Canterbury and Auckland museums.

Wilton, C.L. (Dick). Farmer in Wairarapa, North Island, New Zealand. Became interested in spider fauna. Most of that area had been native forest recently and spider fauna remarkably diverse. After retirement, he began a second career as co-author of the Otago Museum bulletins.

SELECT BIBLIOGRAPHY

This is a list of people whose papers and books have contributed to the knowledge of spiders in New Zealand. Some of them have been mentioned in the text but all have contributed to the information we have brought together here.

Adis Joachim. 1990. Thirty million arthropod species – too many or two few? *Journal of Tropical Ecology*, 6: 115–18.

Aitchison, C.W. 1984. Low temperature feeding by winter-active spiders. *Journal of Arachnology*, 12: 297–306.

Anderson, J.F. 1970. Metabolic rates of spiders. *Comparative Biochemistry and Physiology*, 33: 51–72.

Anderson, J.F., Prestwich, K.N. 1982. Respiratory gas exchange in spiders. *Physiological Zoology*, 55: 72–90.

Andrade, M.C. 1996. Sexual selection for male sacrifice in the Australian redback spider. *Science*, 271: 70–2.

Atkinson, R.K., Wright, L.G. 1991. Studies of the necrotic actions of the venoms of several Australian spiders. *Comparative Biochemistry and Physiology*, 98C 2/3: 441–4.

Atkinson, R.K., Wright, L.G. 1992. The involvement of collagenase in the necrosis induced by the bites of some spiders. *Comparative Biochemistry, Physiology*, C. 102C: 125–8.

Austin, A.D., Anderson, D.T. 1978. Reproduction and development of the spider, *Nephila edulis* (Koch) (Araneidae: Araneae). *Australian Journal of Zoology*, 26: 501–18.

Austin, A.D., Blest, A.D. 1979. The biology of two Australian species of dinopid spider. *Journal of Zoology*, London, 189: 145–56.

Barth, F.G. 1985. (ed.) *Neurobiology of Arachnids*, Springer-Verlag, Berlin, New York, Tokyo.

Berland, L. 1925. Spiders of the Chatham Islands. *Records of the Canterbury Museum*, 2: 295–300.

Berland, L. 1930. Les Araignées des îles avoisinant la Nouvelle Zélande et les relations entre l'Australie et l'Amérique du Sud. *Compte Rendu de la Société de Biogéographie*, 30: 90–4.

Berland, L. 1931. Araignées des îles Auckland et Campbell. *Records of the Canterbury Museum*, 3: 357–65.

Bleckmann, H., Barth, F.G. 1984. Sensory ecology of a semi-aquatic spider (*Dolomedes triton*). II The release of predatory behaviour by water surface waves. *Behavioural Ecology and Sociobiology*, 14: 303–12.

Blest, A.D. 1976. The tracheal arrangement and the classification of linyphiid spiders. *Journal of Zoology*, London, 180: 185–94.

Blest, A.D. 1985. The fine structure of spider photoreceptors in relation to function. In: Barth (ed.) *Neurobiology of Arachnids*, pp. 79–102.

Blest, A.D. 1987. The copulation of a Linyphiid spider, *Baryphyma pratense*. Does a female receive a blood meal from her mate ? *Journal of Zoology*, London, 213(2):

Blest, A.D., Taylor, H.H. 1977. The clypeal glands of *Mynoglenes* and of some other spiders. *Journal of Zoology*, London, 183: 473–93.

Blest, A.D., Pomeroy, G. 1978. The sexual behaviour and genital mechanics of three species of *Mynoglenes* Simon (Araneae: Linyphiidae). *Journal of Zoology*, London, 185: 319–24.

Blest, A.D., Taylor, P.W. 1995. *Cambridgea quadromaculata* n.sp. (Araneae, Stiphidiidae): a large New Zealand spider from wet, shaded habitats. *New Zealand Journal of Zoology*, 22: 351–6.

Blumenthal, H. 1935. Untersuchungen über das 'Tarsalorgan' der Spinne. *Zeitschrift für Morphologie Okologie der Tiere*. 29(5): 667–719.

Bonnet, P. 1945–1961. *Bibliographie Araneorum*. Toulouse. (1) 1945, (2i-v) 1956, 1957, 1958,1959,1961, (3) 1961.

Bristowe, W.S. 1939–41. *The Comity of Spiders* (1), 1939, (2) 1941, London, Ray Society.

Bristowe, W.S. 1958. *The World of Spiders*. Collins, London.

Bryant, E.B. 1933. Notes on the Types of Urquhart's Spiders. *Records of the Canterbury Museum*, 4(1): 1–27.

Bryant, E.B. 1935a. Notes on some of Urquhart's species of spiders. *Records of the Canterbury Museum*, 4(2): 53–70.

Butt, A.G., Taylor, H.H. 1995. Regulatory response of the coxal organs and excretory system to dehydration and feeding in the spider *Porrhothele antipodiana* (Mygalomorphae: Dipluridae). *Journal Experimental Biology*, 198(5): 1137–49.

Bryant, E.B. 1935b. Some new and little known species of New Zealand spiders. *Records of the Canterbury Museum*, 4(2): 71–94.

Carico, J.E., Holt, P.C. 1964. A comparative study of the female copulatory apparatus of certain species in the spider genus *Dolomedes* (Pisauridae: Araneae). *Technical Bulletin* 172, Agricultural Experimental Station, Blacksburg, Virginia (USA).

Catley, K.M. 1993. Courtship, mating and post–oviposition behaviour of *Hypochilus pococki* Platnick (Araneae: Hypochilidae). *Proceedings of the XII International Congress of Arachnology. Memoirs of the Queensland Museum*, 33(2): 469–74.

Chamberlain, G. 1944. Revision of the Araneae of New Zealand, Part 1. *Records of the Auckland Institute Museum*, 3(1): 69–71.

Chamberlain, G. 1946. Revision of the Araneae of New Zealand, Part 2. *Records of the Auckland Institute Museum*, 3(2): 85–97.

Chamberlain, G. 1947. Arachnoidism as applied to New Zealand Spiders. A preliminary note. *Records of the Auckland Institute and Museum*, 3(3): 157–9.

Churchill, T. 1991. Trends in distribution and abundance of spiders in Tasmanian heathland. *Entomology Society, Queensland, News Bulletin*, 19(3): 34–42.

Churchill, T., Raven, R.J. 1992. Systematics of the intertidal trapdoor spider genus Idioctis (Mygalomorphae: Barychelidae) in the Western Pacific with a new genus from the northeast. *Memoirs of the Queensland Museum*, 32: 9–30.

Clausen, I.H.S. 1986. The use of spiders (Araneae) as ecological indicators. *Bulletin of British Arachnological Society*, 7(3): 83–6.

Clyne, Densey. 1967. Notes on the construction of the net and sperm-web of a cribellate spider, *Dinopis subrufus* (Koch) (Araneida: Dinopidae). *The Australian Zoologist*, XIV(2): 189–97.

Clyne, Densey. 1973. Notes on the web of *Poecilopachys australasia* (Griffeth & Pidgeon 1833) (Araneida: Argiopidae). *Australian Entomological Magazine*, 1(3): 23–9.

Coddington, J.A. 1986. The genera of the spider family Theridiosomatidae. *Smithsonian contributions to Zoology*, No. 422: 1–96.

Coddington, J.A. 1989. Spinneret silk spigot morphology: evidence for the monophyly of orbweaving spiders, Cyrtophorinae (Araneae) and the group Theridiidae plus Nesticidae. *Journal of Arachnology*, 17: 71–95.

Coddington, J.A. 1990. Cladistics and spider classification: araneomorph phylogeny and the monophyly of orbweavers (Araneae: Araneaomorphae, Orbiculariae). *Acta Zoologica Fennica*, 190: 75–87.

Coddington, J.A. 1990. Ontogeny and homology in the male palpus of the orbweaving spiders and their relatives, with comments on phylogeny (Araneoclada: Araneoidea: Deinopoidea). *Smithsonian Contributions to Zoology*, 496: 1–52.

Coddington, J.A., Levi, H.W. 1991. Systematics and the evolution of spiders. *Annual Review of Ecology and Systematics*, 22: 565–92.

Colgin, M.A., Lewis, R.V. 1995. Spider Silk: a Biomaterial for the Future. *Journal of Chemistry and Industry*, 24.

Comstock, J.H. 1940. *The Spider Book*. (Revised and edited by W.J. Gertsch) New York.

Cooke, J.A.L., Roth, V.D., Miller, F.H. 1972. The urticating hairs of theraphosid spiders. *American Museum Novitates*, 2498 (1): 1–43

Cooke, J.A.L. 1965. A contribution to the biology of British spiders belonging to the genus *Dysdera*. *Oikos*, 16: 20–5.

Cooke, J.A.L. 1966. Synopsis of the structure and function of the genitalia in *Dysdera crocota* (Araneae, Dysderidae). *Senckenbergiana*, 47(1): 35–43.

Cooke, J.A.L. 1970. Mounting and clearing: notes on some useful arachnological techniques. *Bulletin British Arachnological Society*, 1: 92–5.

Costa, F.G. 1993. Cohabitation and copulation in *Ixeuticus martius* (*Badumna longinqua*). *The Journal of Archnology*, 21(3): 259–60.

Court, D.J. 1974. 'Two-spined' Australian spider, *Poecilopachys australasia* (Griffith & Pidgeon) 1833 in Auckland. *Tane*, 20: 166–8.

Court, D.J. 1982. Spiders from Tawhiti Rahi, Poor Knights Islands, New Zealand. *Journal Royal Society of New Zealand*, 12(4): 359–71.

Crane, Jocelyn 1948–49. *Comparative biology of salticid spiders at Rancho Grande, Venezuela*. Zoologica (Vols 33–35) Pt 1(1–38), Pt 2(139–145, Pt 3(31–51), Pt 4(159–214), Pt 5(253–261).

Dalmas, Comte de. 1917. Araignées de Nouvelle Zélande. *Annales de Société Entomologique de France*, 86: 317–430.

Davies, V. Todd., Raven, R.J. 1981. Spiders and scorpions of medical importance. In: *Animal Toxins and Man*. Ed: John Pearn. Division of Health Education & Information, Queensland Health Department, Brisbane, Australia. pp. 55–62

Dieckmann J., Prebble J., McDonogh A., Sara A., Fisher M. 1989. Efficacy of funnel-web spider antivenom in human envenomation by *Hadronyche* species. *The Medical Journal of Australia*, 151: 706–7.

Duffey, E. 1956. Aerial dispersal in a known spider population. *Journal of Animal Ecology*, 25: 85–111.

Dumpert, K. 1977. Spider odor receptor: Electrophysiological proof. *Experientia*, 34(6): 754–6.

Eberhard, William G. 1985. *Sexual Selection and Animal Genitalia*. Harvard University Press. Cambridge, MA, London, England.

Eberhard, William G. 1990. Function and Phylogeny of spider webs. *Annual Review of Ecological Systematics*, 21: 341–72.

Edwards, E.D., Hurin, A.D. 1995. Annual contribution of terrestrial invertebrates to a New Zealand trout stream. *New Zealand Journal of Marine and Freshwater Research*, 29: 467–77.

Elgar, M.A., Ghaffar, N., Read, A.F. 1990. Sexual dimorphism in leg length among orb-weaving spiders: a possible role for sexual cannibalism. *Journal Zoology*, 222 (Nov. P3): 455–71.

Elgar, M.A. 1992. Sexual cannibalism in spiders and other invertebrates. In: *Cannibalism: ecology and evolution among diverse taxa*, Elgar, M.A., Crespi, B.J. Eds. Oxford University Press. Oxford, New York.

Elgar, M.A. 1995. The duration of copulation in spiders: comparative patterns. *Records Western Australian Museum Supplement*, 52: 1–11.

Elgar, M.A., Bathgate, R. 1995. Female receptivity and mate guarding in the jewel spider *Gasteracantha minas* Thorell (Araneidae). *Journal of Insect Behaviour*, 9(5): 729–38.

Elgar, M.A., Fahey, B.F. 1996. Sexual cannibalism, competition, and size dimorphism in the orb-weaving spider *Nephila plumipes* Latreille (Araneae: Araneoidea). *Behavioural Ecology*, 7(2): 195–8.

Filmer, M.R. 1991. *Southern African Spiders – an Identification Guide*. Struik Publishers.

Foelix, R.F. 1970. Chemosensitive hairs in spiders. *Journal of Morphology*, 132 (313).

Foelix, R.F., Choms, A. 1979. Fine structure of a spider joint receptor and associated synapses. *European Journal of Cell Biology*, 13: 149–159.

Foelix, R.F. 1996. *Biology of Spiders*. Oxford University Press, New York.

Foil, Lane O., Norment, B.R. 1979. Envenomation by *Loxosceles reclusa*. *Journal Medical Entomology*, 16(1):18–25.

Foord, M. 1990. *The New Zealand Descriptive Animal Dictionary*. The Author, Dunedin.

Forster, L.M. 1977a. A qualitative analysis of hunting behaviour in jumping spiders P.C. (Araneae: Salticidae). *New Zealand Journal of Zoology*, 4: 51–62.

Forster, L.M. 1977b. Some factors affecting feeding behaviour in young *Trite auricoma* spiderlings. *New Zealand Journal of Zoology*, 4: 435–43.

Forster, L.M. 1977c. Mating behaviour in *Trite auricoma*, a New Zealand Jumping Spider. *Peckhamia*, 1: 35–6.

Forster, L.M. 1979a. Visual mechanisms of hunting behaviour in *Trite planiceps*, a jumping spider. *New Zealand Journal of Zoology*, 6: 79–93.

Forster, L.M. (and B.J.F. Manly), 1979b. A stochastic model for the behaviour of naive spiderlings (Araneae: Salticidae). *Biometrical Journal*, 21: 115–22.

Forster, L.M. 1982a. Vision and Prey-Catching Strategies in Jumping Spiders. *American Scientist*, 70(2) 165–75.

Forster, L.M. 1982b. Visual Communication in Jumping Spiders (Salticidae). In: *Spider Communication: Mechanisms and Ecological Significance*. Eds: Peter N. Witt and Jerome S. Rovner. Princeton University Press, NJ. pp. 161–212.

Forster, L.M. 1982c. Non-visual prey capture in *Trite planiceps*, a jumping spider (Araneae: Salticidae). *Journal of Arachnology*, 10: 179–83.

Forster, L.M. 1982d. The Australian redback spider – an unwelcome immigrant to New Zealand. *The Weta*, 5(2): 35.

Forster, L.M. 1984. The Australian Redback Spider (*Latrodectus hasselti*): Its Introduction and Potential for Establishment and Distribution in New Zealand. In: *Commerce and the Spread of Pests and Disease Vectors*. Ed: Marshall Laird. Praeger Publishers, New York. pp. 273–89.

Forster, L.M. 1985. Is the redback spider here to stay? *New Zealand Journal of Agriculture*, 150(6): 58–9.

Forster L.M. 1985. Target Discrimination in Jumping Spiders (Araneae: Salticidae). In: *Neurobiology of Arachnids*. Ed: Friedrich

G. Barth. Springer-Verlag. Berlin, Heidelberg, New York, Tokyo. pp. 249–74.

Forster L.M. 1988. Report on the establishment of *Latrodectus hasselti* (redback spider) in Wanaka to the Royal Society of New Zealand, Science House, Wellington, New Zealand.

Forster L.M. 1992a. The stereotyped behaviour of sexual cannibalism in *Latrodectus hasselti* (Araneae: Theridiidae), the Australian redback spider. *Australian Journal of Zoology*, 40: 1–11.

Forster L.M. 1992b. Cannibalism: a growth strategy for *Latrodectus* female spiders (Araneae: Theridiidae). Paper presented at the X111th International Congress of Arachnology, July 1992, Brisbane, Australia.

Forster, L.M. 1992c. The interplay of venom and silk in the predatory behaviour of spiders with some comments on applied arachnology. *Proceedings of the Entomological Society of New Zealand (41st Annual Conference)*, 17–20 May.

Forster L.M. 1995. The behavioural ecology of *Latrodectus hasselti* (Thorell), the Australian redback spider (Araneae: Theridiidae): a review. *Records of the Western Australian Museum Supplement*, 52: 34–24.

Forster, L.M., Kingsford, S. 1983. A preliminary study of development in two *Latrodectus* species (Araneae: Theridiidae). *New Zealand Entomology*, 7: 431–8.

Forster, L.M. & Forster, R.R. 1985. A derivative of the orb-web and its evolutionary significance. *New Zealand Journal of Zoology*, 12: 455–65.

Forster, L.M., Kavale, J. 1989. Effects of food deprivation on *Latrodectus hasselti* Thorell (Araneae: Theridiidae), the Australian redback spider. *New Zealand Journal of Zoology*, 16: 401–08.

Forster, R.R. 1944. The genus *Megalopsis* Roewer in New Zealand with keys to the New Zealand genera of Opiliones. *Records of the Dominion Museum*, 1: 183–92.

Forster, R.R. 1948. The sub-order Cyphophthalmi Simon in New Zealand. *Dominion Museum Records in Entomology*, 1: 1–119.

Forster, R.R. 1952. Supplement to the sub-order Cyphophthalmi. *Dominion Museum Records in Entomology*, 1(9): 180–211.

Forster, R.R. 1954. The New Zealand Harvestmen (Sub-Order Laniatores). *Canterbury Museum Bulletin*, 2: 1–329.

Forster, R.R. 1955. Spiders from the Subantarctic Islands of New Zealand. *Records of the Dominion Museum*, 2 (IV): 167–203.

Forster, R.R. 1955. A New Family of Spiders of the Sub-order Hypochilomorphae. *Paciific Science*, 10(3): 277–85.

Forster, R.R. 1956. Terrestrial environments in New Zealand. *Proceedings of the New Zealand Ecological Society*, 3: 23–4.

Forster, R.R. 1956. New Zealand spiders of the family Oonopidae. *Records of the Canterbury Museum*, VII (II): 89–169.

Forster, R.R. 1959. The spiders of the family Symphytognatidae. *Transactions of the Royal Society of New Zealand*, 86 (3,4): 269–329.

Forster, R.R. 1961. The New Zealand fauna and its origins. *Proceedings of the Royal Society of New Zealand*, 89(1) (9th Science Congress Report).

Forster, R.R. 1962, 1963. A Key to the New Zealand Harvestman. *Tuatara*, 10(3): 129–37, 11(1): 28–40.

Forster, R.R. 1964. The spider family Toxopidae (Araneae). *Annals Natal Museum*, 16: 113–51.

Forster, R.R. 1964. The Araneae and Opiliones of the Sub-Antarctic Islands of New Zealand. *Pacific Islands Monograph*, 7: 58–115.

Forster, R.R. 1965. Harvestmen of the sub-order Laniatores from New Zealand Caves. *Records of the Otago Museum, Zoology* No. 2: 1–18.

Forster, R.R. 1967. *The Spiders of New Zealand, Part I*. Otago Museum Bulletin No.1. Dunedin.

Forster, R.R. 1970. *The Spiders of New Zealand, Part III*. Otago Museum Bulletin No.3. Dunedin.

Forster, R.R. 1975. The Spiders and Harvestman. In: *Biogeography and Ecology in New Zealand*. Ed: G. Kuschel. Dr W. Junk B.V. Publishers, The Hague. pp. 493–505.

Forster, R.R. 1980. Evolution of the tarsal organ, the respiratory system and the female genitalia in spiders. *Proceedings of the 8th International Congress of Arachnology*, (Vienna 1980). pp. 69–84.

Forster, R.R. 1995. The Australasian spider family Periegopidae Simon, 1893 (Araneae: Sicarioidea) *Records of the Western Australian Museum Supplement*, 52: 91–105.

Forster, R.R., Wilton, C.L. 1968. *The Spiders of New Zealand, Part II*. Otago Museum Bulletin No. 2. Dunedin.

Forster, R.R. & Forster, L.M. 1970. *Small Land Animals of New Zealand*. McIndoe, Dunedin, New Zealand.

Forster, R.R. & Forster, L.M. 1973. *New Zealand Spiders: an Introduction*. Collins, Auckland, London.

Forster, R.R., Wilton, C.L. 1973. *The Spiders of New Zealand, Part IV*. Otago Museum Bulletin No.4. Dunedin.

Forster, R.R. & Forster, L.M. 1976. *The Small World: the community life of small land animals*. A Bulletin for Schools. School Publications Branch, Department of Education, Wellington.

Forster, R.R., Platnick, N.I. 1977. A review of the spider family Symphytognathidae (Arachnida, Araneae). *American Museum Novitates*, 2619: 1–29.

Forster, R.R., Blest, A.D. 1979. *The Spiders of New Zealand, Part V*. Otago Museum Bulletin No. 5. Dunedin.

Forster, R.R., Gray, M.R. 1979. *Progradungula*, a new cribellate genus of the spider family Gradungulidae. *Australian Journal of Zoology*, 27: 1051–71.

Forster, R.R., Platnick, N.I. 1984. A review of the Archaeid spiders and their relatives, with notes on the limits of the superfamily Palpimanoidea (Arachnida, Araneae). *Bulletin of the American Museum of Natural History*, 178 (1): 1–106.

Forster, R.R., Platnick, N.I. 1985. A review of the Austral spider family Orsololobidae (Arachnida, Araneae), with notes on the superfamily Dysderoidea. *Bulletin of the American Museum of Natural History*, 181 (1): 1–229.

Forster, R.R., Platnick N.I., Gray, M.R. 1987. A review of the spider superfamilies Hypochiloidea and Austrochiloidea (Araneae, Araneomorphae). *Bulletin of the American Museum of Natural History*, 185 (1): 1–116.

Forster, R.R. Millidge, A.F., Court, D.J. 1988. *The Spiders of New Zealand, Part VI*. Otago Museum Bulletin No. 6. Dunedin.

Forster, R.R., Platnick, N.I, Coddington. J. 1990. A proposal and review of the spider family Synotaxidae (Aranea, Araneoidea), with notes on theridiid interrelationships. *Bulletin of the American Museum of Natural History*, 193: 1–116.

Forster, R.R, *et al.* (in prep.) *The Spiders of New Zealand, Part VII*. Otago Museum Bulletin No. 7. Dunedin.

Franklin, F.E., Field, L.F. 1985. Effect of venom from *Latrodectus katipo* and *Ixeuticus martius* (Arachnida: Araneae) on insect neuromuscular transmission. *New Zealand Journal of Zoology*, 12: 175–80.

Friedrich, Victor L., Langer, Rudolph M. 1969. Fine structure of cribellate silk. *American Zoologist*, 9: 91–6.

Gatenby, J.B. 1912. Nest, life history and habits of *Migas distinctus*, a New Zealand trapdoor spider. *Transactions of the New Zealand Institute*, 44: 234–40.

Gertsch, W.J. 1958. The spider family Hypochilidae. *American Museum Novitates*, 1912: 1–28.

Gertsch, W.J. 1964. A review of the genus *Hypochilus* and a description of a new species from Colorado (Araneae: Hypochilidae), 2203: 1–14.

Gertsch, W.J. 1973. The cavernicolous fauna of Hawaiian lava tubes, 3. Araneae (spiders). *Pacific Insects*, 15(1): 163–80.

Gertsch, W.J. 1979. *American Spiders* (2nd edn), Van Nostrand Reinhold Company. New York. London. Toronto. Melbourne. p. 274.

Gibbs, D.R. 1980. Predation of the spider, *Araneus pustulosa*, by the German wasp, *Vespula germanica*. *The Weta*, 3(2): 13–14.

Gibbs, D.R. 1982. Australian spider in New Zealand: first record of *Araneus heroine* (Koch 1871). *The Weta*, 5: 1–5

Gilbert, C., Rayor, L.S. 1985. Predatory behaviour of spitting spiders (Araneae: Scytodidae) and evolution of prey wrapping. *Journal of Arachnology*, 13: 231–41.

Gillies, R. 1876. Habits of the trapdoor spider. *Transactions of the New Zealand Institute*, 8: 222–9.

Goyen, P. 1888. Descriptions of new spiders. *Transactions of the New Zealand Institute*, 20: 201–12.

Goyen, P. 1891. On New Zealand Araneae. *Transactions of the New Zealand Institute*, 24: 253–7.

Gray, M.R. 1983. The male of *Progradungula carraiensis* Forster and Gray (Araneae: Gradungulidae) with observations on the web and prey capture. *Proceedings of the Linnean Society, New South Wales*, 107: 51–8

Gray, M.R. 1983. The taxonomy of the semi-communal spiders commonly referred to as the species *Ixeuticus candidus* (L. Koch) with notes on the genera *Phryganoporus, Ixeuticus* and *Badumna* (Araneae, Amaurobioidea). *Proceedings of the Linnean Society, New South Wales*, 106(3): 247–61.

Green, C.J. 1989. The golden orbweaver, a spider creating interest in New Zealand. *The Weta*, 12(2): 51–2.

Griswold, Charles. 1987. The African members of the trapdoor family Migidae (Araneae: Mygalomorphae) 2: the genus *Poecilomigas* Simon, 1903. *Annals Natal Museum*, 28(2): 475–97.

Griswold, Charles. 1987. A review of the southern African spiders of the family Cyatholipidae Simon, 1894 (Araneae: Araneomorphae). *Annals Natal Museum*, 28(2): 499–542.

Gruber, Jurgen. 1990. Fatherless spiders. *Newsletter British Arachnological Society*, 58: 3.

Gundermann, J., Horel, A., Roland, C. 1991. Mother-offspring food transfer in *Coelotes terrestris* (Araneae: Agelenidae) *Journal of Arachnology*, 19: 97–101.

Hanemann, W. Michael. 1988. Economics and the preservation of Biodiversity. In: *Biodiversity*, Ed: E.O.Wilson. National Academy of Sciences Press. pp. 193–9.

Hann, S.W. 1990a. A new combination involving *Lithyphantes lepidus* Cambridge 1879 and a new name for *Teutana lepida* Cambridge 1903 (Araneae, Theridiidae). *New Zealand Journal of Zoology*, 17: 283.

Hann, S.W. 1990b. Evidence for the displacement of an endemic New Zealand spider, *Latrodectus katipo* Powell, by the South African species *Steatoda capensis* Hann (Araneae, Theridiidae). *New Zealand Journal of Zoology*, 17: 295–307.

Hann, S.W. 1994. Descriptions of four *Steatoda* species (Araneae, Theridiidae) found in New Zealand. *New Zealand Journal of Zoology*, 21: 225–38.

Harris, Wayne F. 1986. *The breeding ecology of the South Island Fernbird in Otago wetlands*. A thesis submitted for the degree of Doctor of Philosophy at the University of Otago, Dunedin, New Zealand.

Harvey, M.S. 1995. The systematics of the spider family Nicodamidae (Araneae: Amaurobioidea). *Invertebrate Taxonomy*, 9: 279–386.

Hector, J. 1878. Note on a marine spider. *Transactions of the New Zealand Institute*, 10: 300.

Hickman, V.V. 1931. A new family of spiders. *Proceedings Zoological Society, London*, 4: 1321–8.

Hickman, V.V. 1940. The Toxopidae, a new family of spiders. *Papers, Proceedings Royal Society Tasmania*, 1939: 125–30.

Hickman, V.V. 1949. Tasmanian littoral spiders with notes on their respiratory systems. *Papers, Proceedings Royal Society Tasmania*, 1948: 31–43.

Hickman, V.V. 1951. The identity of spiders belonging to the genus *Amaurobioides* Cambridge. *Papers, Proceedings Royal Society, Tasmania*, 1950: 1–2.

Hickman, V.V. 1951. New *Phoroncidia* and the affinities of the New Zealand spider *Atkinsonia nana* Cambridge. *Papers, Proceedings Royal Society, Tasmania*, 1950: 3–25.

Hickman, V.V. 1967. *Some Common Spiders of Tasmania*. Tasmanian Museum and Art Gallery Publication, Hobart.

Hogg, H.R. 1909. Spiders and Opiliones from the Subantarctic Islands of New Zealand. *The Subantarctic Islands of New Zealand*, Vol. 1, Wellington. 1909: 155–181.

Hogg, H.R. 1911. On some New Zealand spiders. *Proceedings Zoological Society, London*, 1911: 297–313.

Homann, H. 1971. Die Augen der Araneae: Anatomie, ontogenie und Bedeutung, für die Systematik (Chelicerata, Arachnida). *Zeitschrift für Morphologie der Tiere*, 69: 201–72.

Hutton, F.W. 1904. *Index Faunae Novae Zealandiae*. Philosophical Institute of Canterbury, New Zealand.

Jackson, R.R., Harding, D.P. 1982. Intraspecific interactions of *Holoplatys* sp. indet., a New Zealand jumping spider (Araneae: Salticidae). *New Zealand Journal of Zoology*, 9: 487–510.

Jackson, R.R., Pollard, S.D., MacNab, A.M., Cooper, K.J. 1990. The complex communicatory behaviour of *Marpissa marina*, a New Zealand jumping spider (Araneae: Salticidae). *New Zealand Journal of Zoology*, 17: 25–38.

Jocqué, R. 1991. A generic revision of the spider family Zodariidae (Araneae). *Bulletin American Museum Natural History*, 201: 1–160.

Jones, Dick. 1984. *Guide to Spiders of Britain and Northern Europe*. The Country Life Books. Hamlyn Publishing Group Limited, Middlesex.

Kaston, B.J. 1935. The slit sense organ of spiders. *Journal of Morphology*, 58(1): 189–207.

Kaston, B.J. 1965. Some little known aspects of spider behaviour. *American Midland Naturalist*, 30(3): 336–56.

Koch, L. 1871. *Die Arachniden Australiens, nach der Natur beschrieben Australiens und abgebildet*. Nurnberg. 1871–1883: 1–1271 by L. Koch. 1272–1489 by E. Keyserling.

Koh, Joseph K.H. 1989. *A guide to common Singapore spiders*. Singapore Science Centre, Singapore.

Korszniak, N., McPhee, C., Story, D. 1994. Australian Spiders: is their publicity worse than their bite? *The Victorian Naturalist*, III(2): 70–3.

Kovoor, J. 1987. Comparative structure and histochemistry of silk-producing organs in arachnids. In: *Ecophysiology of Spiders* (Ed. W. Nentwig), Springer-Verlag, Berlin, New York, pp. 160–86.

Land, M.F. 1972. Mechanisms of orientation and pattern recognition by jumping spiders (Salticidae). In: *Information Processing in the Visual System of Arthropods*, Ed: R. Wehner, Springer Verlag, Berlin, New York. pp. 231–47.

Land, M.F. 1975. A comparison of the visual behaviour of a predatory arthropod with that of a mammal. In: *Invertebrate Neurons and Behaviour*. Ed: C.A.G. Wiersma, 36: 411–18.

Laing, D.J. 1973. Prey and prey capture in the tunnel web spider, *Porrhothele antipodiana*. *Tuatara*, 20(2): 57–64.

Laing, D.J. 1982. Snail-eating behaviour of the tunnel web Spider, *Porrhothele antipodiana*. *Tuatara*, 25(2): 74–81.

Legendre, R. M. 1963. L'audition et l'emission de sons chez les

Araneides. *Annales de biologie*, 2: 371–90.

Lehtinen, Pekka T. 1967. Classification of the cribellate spiders and some allied families. *Acta Zoologica Fennici*, 4: 199–468.

Latro — Levi, H.W. 1959. The spider genus *Latrodectus* (Araneae: Theridiidae). *Transactions of the American Microscopical Society*, 78: 7–43.

Levi, H.W. 1964. The American spiders of the genus *Phoroncidia* (Araneae: Theridiidae). *Bulletin of the Museum of Comparative Zoology*, Harvard, 131(3): 65–86.

Levi, H.W. 1965. Techniques for the study of spider genitalia. *Psyche*, 72(2): 152–8.

Levi, H.W. 1967. Adaptations of respiratory systems of spiders. *Evolution*, 21: 571–83

Levi, H.W. 1978. Orb-weaving spiders and their webs. *The American Scientist*, 66: 734–42.

Levi, H.W. 1980. Orb-webs: primitive or specialised. *Proceedings of 6th International Arachnology, Vienna*, 367–70.

Latro — Levi, H.W. 1983. On the value of genitalic structures and coloration in separating species of widow spiders (*Latrodectus* sp.) (Arachnida: Araneae: Theridiidae). *Verhandlungen Naturwissenschaften*, Hamburg, 26: 195–200.

Levi, H.W., Kirber, W.M. 1976. On the evolution of tracheae in Arachnids. *Bulletin British Arachnological Society*, 3(3): 187–8.

Levi, H.W., Lorna R. Levi. 1968. *A Guide to Spiders and their Kin*. Golden Press, New York. Western Publishing Coy, Inc.

Lincoln, R.J., Boxshall, Clark, P.F. 1982. *A Dictionary of Ecology, Evolution and Systematics*. Cambridge University Press. London, New York, Melbourne, Sydney.

Locket, G.H., Millidge, A.F. 1951, 1953. *British Spiders* Vol 1 (1951), Vol 2 (1953). London, Ray Society.

Locket, G.H. 1973. Two spiders of the genus *Erigone* Audouin from New Zealand. *Bulletin British Arachnology*, 2(8): 158–65.

Latro — McCrone, J.D. 1964. Comparative lethality of several *Latrodectus* venoms. *Toxicon*, 2: 201–03.

McCrone, J.D. 1969. Spider Venoms: Biochemical aspects. *American Zoologist*, 9: 155–6.

Latro — McCutcheon, E.R. 1992. Two species of katipo spiders. *The Weta*, 15(1): 1–2.

McKeown, K.C. 1963. *Australian Spiders*. Sirius Books, Angus & Robertson, Sydney, London.

McLachlan, A.R.G. 1993. Biology of *Spelungula cavernicola* Forster (Gradungulidae), a New Zealand cave-dwelling spider. Unpublished MSc thesis, University of Canterbury, New Zealand.

McQueen D.J., McLay, C.L. 1983. How does the intertidal spider *Desis marina* (Hector) remain under water for such a long time? *New Zealand Journal of Zoology*, 10: 383–92.

McQueen D.J., Pannell, L.K., McLay, C.L. 1983. Respiration rates for the intertidal spider, *Desis marina* (Hector). *New Zealand Journal of Zoology*, 10: 393–400.

Magni, F., Papi, F., Savely, H.E., Tongiorgi, P. 1964. Research on the structure and physiology of the eyes of a lycosid spider. II. The role of different pairs of eyes in astronomical navigation. *Archives Itallienes de Biologie*, 103: 146–58.

Main, Barbara. 1967. *Spiders of Australia*. The Jacaranda Press.

Latro — Maretic, Z. 1983. Latrodectism: variations in clinical manifestations provoked by *Latrodectus* spiders. *Toxicon*, 21(4): 457–66.

Maretic, Z., Levi, H.W., Levi, L.R. 1964. The theridiid spider, *Steatoda paykulliana*, poisonous to mammals. *Toxicon*, 2: 149–54.

Marples, B.J. 1948. An unusual type of web constructed by a Samoan spider of the family Argiopidae. *New Zealand Science Congress*, 1947: 232–3.

Marples, B.J. 1955. A new type of web spun by spiders of the genus *Ulesanis* with the description of two new species. *Proceedings of the Zoological Society, London*, 125: 751–60.

Marples, B.J. 1962. Notes on spiders on the family Uloboridae. *Annals of Zoology*, 4: 1–10.

Marples, B.J. 1962. The Matachiinae, a group of cribellate spiders. *Journal of Linnean Society (Zoology)*, London, 301: 701–720.

Marples, B.J. 1967. The spinnerets and epiandrous glands of spiders. *Journal of the Linnean Society (Zoology)*. 46(310): 209–22.

Marples, R.R. 1959. The Dictynid spiders of New Zealand. *Transactions of the Royal Society of New Zealand*, 87: 333–61.

Martendale, C.B., Newlands, G. 1982. The widow spiders: A complex of species. *South African Journal of Science*, 78(2): 78–9.

Mascord, R.E. 1966. The mating behaviour of *Gasteracantha minax* Thorell, 1859. *Journal of the Entomological Society of Australia (NSW)*, 3: 44–7.

Mascord, R.E. 1980. *Spiders of Australia: a field guide*. A.H. & A.W. Reed Pty Ltd., London, Auckland, Christchurch.

May, B.M. 1963. New Zealand cave fauna: 2. The limestone caves between Port Waikato and Piopio District. *Transactions of the Royal Society of New Zealand (Zoology)*, 3(19): 181–204.

May, B.M., Gardiner, D.I. 1995. Observations of the Australian 'two-spined' spider *Poecilopachys australasia* in an Auckland garden. *The Weta*, 18: 1–5.

Merrett P., Rowe, J.J. 1961. A New Zealand spider *Achaearanea veruculata* (Urquhart), established in Scilly, and new records of other species. *Annals Magazine Natural History*, 13(4): 89–96.

Millidge, A.F. 1951. Key to the British genera of the subfamily Erigoninae (family Linyphiidae: Araneae) including the description of a new genus (*Jacksonella*). *Annals and Magazine of Natural History*, 12(4): 545–62.

Murphy, Frances. 1980. *Keeping Spiders, Insects and Other Land Invertebrates in Captivity*. John Bartholomew & Son. Edinburgh.

Nelson, G., Platnick, N.I. 1981. *Systematics and Biogeography: cladistics and vicariance*. Columbia University Press. New York.

Nelson, G., Platnick, N.I. 1984. *Biogeography*. Carolina Biological Supply Company. 119: 1–16.

Nentwig, W. 1987. *Ecophysiology of Spiders*. Springer. Berlin, New York, London, Paris and Tokyo.

O'Donnell, M. 1983. A review of records of spider bites on humans in New Zealand including some previously unpublished records. *The Weta*, 6(2): 72–4.

Opell, Brent D. 1979. Revision of the genera and tropical American species of the spider family Uloboridae. *Bulletin of the Museum of Comparative Zoology*, 148: 443–549.

Opell, Brent D. 1994. Factors governing the stickiness of cribellar prey capture threads in the spider family Uloboridae. *Journal of Morphology*, 221: 111–19.

Parrott, A.W. 1946a. A systematic catalogue of New Zealand spiders. *Records Canterbury Museum*, 5(2): 51–93.

Parrott, A.W. 1946b. The eyes as taxonomic characters in spiders. *Records Canterbury Museum*, 5(2): 95–103.

Parrott, A.W. 1952. The banana spider (*Heteropoda venatoria* Linnaeus) recorded from New Zealand. *New Zealand Science Review*, 18: 8.

Parrott, A.W. 1960. Notes on New Zealand mygalomorph spiders, with a description of new species. *Records of the Canterbury Museum*, 7(3): 107–2202.

Parry, D.A. 1960. The small leg–nerve of spiders and a probable mechanoreceptor. *Quarterly Journal Microscopical Science*, 108: 1–8.

Parry, D.A. 1960. Spider hydraulics. *Endeavour*, July: 156–62.

Parry, D.D. 1965. The signal generated by an insect in a spider's web. *Journal of Experimental Biology*, 43: 185–92.

Parry, D.A., Brown, R.H.J. 1959. The jumping mechanism of salticid spiders. *Journal of Experimental Zoology*, 36: 654.

Patterson, Peter J. 1995. *Spider Relatives*. Shortland Publications Ltd., Auckland, New Zealand.

Paul, Rudiger, Till, Fincke & Bernt, Linzen. 1989. Book lung function in arachnids. *Journal of Comparative Physiology*, B. 159: 409–18.

Peakall, D.B. 1971. Conservation of web proteins in the spider *Araneus diadematus*. *Journal Experimental Zoology*, 176: 257.

Peakall, D.B. 1968. The spider's dilemma. *New Scientist*, 37: 28–9.

Peckham, G.W., Peckham, E.G. 1889. Observations on sexual selection in spiders of the family Attidae. *Occasional papers*, Natural History Society, Wisconsin, 1: 1–60.

Peters, Hans M. 1990. On the structure and glandular origin of bridging lines used by spiders for moving to distant places. *Acta Zoologie Fennica*, 190: 309–14.

Petrunkevitch, A. 1933. An inquiry into the natural classification of spiders based on a study of their internal anatomy. *Transactions Connecticut Academy of Arts and Science*, 31: 303–89.

Petrunkevitch, A. 1942. A study of amber spiders. *Transactions Connecticut Academy of Arts and Science*, 34: 119–464.

Pickard-Cambridge, F.O. 1901. A revision of the genera of Araneae, or spiders, with reference to their type species. *Annals and Magazine of Natural History*, 7(7): 51–65.

Platnick, N.I. 1971. The evolution of courtship behaviour in spiders. *Bulletin British Arachnology Society*, 2: 40.

Platnick, N.I. 1989. *Advances in Spider Taxonomy 1988–1987. A supplement to Brignoli's – A Catalogue of the Araneae described between 1940 and 1981*. Manchester University Press, Manchester and New York.

Platnick, N.I. 1990. Spinneret morphology and the phylogeny of ground spiders (Araneae: Gnaphosidae). *American Museum Novitates*, 2978: 1–42.

Platnick, N.I. 1993. *Advances in Spider Taxonomy 1988–1991, with synonymies and transfers 1940–1980*. New York Entomological Society, New York.

Platnick, N.I. 1995. An abundance of spiders. *Natural History*, 104(3): 50–3.

Platnick, N.I. 1997. *Advances in Spider Taxonomy 1992–1995, with redescriptions 1940–1980*. New York Entomological Society, in association with The American Museum of Natural History.

Platnick, N.I. (in press) A relimitation of the Australasian ground spider family Lamponidae (Araneae: Gnaphosidae). *Bulletin American Museum of Natural History*.

Platnick, N.I., Forster, R.R. 1982. On the Micromygalinae, a new sub-family of Mygalomorph Spiders (Araneae: Microstigmatidae). *American Museum of Natural History Novitates*, New York, 2734: 1–13.

Platnick, N.I., Forster, R.R. 1989. A revision of the temperate South American and Australasian spiders of the family Anapidae (Araneae, Araneoidea). *Bulletin of the American Museum of Natural History*, 190: 1–139.

Pocock, R.I. 1902. On the marine spiders of the genus *Desis*, with description of a new species. *Proceedings Zoological Society, London*, 1902(2): 98–106.

Pollard, S.D., Jackson, R.R. 1982. The biology of *Clubiona cambridgea* (Araneae: Clubionidae): intraspecific interactions. *New Zealand Journal of Ecology*, 5: 44–50.

Pollard, S.D., Jackson, R.R, van Olphen, A., Robertson, M.W. 1995. Does *Dysdera crocota* (Araneae: Dysderidae) prefer woodlice as prey? *Ethology, Ecology & Evolution*, 7: 271–5.

Powell, L. 1871. Katipo (*Latrodectus*), the poisonous spider of New Zealand. *Transactions New Zealand Institute*, 3: 56–9.

Powell, L. 1879. On *Desis robsoni*, a marine spider from Cape Campbell. *Transactions New Zealand Institute*, 11: 263–8.

Prestwich, K.N. 1983. The roles of aerobic and anaerobic metabolism in active spiders. *Physiological Zoology*, 56: 122–32.

Quicke, Donald. 1988. Spiders bite their way towards safer insecticides. *New Scientist*, November: 38–41.

Ralph, T.S. 1855. On the katipo, a supposed poisonous spider of New Zealand. *Journal Proceedings Linnean Society*, 1: 1–12.

Raven, R.J. 1978. Systematics of the Spider Subfamily Hexathelinae (Dipluridae: Mygalomorphae: Arachnida). *Australian Journal of Zoology Supplement*, 65: 1–75.

Raven, R.J., Churchill, T.B. 1988. Funnel-web spiders. In: *Venoms and Victims*, Pearn J., Covacevich. J. (eds), Queensland Museum. pp. 67–72.

Raven, R.J. 1985. The spider infraorder Mygalomorphae: cladistics and systematics. *Bulletin American Museum Natural History*, 182: 1–180.

Raven, R.J., Gallon, J. 1987. The redback spider. In: *Toxic plants and animals; a guide for Australia*. Queensland Museum, Brisbane, Australia. pp. 313–21.

Reichert, S.E., Lockley, T. 1984. Spiders as biological control agents. *Annual Review of Entomology*, 29: 299–320.

Roberts, N.L. 1936–37. Some notes on the bird-dung spider (*Celaenia excavata*). *Proceedings of the Royal Society of New South Wales*, pp. 23–8.

Robinson, M.H., Robinson, B. 1972. The structure, possible function and origin of the remarkable ladder-web built by a New Guinea orb-web spider. *Journal of Natural History*, 6: 687.

Robinson, M.H., Robinson B. 1976. Discrimination between prey types: an innate component of the predatory behaviour of araneid spiders. *Z. Tierpsychol.*, 41: 266–76.

Robinson, M.H. 1982. Courtship and mating behaviour in spiders. *Annual Review of Entomology*, 27: 1–20.

Robson, C.H. 1878. Notes on a marine spider found at Cape Campbell. *Transactions New Zealand Institute*, 10: 299–300.

Roewer, C.F. 1942. *Katalog der Araneae*, Bremen 1: 1–1040.

Roewer, C.F. 1954. *Katalog der Araneae*, Bremen. 2: 1–1951.

Rovner, J.S. 1977. Spider behaviour. In: *International Encyclopedia of Psychiatry, Psychology, Psychoanalysis, and Neurology*. Aesculapius Publishers, Inc. pp. 419–23.

Rovner, J.S. 1980. Morphological and ethological adaptations for prey capture in wolf spiders (Araneae: Lycosidae). *Journal of Arachnology*, 8: 201–15.

Rovner, J.S. 1987. Nests of terrestrial spiders maintain a physical gill: flooding and the evolution of silk constructions. *Journal of Arachnology*, 14: 327–37.

Rovner, J.S. 1993. Visually mediated responses in the lycosid spider *Rabidosa rabida*: the roles of different pairs of eyes. *Proceedings of the XII International Congress of Arachnology, Memoirs of the Queensland Museum*, 33(2): 635–44.

Sauer, Frieder., Wunderlich, Jörg. 1987. *Die Schönsten Spinnen Europas* (in German). Fauna-Verlag.

Savory, T.H. 1961. *Spiders, Men and Scorpions*. University of London Press, London.

Schultz, Stefan. 1997. The chemistry of spider toxins and spider silk. *Angewandte Chemie International Edition in English*, 36: 314–26.

Simon, E. 1892–1903. *Histoire naturelle des Araignées*. Paris. 1(i–iv): 1–1084, 2(v–viii): 1–1080.

Shear, W.A. 1969. Observations on the predatory behaviour of the spider, *Hypochilus gertschi* Hoffman (Hypochilidae). *Psyche*, 76(4): 407–17.

Shear, W.A. 1986. (Ed.) *Webs, Behaviour and Evolution*. Stanford University Press. Stanford, CA., USA.

Shear, W.A, Palmer, J.M., Coddington, J.A., Bonamo, P.M. 1989. A Devonian Spinneret: Early evidence of spiders and silk use. *Science*, 246: 479–81.

Smith, D.J. 1971. The habitat and distribution of the katipo spider at South Brighton beach, Christchurch, New Zealand. *New Zealand Entomologist*, 5: 96–100.

Spiller, D., Turbott, E.G. 1944. The occurrence of some Australian insects and a spider in New Zealand. *Records Auckland Institute Museum*, 3(1): 79–83.

St George, Ian, Forster, Lyn. 1991. Skin necrosis after white-tailed spider bite? *New Zealand Medical Journal*, 104: 207–08.

Sunde, R.G. 1980. *Lampona cylindrata* (Araneae: Gnaphosidae): a note. *New Zealand Entomologist*, 7(2): 175–6.

Sutherland, Struan K. 1990. Treatment of arachnid poisoning in Australia. *Australian Family Physician*, 19(11): 47–64.

Tenquist, J.D. 1984. A suspected bite by *Latrodectus katipo*. *The Weta*, 7(3): 60–1.

Todd, V. 1945. Systematic and biological account of the New Zealand Mygalomorphae (Arachnida). *Transactions of the Royal Society of New Zealand*, 74(4): 375–407.

Todd-Davies, Valerie. 1986. Australian spiders: Araneae. *Queensland Museum Booklet*, 14: 1–60.

Toft, R., Rees, J.S. 1998. Reducing predation of orbweb spiders by controlling common wasps (*Vespula vulgaris*) in a New Zealand beech forest. *Ecological Entomology*, 23: 90–5.

Turnbull, A.L. 1973. Ecology of the true spiders (Araneomorphae). *Annual Review of Entomology*, 18: 305.

Urquhart, A.T. 1883. Protective resemblances in the Araneidae in New Zealand. *Transactions New Zealand Institute*, 15: 174–8.

Urquhart, A.T. 1884. On the spiders of New Zealand. *Transactions New Zealand Institute*, 17: 31–53.

Urquhart, A.T. 1885. On the spiders of New Zealand. *Transactions New Zealand Institute*, 18: 184–205.

Urquhart, A.T. 1887. On new species of Araneidea. *Transactions New Zealand Institute*, 19: 72–118.

Urquhart, A.T. 1888. On new species of Araneidea. *Transactions New Zealand Institute*, 20: 109–25.

Urquhart, A.T. 1890. On two species of Araneae new to science from the Jenolan Caves, New South Wales. *Transactions New Zealand Institute*, 22: 236–9.

Urquhart, A.T. 1892. Catalogue of the described species of New Zealand Araneidae. *Transactions New Zealand Institute*, 24: 220–30.

Vink, C.J., Sirvid, P.J. 1999. The Oxyopidae (lynx spiders) of New Zealand. *New Zealand Entomologist*, 21: 1–9.

Vollrath, Fritz. 1979. Behaviour of the kleptoparasite spider *Argyrodes elevatus* (Araneae: Theridiidae). *Animal Behaviour*, 27: 515–21.

Walcott, C. & Van der Kloot, W.G. 1959. The physiology of the spider vibration receptor. *Journal of Experimental Zoology*, 14(12): 191–244.

Watt, J.C. 1971. The toxic effects of the bite of a clubionid spider. *New Zealand Entomologist*, 5(1): 87–90.

Whitehouse, Mary E.A. 1986. The foraging behaviours of *Agyrodes antipodiana* (Theridiidae), a kleptoparasitic spider from New Zealand. *New Zealand Journal of Zoology*, 13(2): 151–68.

Whitehouse, Mary E.A. 1987. 'Spider eat spider': The predatory behaviour of *Romphaea* sp. from New Zealand. *Journal of Arachnology*, 15: 355–62.

Williams, D.S. 1979. The feeding behaviour of New Zealand *Dolomedes* species (Araneae: Pisauridae). *New Zealand Journal of Zoology*, 6: 95–105.

Wilson, R.S. 1962. The control of dragline spinning in the garden spider. *Quarterly Journal of Microscopic Science*, 103: 557–71.

Wilson, R.S. 1965. The pedicel of the spider, *Heteropoda venatoria*. *Journal of Zoology*, 147: 38–45.

Wilton, C.L. 1946. A new spider of the family Archaeidae from New Zealand. *Dominion Museum Records of Entomology*, 1 (3): 19–26.

Wise, D.H. 1993. *Spiders in Ecological Webs*. Cambridge University Press, Cambridge.

Witt, P.N., Reed, Ch.F., Peakall, D.B. 1968. *A spider's Web: Problems in regulatory Biology*, Springer-Verlag, Berlin, Heidelberg, New York.

Witt, P.N., Scarboro, M.B., Daniels, R., Peakall, D.B., Gause, R.L. 1977. Spider web-building in outer space: evaluation of records from the Skylab spider experiment. *Journal of Arachnology*, 4(2): 115–24.

Witt, P.N. & Rovner, J.S. (Eds) 1982. *Spider Communication: Mechanisms and Ecological Significance*. Princeton University Press, NJ.

Wunderlich, J. 1976. Uloboridae, Theridiosomatidae und Symphytognathidae (Arachnida: Araneidae). *Senckenbergiana biologie*, 57(1–3): 113–24.

Wunderlich, J. 1987. *The Spiders of Canary Islands and Madeira* (in German). Tropical Scientific Books.

Yaginuma Takeo. 1986. *Spiders of Japan in colour* (in Japanese). Hoikusha Publishing Co., Ltd.

INDEX

Numerals in **bold** refer to illustrations. Where the name of a spider is followed by another name in parentheses, the second name is identical in meaning to the first, and may be the name used in the text.

Acari (mites), 47-8, **48**, 49
Achaearanea ampliata, 173, **173**
Achaearanea tepidariorum, 172, **172**
Achaearanea veruculata, 172, **172**, 173
Achaearanea, 172, **173**
Acropsopilionidae, 58
Africa 119, 136, 180, 191 – *see also* South Africa, Madagascar
Agelena, 223
Agelenidae, 45, 213, 221, 223-6
Akatorea otagoensis, 232, **232**
Akatorea, 232
Algidia viridata, 56
Algidia, **55**, 56
Allotrochosina schauinslandi, 83, 85
Amarara fera, 183
Amaurobiidae, 221, 223-7, 229
Amaurobioides maritimus, 212, **213**
Amaurobioides picunus, **212**
Amaurobioides, 212, 251
Amaurobioididae, 212-13, 251
Amaurobius, 223
Amblypygi (tail-less whip scorpions), **47**
America, 42 – *see also* Mexico, Panama, South America, United States
Amphinecta, 229-31
Amphinecta pika, **229**
Amphinectidae, 221, 223, 226, 229-32
Anapidae, **26**, 27, 197-199, 201
Anaua unica, 116, **116**
Andrade, Maydianne 179
Antarctica, 48
Anyphaenidae, 212
Anzacia gemmea, 109, **109**
Aorangia ansa, **225**
Aorangia, 224
Aotearoa magna, **43**, 204, **204**
Aparua kaituna, **66**
Aparua, 60-1, 66
Aphonopelma chalcodes, **7**
Apneumonomorphae, 199
Arachnida, 47
Arachnura feredayi, 158, **158**
Arachnura higginsi, 158
Arachnura, 40, **40**, 159
Araeoncus humilis, **206**
Aranea transmarina, 239
Araneae, 4, 5, 47
Araneidae (orb-weaving spiders), 4, **19**, 26, **26**, **45**, 114, 145-6, 151-6, 156-163, **160**, 197, 202, 238

Araneinae, 151
Araneoidea, 151, 168-9
Araneomorphae (True Spiders), 5, 10, 25, 26, 28, 41, 69-80, 199, 224, 227
Araneophages, 111, 204
Araneus diadematus, 40, 150
Arangina cornigera, 222
Arangina pluva, **222**
Arangina, 222
Archaea gracilicollis, **202**
Archaeidae, 203-4
archaeids, true, 202-3
Argentina, 72-3
Argiope protensa, **28**, 159-60
Argiope, 238
Argyrodes antipodiana, 183, **183**
Argyrodes, 154, 182-3
Argyroneta marina, 210
Ariadna septemcinta, **41**
Ariadna, 41, **41**, 116, 136, 251
Arizona, 7 – *see also* United States
Asia, 4, 69, 210 – *see also* China, Japan
Astrolabe (French corvette), 61
Atkinson, R.K., 239
Atrax robustus, 238-9
Atrax (funnelweb spider), 238
Auckland Islands, 56, 136, 200, 206, 212, 225 – *see also* subantarctic islands
Auckland, 6, 96, 101, 105, 107, 117, 154, 184, 242 – *see also* Avondale, Hauraki Gulf
Ausserer, Anton, 61
Austin, A.D., 170
Australia, 4, 6, 7, 14, 16, 25, 42-3, 47, 55, 58, 60, 70-2, 74-5, 80, 87, 96-7, 106-9, 110-11, 113-14, 117-19, 136, 145, 152-5, 162, 164, 166-9, 177, 179, 183-4, 190-1, 193-4, 199, 200, 202-5, 210, 212, 215, 217-9, 229, 233-4, 238-43, 245 – *see also* Carrai Bat Cave, New South Wales, Queensland, Sydney
Australian cave spiders, 71-5
Australian grey house spider (*Badumna longinqua*), 6, **28-9**, 41, **112**, 218, **218**, 219-20, **221**, 240, 246
Australian white-tailed spider (*Lampona*), 6, 105, 111-12, 245-6
Austrochilidae, 5, 69, 72-3
Austrochilus, 73
Avella, 169, 170
Avondale spider (*Delena cancerides*), 6, 96, **96**, 97, **97**, 241, **241**

Badumna robusta (black house spider), 219-20
Badumna insignis (black window spider), 240, **240**
Badumna longinqua (Australian grey house spider), 6, **28-9**, 41, **112**, 218, **218**, 219-20, **221**, 240, 246
Badumna, **15**, 111, 219, **219**, 246
banana spider (*Heteropoda venatoria*), 6, 47, 95, **95**
Bell Block, 176
big-jawed spiders (Tetragnathidae), 6, 145, 151, 154-5, 160, 164-5, **165**, 166, 186
bird-catching spider (*Nephila edulis*, bird-eating spider), 154, **154**, **238**, 239-40
bird-dropping spider (*Celaenia*), 13, **13**, 14, 35, **35**, 42, **42**, 160-2, **163**, 197
bird-eating spiders – *see* bird-catching spiders
bites – *see* venom
black house spider (*Badumna robusta*), 219-20
black katipo (*Latrodectus atritus*), 173-4, **174**, 176-7, 242-3
black widow spider (*Latrodectus mactans*), 179, 242
black window spider (*Badumna insignis*), 240, **240**
Bleckmann, H., 93
Blest, A.D., 170, 190
Bonnet, Pierre, 39, 92, 189
Bristowe, W. S., 8, 35, 113, 187, 205
brown widow spider (*Latrodectus geometricus*), 243, **243-4**
Bryant, Elizabeth, 99, 115-16, 193, 195

California, 220 – *see also* United States
Cambridgea antipodiana, **29**, 38-9
Cambridgea arboricola, 185, **185**
Cambridgea foliata, 171, 185, **185**, 186, 240
Cambridgea, 17, **20**, 42, 169, **171**, 171, 184, **184**, 185-6, 192, 225, 240
Campbell Island, 56, 200, 212, 225 – *see also* subantarctic islands
Canterbury, 61, 64, 66, 107-8, 195, 242 – *see also* Cass, Christchurch, Hanging Rock, Kaituna Valley
Cantuaria, 60
Carrai Bat Cave (Australia), 71, 74, 75
Cass, 195
Catley, Kefyn, 73

cave spiders, 71-6; Australian, 71-6; New Zealand, 71, 75, 77, **76-77**, 78, **76**; Tasmanian, 73
Celaenia (bird-dropping spider), 13, **13**, 14, 35, **35**, 42, **42**, 160-2, **163**, 197
Central Otago, 39, 51, 86-8, 100, 105, 107 – *see also* Otago
Chalmers, Barry, 78
Chamberlain, G., 240
Chatham Islands, 156, 184, 210, 213
Cheiracanthium stratioticum, 108, **108**, 241
Cheiracanthium, 44, **45**, 108, 241, **241**
Chelonethi (false scorpions), 47-9, **49**, 50-2
Chile, 7, 72, 73, 136, 192, 203-4 – *see also* South America
China, 4, 70, 72 – *see also* Asia
Chiracanthium – *see Cheiracanthium*
Christchurch, 116, 127-8, 187, 243 – *see also* Canterbury, Port Hills
classification, 3-8, 60, 72, 151, 209, 233, 235
Clubiona cambridgea, 106, **106**
Clubiona, **15**, 23-4, **28**, 106, **106**
Clubionidae (hopping spiders), **24**, 43-5, 105-6, **106**, 108, 216, 242
Clyne, Densey, 155
cobweb spiders (Theridiidae), 4, 6, **14**, 30, 42, 44, **45**, 72, 83, 112-3, 154, 171-183, 192-3, 197, 202, 208, 241
Coelotes terrestris, 35
Colaranea melanoviridis, 157, **157**
Colaranea verutum, 157, **157**
Colaranea viriditas (green orbweb spider), **146**, **149**, 150, **150**
Colaranea, 157
Colgin, M.A., 40
collecting spiders, 248-51
comb-footed spiders, 171-2, **171**
Corinnidae (fleet-footed spiders), 6, 105-7
courtship – *see* mating
crab spiders, 81, 95-104
cribellate spiders (hackled-silk spiders), 13, 69, 72, 74, 169, 183-4, 209-10, 213-14, 216, 219, 221-232
Cryptaranea albolineata, **151**
Cryptaranea atrihastula (ladderweb spider), 160, **160**, 161
Cryptaranea subalpina, **23**
Cryptaranea subcompta, 161, **161**
Cryptaranea venustula, **145**
Ctenidae, 233-4
Ctenizidae, 60
Cuvier Island, 107
Cyatholipidae, 26, **26**, 171, 193-195
Cyatholipus, 193
Cycloctenidae (scuttling spiders), **19**, 105, 114-16, 209
Cycloctenus fugax, **115**
Cycloctenus, **19**, 114, **114**, 115-16
Cyclosa trilobata, 159, **159**
Cyclosa, **28**
Cymbachina albobrunnea, 101, **101**
Cymbachina, 98-9
Cyphophthalmi (mite-like harvestmen), 47-9, 54

daddy-long-legs (*Pholcus* spp.), 6, 21, 33, **34**, 53, 187-8, **188**
Dahl, F., 227
Dalmas, Compte de, 110, 222
Davies, Valerie, 113, 193
de Dalmas, Compte, 110, 222
Deinopidae, 6, 145, 169-170
Deinopis (ogre-faced spider, net-casting spider), 6, 151, 169
Deinopis subrufa, 170
Deinopoidea, 151
Delena cancerides (Avondale spider), 6, 96, **96**, 97, **97**, 241, **241**
Delena, 98
Desidae, 6, **26**, 112, 209-11, 213-7, 219, 221, 223
Desis marina (intertidal spider), 209-11, **210**, 213, 251
Desis, 210
Diaea albomaculata, 101, **101**
Diaea ambara, 99,100
Diaea, 44, 98-9, **99**, **100**, 101, 112
Dichrostichus, 42
Dictyna cornigera, 222, **222**,
Dictyna, 222
Dictynidae, 214, 221-3,
Dieckmann, J., 238
Dinopis – *see Deinopis*
Diplocephalus cristatus, 205, **206**
Diplocephalus, 206-7
Dipluridae, 60, 66
Dipoena, 181, **182**
Diptera, 129
Dolomedes aquaticus (water spider), 88, 90, **90**, 91-2, **92**, 93-4, 211
Dolomedes minor, **28**, 87, **87**, 88, **88**, 92-3, **94**
Dolomedes triton, 91, 93
Dolomedes, **89**, 93, **93**, 94
Drassidae, 45
Drassus formicarius, 111
Dry River (Nelson), 78
Dunedin, 36, 82, 87, 110, 116, 235 – *see also* Otago
Dysdera crocota (slater-eating spider), 6, **24**, 135, 241
Dysdera hungarica, 18
Dysdera, **18**, 188
Dysderidae, 6, 16, **19**, **24**, 26, **26**, 30, 135-6, 139-41

East Coast, 107
ecribellate spiders, 74, 77, 184, 190, 209-10, 221
eight-eyed spiders, **19**, 115, 135, 197, 201
England, 6, 52, 94, 173, 189, 210
entelegyne spiders, 16, 22, **22**, 44, 46
Eomatachia, 215
Epeira brounii, 154
Epeira orientalis, 154
Episinus, 112, 182, **182**
Erigone wiltoni, 206, **206**
Erigoninae (money spiders), 6, 36, 191, 197, 205, **206**, 207
Eriophora decorosa, 152, **152**, 153

Eriophora heroine, 154
Eriophora transmarina, 239
Eriophora pustulosa, 6, **34**, 37, **42**, **150**, 152, **152**, 153, **153**, 156, 183, 239, 252
Eriophora, **19**, 152
Ero furcata, 113
Eryciniolia purpurapunctata, 168, **168**
Eryciniolia, 168-9
eucalyptus spider (*Hemicloea rogenhoferi*), 110, **110**
Euophrys parvula (house hopper), 119, 127-9
Euophrys, **127**, 128-9, **129**
Europe, 58, 179 – *see also* England, France, Italy
European harvestman (*Phalangium opilio*), 52, **52**, 54
Euryopis, 182, **182**

false katipo (*Steatoda capensis*), 180, **180**, 241
false scorpions (Chelonethi), 47-9, **49**, 50-2
fiddleback spider (*Loxosceles reclusa*), 242
Fiji, 117
Filistatidae, 26, **26**, 40
Fiordland, 78, 116, 187, 203-4, 226, 234 – *see also* Hauroko, Southland, Te Anau
fisher spiders, 88 – *see* Pisauridae
fleet-footed spiders (Corinnidae), 6, 105-7
Forster, R.R, 70, 75, 80, 136
Forsterella faceta, 233, **233**
Forsterella, **16**, 233
Fox Glacier, 88
France, 66, 242 – *see also* Europe
Franz Josef Glacier, 99
funnelweb spider (*Atrax*), 61, 238

Gasparia, 216
Gertsch, Willis, 46, 72
giant crab spiders (Sparassidae), 6, 95-98
Gippsicola, 41, 116
Given, Bruce, 180
Gnaphosidae (stealthy spiders), 6, **19**, 43, 105-6, 109-10
Gondwana, 7, 136, 189, 192, 197, 212
Goyen, P., 64, 87-8, 129, 211
Goyenia fresa, **216**
Goyenia, 216
Gradungula sorenseni, **24**, 70, **71**, 71, 78-9
Gradungula woodwardi, 70
Gradungula, 73-4, 77
Gradungulidae, 5, **13**, 16, **16**, **24**, 25, 30, 35, 44, 69-71, **71**, 72-81, 214
Gray, Mike, 75, 80
green orbweb spider (*Colaranea viriditas*), **146**, **149**, 150, **150**
grey house spider, Australian (*Badumna longinqua*), 6, **28-9**, 41, **112**, 218, **218**, 219-20, **221**, 240, 246
Griswold, Charles, 136, 193

Habronestes marinus, 211
hackled-silk spiders (cribellate spiders), 13, 69, 72, 74, 169, 183-4, 209-10, 213-14, 216, 219, 221-232

Hadronyche infensa, 239, **239**
Hadronyche, 238
Hahniidae, 197, 205, 207
Hall, Grace, 96
Hamilton, 6
Hanea paturau, 194
Hanea, 194
Hanging Rock (Canterbury), 195
haplogyne spiders, 16, 22, **22**, 44, 46, 109, 135
Hapona otagoa, **218**
Hapona, 218
Harris, Anthony, 68, 71
harvestman, European (*Phalangium opilio*), 52, **52**, 54
harvestmen (Opiliones), 47-9, 52, **52**, 53-4, **54-6**, 56-8
Hauraki Gulf, 107
Hauroko, Lake, 187, 234
Haurokoa filicicola, 234, **235**
Haurokoa, 234
Hawaii, 77, 220 – see also United States
Hector, James, 210
Helpis minitabunda, 6, 117, **117**
Hemicloea alacris, 110
Hemicloea celerrima, 110
Hemicloea plautus, 110
Hemicloea rogenhoferi (eucalyptus spider), 110, **110**
Hendea spina, 56
Hendea, 56
Heptathela kimurai, **4**
Heptathela, 4, 5
Hesperauximus sternitzkii, 220
Heteropoda venatoria (banana spider), 6, 47, 95, **95**
Hexathele hochstetteri, 61, 63, **63**, 239
Hexathele huka, 63, **63**, 64
Hexathele maitaia, 63
Hexathele petriei, 61, 64, **64**
Hexathele rupicola, 61, 64
Hexathele, 60-3, 66
Hexathelidae (tunnelweb spiders), **10**, **19**, **24**, 41, 59, 60-1, **61**, 62, **62**, 63, **63**, 64, **64**, 66-8, 79, 239
Hickman, V.V., 200, 202-3, 208
Hickmania troglodytes, 72, 73
Hickmania, 73
Hochstetter, Ferdinand von, 61
Holarchaea novaseelandiae, 203, **203**
Holarchaeidae, 197, 202-3
Holoplatys 132, **133**, 134
Homann, Heinrich, 118
Honeycomb Cave system (Karamea), 75, 77-8
hopping spiders (Clubionidae), **24**, 43-5, 105-6, **106**, 108, 216, 242
Horioctenoides, 234, **234**
house cobweb spider (*Steatoda grossa*), 179, **179**, 241
house hopper (*Euophrys parvula*), 119, 127-9
house spider, Australian grey (*Badumna longinqua*), 6, **28-9**, 41, **112**, 218, **218**, 219-20, **221**, 240, 246; black (*Badumna robusta*) 219-20; European (*Tegenaria domestica*), 41, 224
hunting spiders, 43, **45**, 70, **71**, 81-2, 95, 105-116, 241-2
huntsman spiders, 241
Hutton, Captain F.W., 235
Hutton, Russell, 166
Huttonia palpimanoides, 235-6, **236**
Huttonia, 13, **14**
Huttoniidae, 6, 81, 233, 235-6
Hypochilidae, 5, 69-73
Hypochilus gertschi, 72
Hypochilus pococki, 72, **72**, 73
Hypochilus, 72
Hypodrassodes, **19**, **29**, 109
Hyptiotes, 160, 208

Idiopidae (true trapdoor spider), 37, 59-61, 64-6
Ihurakius forsteri, 227
intertidal spider (*Desis marina*), 209-11, **210**, 213, 251
introduced spiders, 6, 37, 96, 107, 110, 152, 153-6, 177, 179, 183, 187, 205, 240-1, 245
Ischalea spinipes, 186-7, **187**
Isopeda immanis, 239
Isopeda montana, 241
Italy, 7, 237, 242
Ixeuticus dalmasi, 195
Ixeuticus janus, 195
Ixeuticus martius, 218
Ixeuticus robustus, 240

Japan, 4, 58, 210, 220 – see also Asia
Joque, R., 233
jumping spiders (Salticidae), 6, 18, **19**, 30, 39, **39**, 43-5, 81, 117-134, **118**, **124**, 129, **130-1**, 209

Kaituna Valley, 66 – see also Canterbury
Kapitia obscura, 144
Karamea, 56, **56**
Karamea, 75, 77, 78 – see also West Coast
katipo (*Latrodectus katipo*), **45**, 173-4, **174**, 175, **175**, 176, **176**, 177-80, 238-9, 241-3, **243**, 244-5
kleptoparasites, 182-3
Koch, L., 111
Kohukohu, 216
Korszniak, N., 241

ladderweb spiders (*Cryptaranea atrihastula*), 160, **160**, 161
Laestrygones otagoensis, **217**
Laestrygones, 216
Laetesia trispathulata, **190**
Laetesia, 190
Laing, Don, 62, 68
Lamina minor, 216, **217**
Lamina montana, **217**
Lamina, 216
Lampona (white-tailed spider), 111-12, 245-6
Lampona cylindrata, 6, 111, **111**, 112, 245, **246**
Lampona murina, 6, 111, 246
Lamponidae (white-tailed spiders), 6, 105, 109, 111-12, 245-6
lampshade webs, 72-73
Land, Mike, 118
Laniatores (short-legged harvestmen), 54-7, 58
Latrodectus atritus (black katipo), 173-4, **174**, 176-7, 242-3
Latrodectus geometricus (brown widow spider), 243, **243-4**
Latrodectus hasselti (redback spider), 6, **43**, 177, **177-8**, 178-9, 242, 244-5
Latrodectus indistinctus, 242
Latrodectus katipo (katipo), **45**, 173-4, **174**, 175, **175**, 176, **176**, 177-80, 238-9, 241-3, **243**, 244-5
Latrodectus mactans (black widow spider), 179, 242
Latrodectus tredecimguttatus, 242
Latrodectus, 179, 238, 242-5
Lauder, Stewart, 162
Lehtinen, Pekka, 195, 203, 227
Lepthyphantes tenuis, 189, **189**
Leucage dromedaria, 155, **155**, 156
Leucage, 155
Lewis, R.V., 40
Linyphia, 194
Linyphiidae (sheetweb spiders), 6, 36, 44, **45**, 171, 189-192, 197, 240, 205-7
Liphistiidae, 5, 15, **15**, 25, 79
Liphistiomorph spiders, 4
Liphistius, 4, 5, 79
Lockett, G.H., 207
long-legged harvestmen (Palpatores), 54, **54**, 55, 57-8
Longwood Forest (Southland), 70
Loxosceles reclusa, 242
Lycosa arenivaga, 86
Lycosa bellicosa, 87
Lycosa erythrognatha, 237
Lycosa hilaris, 81, **81-2**, 84
Lycosa rabida, 45
Lycosa tarantula, 237
Lycosa, **19**, 77, 86
Lycosidae (wolf spiders), 7, 18, **19**, 26, **26**, 33, 35, **35**, 43-4, **45**, 70, 81-2, **82**, 83-5, **85-6**, 86-7, 94, 237-8
lynx spiders (Oxyopidae), 105, **105**, 106

Macrogradungula, 74, 79
Madagascar, 55, 202-3, 242 – see also Africa
Magni, F., 82
Mahura, 226
Makarora Valley, 129
Malkara loricata, 113
Malkaridae (shield spiders), 105, 113-4, **113**
Mamoea rufa, **230**
Mamoea, 229-30
Mangua medialis, **193**
Mangua, 193
Maniho gracilis, **231**
Maniho, 231-2
Maretic, Z., 241

Marlborough, 107, 234
Marpissa marina, 119, 129, **129**, 209, **209**
Marpissa, 129
Marples, Brian and Molly, 66, 208
Marples, Richard, 195, 227
Marplesia dugdalei, 195
Marplesia janus, 195
Marplesia pohara, **195**
Marplesia, 195-6
Matachia australis, **213**
Matachia, 211, 213-15, 219
Matamata, 6
Matilda, 194
mating, 43-46, 51, 54, 62, 63, 86, 91-2, 100, 104, 106, 123-6, 128-9, 131, 134, 151, 154, 164-5, 176, 178-9, 185, 188, 191, 200, 207
McCrone, J.D., 244
McCutcheon, Eddie, 176
McLachan, A., 77
McLay, C.L., 211
McQueen, D.J., 211
Mecysmaucheniidae, 16, 28, **43**, 46, 197, 202-4
Mecysmauchenius segmentatus, 203
Mecysmauchenius, 204
Megadictyna, 228
Megadictyna thileniusi, 227, **227,** 228
Megadictynidae, 227-8
Megalopsalis, **54, 57**, 58, **58**
Menemerus, 117, **117**
Meringa otagoa, **193**
Mesothelae, 4, 5
Meta arborea, 166, **166**
Meta lautiuscula, 165, **165**
Meta segmentata, 6, 166
Metidae, 6, 165, 166
Mexico, 47 – *see also* America
Microctenonyx subitaneus, **206**
Micromygale diblemma, 59
Micropholcomma caeligenus, 199
Micropholcomma longissima, 199, **199**
Micropholcommatidae, 197, 199-200
midget spiders, 197-208
Migas cantuarius, **66**, 67
Migas goyeni, **66**
Migas saxatilis, **67**, 68
Migas, 61, 67-8, 251
Migidae (tree-trapdoor spider), 61, 67-8, 251
Millar, Ian, 77-8
Millidge, Frank, 190
Mimetidae (pirate spiders), 30, 105, 112-14, 203
Mimetus, **30**, 112-13
Misgolas assimilis, **65**
Misgolas huttoni, 60
Misgolas napua, **65**
Misgolas toddae, **64**, 66
Misgolas vellosa, 37
Misgolas, 61, 64, 66, 68
mite-like harvestmen (Cyphophthalmi), 47-9, 54
mites (Acari), 47-8, **48**, 49
Miturga, 107-8, **108**, 242

Miturgidae (prowling spiders), 105, 107-8, 241-2
Mole Creek (Tasmania), 72
Moneta, **14**, 182, **182**
money spiders (Erigoninae), 6, 36, 191, 197, 205, **206**, 207
Monrovia moth, 116
Motupipi Cave (Nelson), 78
Mount Pisa (Otago), 130
mountain salticid, 130-1, **131-2**
mountain wolf spider, **86**
Mygalomorphae (trapdoor spiders), 4, 5, 7, 10, 15-16, 25, 28, 37, **37**, 40, 44-5, 59-70, 135, 204, 238
Mynoglenes major, 190
Mynoglenes titan, 190, **191**
Mynogleninae, 36, 190
Myro marinus, 211, **211**
Myro, 30
Mysmena tasmaniae, 200
Mysmena, 201
Mysmenidae, 197, 200-2

naming spiders, 3-8, 60, 72, 99, 110, 151, 209-10, 233, 235
Nanocambridgea gracilipes, 186, **186**
Nanocambridgea, 17
Nanometa gentilis, 167
Nanometa, 167, **167**, 168
Nanometidae, 145, 166-8
Napier, 166
Nelson Lakes National Park, 153
Nelson, 56, 63, 71, 74, 78, 107, 165, 186, 194, 234 – *see also* Takaka
Nemesiidae, 60-1, 66-67
Nemoctenus aurens, 234, **234**
Nemoctenus, 234
Neolana dalmasi, **196**
Neolana, **16**, 195-6
Neolanidae, 6, 171, 195-6
Neopurcellia, 55
Neoramia janus, 195
Neoramia alta, 225, **225**
Neoramia, 225-6
Neoscona orientalis, 154, **154**
Nephila argentata, 155
Nephila clavipes, 40
Nephila edulis (bird-catching or bird-eating spider), 154, **154, 238, 239**-40
Nephila, 154
Nesticodes rufipes, 172, **172**
net-casting spider (*Deinopes*, ogre-faced spider), 6, 151, 169
New Caledonia, 109, 200
New Guinea – *see* Papua New Guinea
New Plymouth, 176
New South Wales, 71, 74-5, 168 – *see also* Australia
New Zealand cave spider, 71, 75, 77, **76-77**, 78
Nicodamidae, 221, 227-8
Nicodamus bicolor, **228**
Nicodamus, 228
Nomaua, 192

Northland, 87, 216
Notiodrassus distinctus, 110, **111**
Notomatachia, **29**, 213
Novanapis spinipes, 198
Novaranea laevigata, 157, **157**
Nuisiana arboris, 215, **215**
Nuncia, **53**, 56
nurseryweb spiders (Pisauridae), 43, 81, 87-94

observing spiders, 247, 251-3
Oecobiidae, 6, 221, 228,
Oecobius navus, 228, **229**
Oecobius, 6
ogre-faced spider (*Deinopis*, net-casting spider), 6, 151, 169
Oonopidae, 135, 144, 136
Oparara, 75, 77
Opiliones (harvestmen), 47-9, 52, **52**, 53-4, **54-6**, 56-8
Opisthothelae, 5
Opsochernes carbophilus, **50**
Oramia littoralis, **213**
Oramia, 213, 226
orb-weaving spiders (Araneidae), 4, **19**, 26, **26**, **45**, 114, 145-6, 151-6, 156-163, **160**, 197, 202, 238
orbweb spiders, 13, **13**, 15, **28**, 33, **34**, **35**, 37, 40, 42, 75, 114, 145-170, 198, 202, 221, 239 – *see* (orbweb families) Araneidae, Deinopidae, Nanometidae, Tetragnathidae, Theridiosomatidae, Uloboridae
Orepuki (Southland), 70
Orepukia, **20**, 225-6
Orsiella lagenifera, 167, **167**
Orsolobidae, **13**, 16, **39**, 135-138
Orsolobus, 136
Ostearius melanopygius, **45**, 189, **189**
Otago, 64, 119, 129-30, 193, 208, 212, 226; Central **39**, 51, 86-8, 100, 105, 107; Harbour 211 – *see also* Dunedin, Mount Pisa, Palmerston, Piano Flat, Queenstown, Remarkables, Rock and Pillar Range, Saddle Hill, Taieri River, Wanaka
Otagoa nova, **212**
Otagoa, 212
Otira, 226-7
Oxyopes gregarius, 105, **105**
Oxyopidae (lynx spiders), 105, **105**, 106

Pahora murihiku, **192**
Pahora, 192, **192**
Pahorinae, 192
Pahoroides, 192
Pakeha subtecta, **226**
Pakeha, 226
Palmerston (Otago), 65
Palmerston North, 117
Palpatores (long-legged harvestmen), 54, **54**, 55, 57-8
Palpigradi, 47-8
Palpimanidae, 235
Panama, 59, 75 – *see also* America
Pantopsalis, 58

Papua New Guinea, 160, 200, 219
Paradictyna rufoflava, **223**
Paradictyna, 222-3
Paralinyphia, **190-1**
Paramamoea parva, **230**
Paramamoea waipoua, **230**
Paramamoea, 230
Paramatachia, 215
Parapua punctata, 200, **200**
Parapua, 200
Pararchaea alba, 204
Pararchaea rubra, 204, **205**
Pararchaea, 204
Pararchaeidae, 197, 202-5
Paratupua, 192
Paravoca, 226
Paravoca otagoensis, **226**
Pardosa goyeni, 86
Patu marplesi, 201
Paul, Rudiger, 82
Periegopidae, **19**, 135, 143-4
Periegops, **19**
Phalangium opilio (European harvestman), 52, **52**, 54
Philodromus ambarus, 99
Pholcidae, 6, 21, 27, 33, 34, 171, 187-189
Pholcomma, 173
Pholcus (daddy-long-legs), 6, 188
Pholcus phalangioides (daddy-long-legs), 21, 33, **34**,187, **188**
Phoroncidia puketoru, **208**
Phoroncidia quadrata, **208**
Phoroncidia, 41, 180, 197
Phoroncidiinae, 208
photographing spiders, 253-4
Physoglenes, 192
Physogleninae, 192
Piano Flat (Southland), 71
Pianoa isolata, **16**, **32**, **35**, **43**, 71, 78, **78**, 79, **79**
Pianoa, 73-4, 77, 79-80
Pickard-Cambridge, Rev. F.O., 202, 212, 235
pirate spiders (Mimetidae), 105, 112-14, 203
Pisaura mirabilis (nurseryweb spider), 92
Pisauridae (nurseryweb spiders, swamp spiders, fisher spiders), 81, 87-94
Platnick, N.I., 7, 80, 136, 204, 245
Plectophanes, 115, **115**, 116, **116**
Plectreuridae, 115
Plectreurys, 116
Poaka graminicola, 234, **235**
Poaka, 234
Poecilopachys australasia (two-spined spider), 154, **155**
Poecilopachys (two-spined spider), **14**
poison – *see* venom
Porrhothele (tunnelweb spider), **19**, 41, 60-3, 66, 79
Porrhothele antipodiana (tunnelweb spider), **10**, **24**, **59**, 61-2, **62**, 68, 239
Port Hills, 116 – see also Christchurch
Portia, 119, **119**
Pounamuella australis, 136

Pounamuella, 4
Prasma, 56
preserving spiders, 247
Priocnemis (Trichocurgus) monachus, 68
Priocnemis nitidiventris, 86, **86**
Pristobunus, 56, **56**
Progradungula carraiensis, 71, 74, **74**, 75, **75**, 214
Progradungula, 74-5, 79-80
prowling spiders (Miturgidae), 105, 107-8, 241-2
Pschridae, 233-4
Pua, 200

Queensland, 71, 74, 168 – *see also* Australia
Queenstown, 131
Quicke, Donald, 238

rain spiders, 205
Rakaia pauli, **53**
Rakaia, **54**, 55
Ralph, T.S., 242
Rapua australis, 216, **216**
Raven, Robert, 60, 238
ray spiders (Theridiosomatidae), 42, 145, 151, 168, **168**
redback spider (*Latrodectus hasselti*), 6, **43**, 177, **177-8**, 178-9, 242, 244-5
Remarkables, The, **39**, 130-1 – *see also* Otago
Rhomphaea, 180-1, **181**, 193
Ricinulei, 47, **47**
Ricinuleids 47
Robinson, Mike and Barbara, 75
Robson, C.H., 210-11
Rock and Pillar Range (Otago), 130-1
Rotorua, 195
Rovner, Jerome, 82
Runga, 192

Saddle Hill (Otago), 116
Salticidae (jumping spiders), 6, 18, **19**, 30, 39, **39**, 43-5, 81, 117-134, **118**, **124**, 129, **130-1**, 209
Samoa, 201, 208
Schultz, Stefan, 238
Scilly Islands, 173
scorpions, 47, **47**
Scotophaeus pretiosus, 110, **111**
scuttling spiders (Cycloctenidae), **19**, 105, 114-16, 209
Scytodes (spitting spider), 43-4, 204
Scytodes thoracica, **45**, 144
Scytodidae, 43-4, **45**, 135-6, 144, 204
seashore spiders, 209-220
Segestriidae, 16, 135-6, 141-3
sexual cannibalism, 178-9
Shear, Bill, 73
sheetweb spiders (Linyphiidae), 6, 36, 44, **45**, 171, 189-192, 197, 240, 205-7
shield spiders (Malkaridae), 105, 113-4, **113**
shoebutton spider, 242
short-legged harvestmen (Laniatores), 54-7
Sicariidae, **26**, 27

Sidymella, 98, **98**, 101, **101**, 102, **103**, **104**, 112
silk, 4, 8-9, 27-35, **28-32**, **34-35**, 40-3, **50**, 50, **51**, 52-3, 59-60, 62-3, 69, 72, 74, 78-9, 83-5, 92, 98, 110, 119, **126**, 131, 146-150, 157-8, 160-161, 168, 171, 175, 184-5, 188-9, 192, 194, 198, 207-8, 210, 213-6, 219, 221-2, 238
Simon, Eugene, 119, 193, 203, 218
six-eyed spiders, 4, 6, **19**, 26, 41, 44-5, 135-144, 188, 204, 251
Skylab experiment, 150
slater-eating spider (*Dysdera crocota*), 6, **24**, 135, 241
small crab spiders (Thomisidae), 43-4, **45**, 95, 98-104, 112
Soerensenella prehensor, **55**
Soerensenella, 55-6
Sorensen's Gradungula, 70
Sorensen, Jack, 70
South Africa, 7, 42, 55, 58, 113, 180, 193, 202-3, 210, 212, 242 – *see also* Africa
South America, 55, 58, 60, 70, 72, 113, 192, 202-3, 212, 237 – *see also* America, Argentina, Chile
Southland, 70-1, 73, 107, 226 – *see also* Fiordland, Te Anau, Longwood Forest, Orepuki
spaceweb spiders, 171-196
Sparassidae (giant crab spiders), 6, 95-98
Spelungula cavernicola (cave spider), 71, 75, 77, **76-77**, 78
Spelungula, 74-5, 77-8
spitting spider (*Scytodes*), 43-4, 204
Stanwellia, 60
stealthy spiders (Gnaphosidae), 6, **19**, 43, 105-6, 109-10
Steatoda capensis (false katipo), 180, **180**, 241
Steatoda grossa (house cobweb spider), 179, **179**, 241
Steatoda lepida, 180
Steatoda paykulliana, 241
Steatoda truncata, 180
Steatoda, 17, 46, 241, 243
Stewart Island, 66, 70, 73, 78, 87, 203, 211
Stiphidiidae, 171, 183-187, 192, 223, 226
Stiphidion facetum, 183-4, **183**
Stiphidion, 183-4
Storena, 233
Subantarctia trina, **39**
Subantarctia turbotti, 136
Subantarctia, **16**, 136, 144, 183
subantarctic islands, 37, 56, 152, 191, 193, 200, 211, 225 – *see also* Auckland Islands, Campbell Island
Supunna picta, 106-7, **107**
Supunna, 6
Sutherland, Struan, 239
swamp spiders, 88 – *see* Pisauridae
Syctodes, 6
Syctodidae, 6
Sydney, 61, 218, 238-9 – *see also* Australia
Symphytognathids, 197, 201
Synotaxidae, 83, 171, 192-3

Synotaxus, 192
Synthetonychia, **57**
Synthetonychidae, 55, 57

Taieri River, 193
tail-less whip scorpions (Amblypygi), **47**
tailed spider, **40**
Takaka, 78
Taradale, 166
tarantula, 6, **7**, 47, 237-8
Tararua, 226
tarentella, 237
Tarento, 237
Tarlina woodwardi, 70
Tasmania, 70, 72-3, 152, 158, 193, 200, 203, 208-9, 215, 227 – see also Australia
Tasmanian cave spider, 73
Te Anau, Lake, 116
Te Karaka, 105, 110, 154
Tegenaria domestica, 41, 224-5, **224**
Tegenaria, 224-5
Tekella absidata, 194
Tekella, 193, 194-5
Tekelloides australis, **195**
Tekelloides, 194, **194**, 195, **195**
Tetragnatha, **164**, 165, 167
Tetragnathidae (big-jawed spiders), 6, 145, 151, 154-5, 160, 164-5, **165**, 166-7, 186
Textricella salmoni, **199**
Textricella, 199, 200
Thaida, 73
Thelassochernes pallipes, **52**
Theridiidae (cobweb spiders), 4, 6, **14**, 30, 35, 42, 44, **45**, 72, 83, 112-3, 154, 171-183, 192-3, 197, 202, 208, 241
Theridion ampliatum, 173
Theridion sisyphium, 35
Theridion troglodytes, 72
Theridion, 113, 172
Theridiosomatidae (ray spiders), 42, 145, 151, 168, **168**
Thomisidae (small crab spiders), 43-4, **45**, 95, 98-104, 112
Three Kings Islands, 66, 203, 230
ticks, 48, **48**
Todd, Valerie, 66
Toft, Richard, 153
Toxopidae, 114, 209
Toxops montanus, 209

Toxopsiella minuta, 115, **115**
Toxopsiella, **15**, **16**, 114, 209
trapdoor spiders (Mygalomorphae), 4, 5, 7, 10, 15-16, 25, 28, 37, **37**, 40, 44-5, 59-70, 135, 204, 238
trapdoor spiders, true (Idiopidae), 37, 59-61, 64-6
tree-trapdoor spider (Migidae), 67
Triaenonychidae, **53**, 55
Triantelope, 6
Trite auricoma, 119-21, **120**, **122-3**, **125-6**, 132
Trite planiceps, 39, 119, **121**, 125, **125**, 126, **126**
Trite, **19**, **118**, 130
Trogloneta, 200-1, **201**, 202, 208
trogloxenes, 232
Trombidiid mites, **48**
true archaeids, 202-3
True Spiders (Araneomorphae), 5, 10, 25, 26, 28, 41, 69-80, 199, 224, 227
true trapdoor spider (Idiopidae), 37, 59-61, 64-6
Tuata insulata, **207**
tunnelweb spiders (Hexathelidae) **10**, **19**, **24**, 41, **59**, 60-1, **61**, 62, **62**, 63, **63**, 64, **64**, 66-8, 79, 239
Tupua, 192
Turbott, Graham, 136
Turnbull, A.L., 94
two-spined spider (*Poecilopachys* spp.), **14**, 154, **155**
Tylorida ventralis, 160

Ulesanis, 208
Uloboridae, 145, 151, 160, 169, 208, 221
United States, 6, 7, 58, 69-70, 72, 80, 179, 220 – Arizona 7; California 220; Hawaii 77, 220
Uropygi (whip scorpions), **47**
Urquhart, A.T., 99, 101, 105, 110-11, 151, 154-5, 167-8, 172, 194

Vachon, Max, 51
vagrant spiders – see hunting spiders
venom, 8, 10, 11, 50, 52-3, 108, 185, 237, 239-46
vinegaroon (whip scorpion), **47**
Viridictyna kikkawai, **222-3**
Viridictyna, 222
von Hochstetter, Ferdinand, 61

Waikaia Forest, 73, 78
Wairarapa, 63, 116, 203
Wairua, 192
Waitkera waitakeriensis, 169, **169**, **221**
Waitkera, 15
Walckenaer, C., 152
Wanaka, 177, 244
Wanganui, 66
wasps, 68, 86, 106, 153, 161
water spider (*Dolomedes aquaticus*), 88, 90, **90**, 91-2, **92**, 93-4, 211
Waterea, 196
webs – see silk
Wellington, 183
West Coast (South Island), 66, 73-4, 78, 88, 99, 107, 115, 176, 186 – see also Fox Glacier, Franz Josef Glacier, Karamea, Oparara
Whangarei Harbour, 99
whip scorpions (*Uropygi*), **47**
white-tailed spider (*Lampona*), 6, 105, 111-12, 245-6 – see also *Lampona* spp.
Whitehouse, Mary, 181, 183
Williams, David, 93
Wilton, C.L., 116, 203, 206-7
Wiltonia, **20**
Witt, Peter, 150, 166
wolf spiders (Lycosidae), 7, 18, **19**, 26, **26**, 33, 35, **35**, 43-4, **45**, 70, 81-2, **82**, 83-5, **85-6**, 86-7, 94, 237-8
Wood, Phil, 78
Woodward, T.E., 70
Worthy, Trevor, 77
Wright, L.G., 239
Wunderlich, Jörg, 168, 193

Xysticus, 44, **45**

Zealaranea crassa, 156, **156**, 157
Zealoctenus cardronaensis, 234
Zealoctenus, 234
Zearchaea clypeata, 203-4, **203**
Zearchaea globosa, 203
Zearchaea magna, 204
Zearchaea, 204
Zeopsopilio novaezealandiae, 58
Zodariidae, 105, 226, 233
Zodarion, 233
Zygiella x-notata, 166
Zygiella, 166